11 - 19 PROGRESSION

Edexcel GCSE (9-1)

Mathematics

Higher

Practice, Reasoning and Problem-solving Book

Confidence • Fluency • Problem-solving • Reasoning

ALWAYS LEARNING

Published by Pearson Education Limited, 80 Strand, London WC2R 0RL.

www.pearsonschoolsandfecolleges.co.uk

Copies of official specifications for all Edexcel qualifications may be found on the website: www.edexcel.com

Edited by ProjectOne Publishing Solutions, Scotland
Typeset and illustrated by Tech-Set, Gateshead

First published 2015

18 17 16 15
10 9 8 7 6 5 4 3 2 1

British Library Cataloguing in Publication Data
A catalogue record for this book is available from the British Library

ISBN 978 1 447 98360 6

Printed in Slovakia by Neografia

Acknowledgements
We would like to thank Glyn Payne for his work on this book.

The publisher would like to thank the following for their kind permission to reproduce their photographs:

Cover images: Front: Created by **Fusako**, Photography by NanaAkua

Every effort has been made to contact copyright holders of material reproduced in this book. Any omissions will be rectified in subsequent printings if notice is given to the publishers.

A note from the publisher
In order to ensure that this resource offers high-quality support for the associated Pearson qualification, it has been through a review process by the awarding body. This process confirms that; this resource fully covers the teaching and learning content of the specification or part of a specification at which it is aimed. It also confirms that it demonstrates an appropriate balance between the development of subject skills, knowledge and understanding, in addition to preparation for assessment.

Endorsement does not cover any guidance on assessment activities or processes (e.g. practice questions or advice on how to answer assessment questions), included in the resource nor does it prescribe any particular approach to the teaching or delivery of a related course.

While the publishers have made every attempt to ensure that advice on the qualification and its assessment is accurate, the official specification and associated assessment guidance materials are the only authoritative source of information and should always be referred to for definitive guidance.

Pearson examiners have not contributed to any sections in this resource relevant to examination papers for which they have responsibility.

Examiners will not use endorsed resources as a source of material for any assessment set by Pearson.

Endorsement of a resource does not mean that the resource is required to achieve this Pearson qualification, nor does it mean that it is the only suitable material available to support the qualification, and any resource lists produced by the awarding body shall include this and other appropriate resources.

Contents

Welcome to Edexcel GCSE (9-1) Mathematics Higher Practice, Reasoning and Problem-solving Book

This Practice Book is packed with extra practice on all the content of the Student Book – giving you more opportunities to practise answering simple questions as well as problem-solving and reasoning ones.

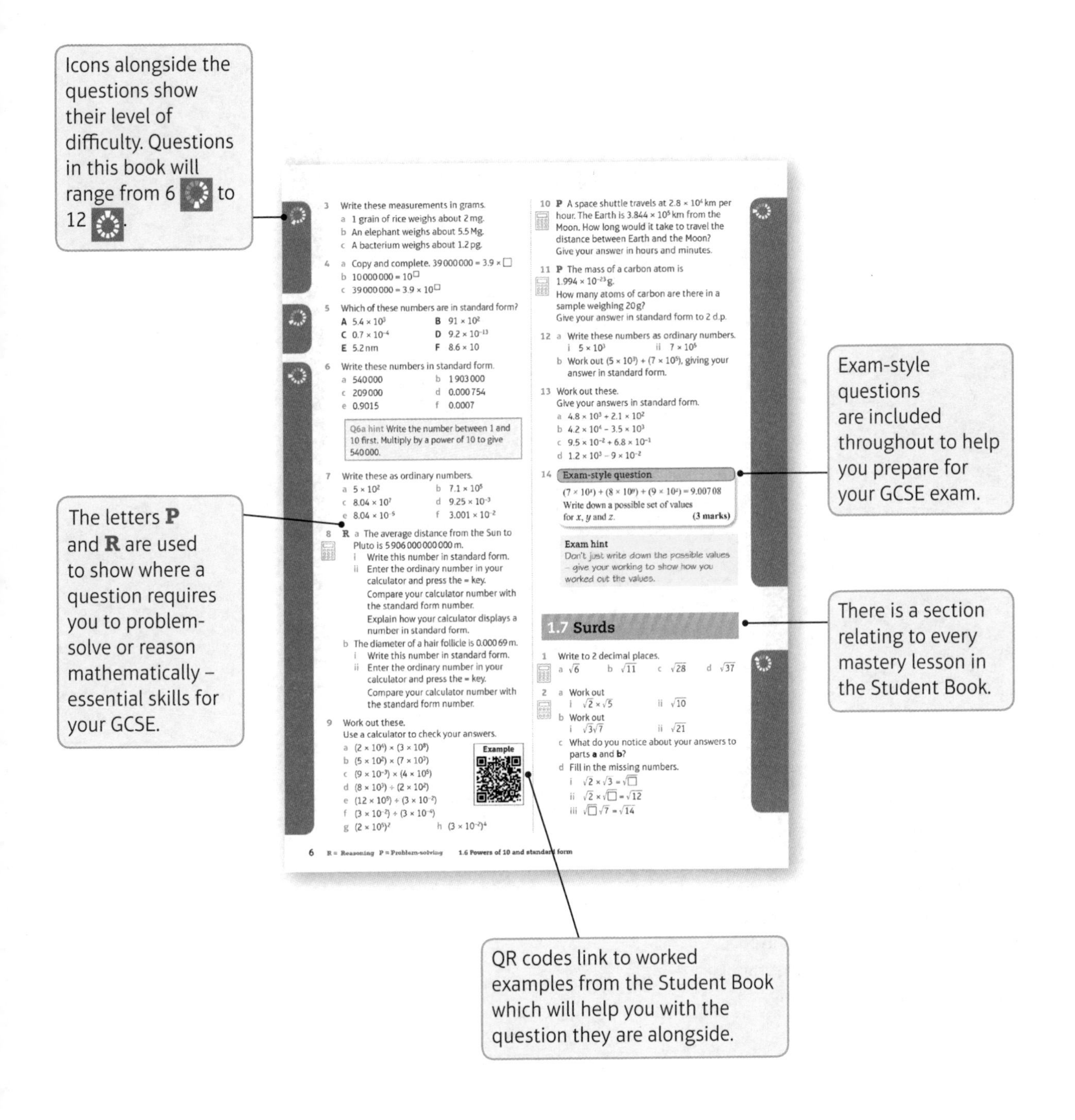

3 Write these measurements in grams.
a 1 grain of rice weighs about 2 mg.
b An elephant weighs about 5.5 Mg.
c A bacterium weighs about 1.2 pg.

4 a Copy and complete. $39000000 = 3.9 \times \square$
b $10000000 = 10^{\square}$
c $39000000 = 3.9 \times 10^{\square}$

5 Which of these numbers are in standard form?
A 5.4×10^3 **B** 91×10^2
C 0.7×10^{-4} **D** 9.2×10^{-13}
E 5.2 nm **F** 8.6×10

6 Write these numbers in standard form.
a 540000 b 1903000
c 209000 d 0.000754
e 0.9015 f 0.0007

Q6a hint Write the number between 1 and 10 first. Multiply by a power of 10 to give 540000.

7 Write these as ordinary numbers.
a 5×10^2 b 7.1×10^5
c 8.04×10^7 d 9.25×10^{-3}
e 8.04×10^{-5} f 3.001×10^{-2}

8 **R** a The average distance from the Sun to Pluto is 5906000000000 m.
i Write this number in standard form.
ii Enter the ordinary number in your calculator and press the = key.
Compare your calculator number with the standard form number.
Explain how your calculator displays a number in standard form.
b The diameter of a hair follicle is 0.00069 m.
i Write this number in standard form.
ii Enter the ordinary number in your calculator and press the = key.
Compare your calculator number with the standard form number.

9 Work out these.
Use a calculator to check your answers.
a $(2 \times 10^4) \times (3 \times 10^8)$
b $(5 \times 10^2) \times (7 \times 10^3)$
c $(9 \times 10^{-3}) \times (4 \times 10^5)$
d $(8 \times 10^3) \div (2 \times 10^2)$
e $(12 \times 10^5) \div (3 \times 10^{-2})$
f $(3 \times 10^{-2}) \div (3 \times 10^{-4})$
g $(2 \times 10^5)^2$ h $(3 \times 10^{-2})^4$

Example

10 **P** A space shuttle travels at 2.8×10^4 km per hour. The Earth is 3.844×10^5 km from the Moon. How long would it take to travel the distance between Earth and the Moon?
Give your answer in hours and minutes.

11 **P** The mass of a carbon atom is 1.994×10^{-23} g.
How many atoms of carbon are there in a sample weighing 20 g?
Give your answer in standard form to 2 d.p.

12 a Write these numbers as ordinary numbers.
i 5×10^3 ii 7×10^5
b Work out $(5 \times 10^3) + (7 \times 10^5)$, giving your answer in standard form.

13 Work out these.
Give your answers in standard form.
a $4.8 \times 10^3 + 2.1 \times 10^2$
b $4.2 \times 10^4 - 3.5 \times 10^3$
c $9.5 \times 10^{-2} + 6.8 \times 10^{-1}$
d $1.2 \times 10^3 - 9 \times 10^{-2}$

14 **Exam-style question**
$(7 \times 10^x) + (8 \times 10^y) + (9 \times 10^z) = 9.00708$
Write down a possible set of values for x, y and z. **(3 marks)**

Exam hint
Don't just write down the possible values – give your working to show how you worked out the values.

1.7 Surds

1 Write to 2 decimal places.
a $\sqrt{6}$ b $\sqrt{11}$ c $\sqrt{28}$ d $\sqrt{37}$

2 a Work out
i $\sqrt{2} \times \sqrt{5}$ ii $\sqrt{10}$
b Work out
i $\sqrt{3}\sqrt{7}$ ii $\sqrt{21}$
c What do you notice about your answers to parts **a** and **b**?
d Fill in the missing numbers.
i $\sqrt{2} \times \sqrt{3} = \sqrt{\square}$
ii $\sqrt{2} \times \sqrt{\square} = \sqrt{12}$
iii $\sqrt{\square}\sqrt{7} = \sqrt{14}$

6 R = Reasoning P = Problem-solving 1.6 Powers of 10 and standard form

These are strategies that you have learned so far in the Student Book. They build up as you work through the book. Consider whether they could help you to answer some of the following questions.

A QR code is given in the problem-solving section where you learn a new strategy in that unit in the Student Book. Scan it to see the worked example and remind yourself of the strategy.

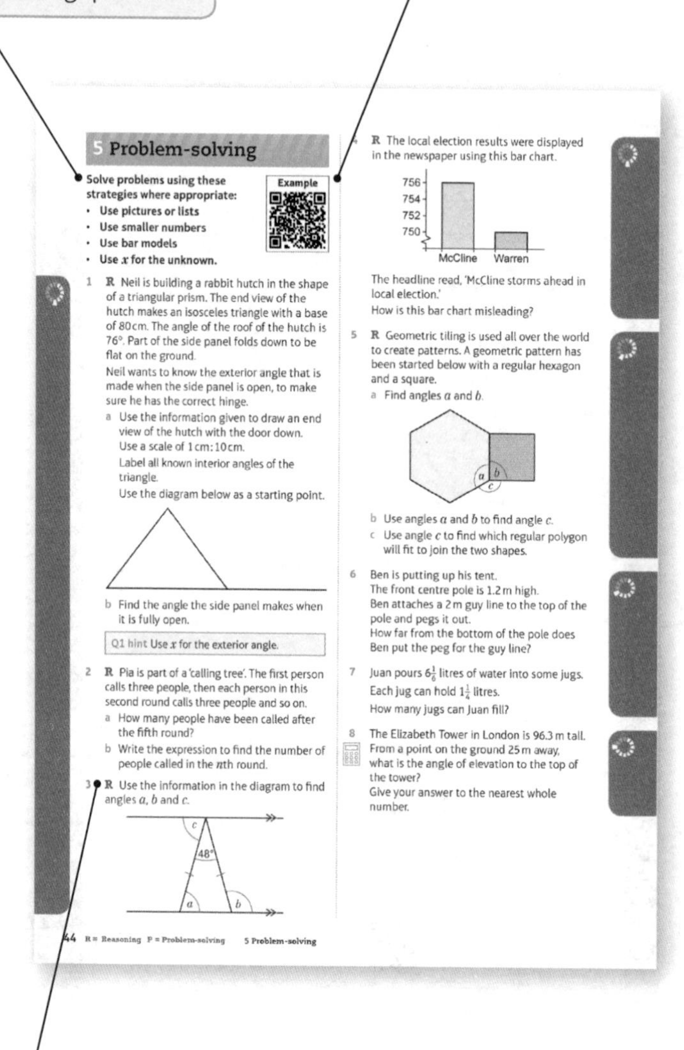

5 Problem-solving

Solve problems using these strategies where appropriate:
- **Use pictures or lists**
- **Use smaller numbers**
- **Use bar models**
- **Use x for the unknown.**

Example

1 **R** Neil is building a rabbit hutch in the shape of a triangular prism. The end view of the hutch makes an isosceles triangle with a base of 80 cm. The angle of the roof of the hutch is 76°. Part of the side panel folds down to be flat on the ground.
Neil wants to know the exterior angle that is made when the side panel is open, to make sure he has the correct hinge.
  a Use the information given to draw an end view of the hutch with the door down.
  Use a scale of 1 cm : 10 cm.
  Label all known interior angles of the triangle.
  Use the diagram below as a starting point.
  b Find the angle the side panel makes when it is fully open.

Q1 hint Use x for the exterior angle.

2 **R** Pia is part of a 'calling tree'. The first person calls three people, then each person in this second round calls three people and so on.
  a How many people have been called after the fifth round?
  b Write the expression to find the number of people called in the nth round.

3 **R** Use the information in the diagram to find angles a, b and c.

4 **R** The local election results were displayed in the newspaper using this bar chart.

The headline read, 'McCline storms ahead in local election.'
How is this bar chart misleading?

5 **R** Geometric tiling is used all over the world to create patterns. A geometric pattern has been started below with a regular hexagon and a square.
  a Find angles a and b.
  b Use angles a and b to find angle c.
  c Use angle c to find which regular polygon will fit to join the two shapes.

6 Ben is putting up his tent.
The front centre pole is 1.2 m high.
Ben attaches a 2 m guy line to the top of the pole and pegs it out.
How far from the bottom of the pole does Ben put the peg for the guy line?

7 Juan pours $6\frac{1}{6}$ litres of water into some jugs.
Each jug can hold $1\frac{1}{4}$ litres.
How many jugs can Juan fill?

8 The Elizabeth Tower in London is 96.3 m tall.
From a point on the ground 25 m away, what is the angle of elevation to the top of the tower?
Give your answer to the nearest whole number.

44 R = Reasoning P = Problem-solving 5 Problem-solving

The **R** symbol indicates questions where you are required to reason mathematically. Questions in this section do not have the **P** symbol as they are all problem-solving.

1 NUMBER

1.1 Number problems and reasoning

1 A T-shirt manufacturer offers a choice of three different coloured shirts – red, green or blue, and two different patterns – spots or stripes.

a Write down all possible combinations of colours and patterns.

b How did you order your list to make sure that you didn't miss any combinations?

The manufacturer decides to offer yellow T-shirts as well.

c How many possible combinations are there now?

d Copy and complete.

3 colours and 2 patterns:
☐ combinations

4 colours and 2 patterns:
☐ combinations

n colours and m patterns:
☐ combinations

Another T-shirt company offers 4 colours, 2 patterns and 3 different sizes.

e How many possible combinations are there now?

2 **Exam-style question**

A bank asks for a 3-character PIN code.
The first character is a letter.
The other two characters are digits between 0 and 9 inclusive.

a How many choices are possible for each character of the code?

b What is the total number of 3-character PIN codes possible?

Amy chooses a PIN code beginning with the letter A.

c How many different PIN codes are possible for Amy? **(5 marks)**

Q2 hint **Inclusive** means that the end numbers are also included.

3 Four people, Anne, Brian, Colin and Danni enter a competition. Between the four of them, they win first, second, third and fourth prizes.

a Write down the different orders in which they can finish first, second, third and fourth.

b How many different ways can the prizes be awarded if there are

i 5 people and 5 prizes

ii 10 people and 10 prizes

iii 12 people and 12 prizes?

4 **P** A shop alarm needs a 5-character code.

a How many different codes can be created using

i 5 numbers

ii 5 letters

iii 4 letters followed by 1 number?

The shop owners decide not to repeat a digit or letter.

b How many ways are possible in parts **i** to **iii** now?

1.2 Place value and estimating

1 Work out

a 56×12

b 28×24

c 14×48

d 7×96

2 $5.4 \times 7.39 = 39.906$

Use this fact to work out the calculations below.

Check your answers using an approximate calculation.

a 54×7.39

b $5.4 \times 0.007\,39$

c 0.0054×0.729

d $39.906 \div 7.39$

e $3990.6 \div 73.9$

f $399.06 \div 739$

3 **R** $18.4 \times 9.07 = 166.888$

a Write down three more calculations that have the same answer.

b Write down a division that has an answer of 18.4.

c Write down a division that has an answer of 0.907.

d Prue says that $18.4 \times 907 = 16\,888.08$. Explain why Prue must be wrong.

4 a Write down the value of $\sqrt{9}$ and $\sqrt{16}$.
b Estimate the value of $\sqrt{10}$, $\sqrt{11}$, $\sqrt{12}$ and $\sqrt{13}$.
Round each estimate to 1 decimal place.
c Use a calculator to check your answers to part **b**.

Q4b hint Use a number line to help.

$\sqrt{9}$ $\sqrt{10}$ $\sqrt{16}$

□ □ □

5 Estimate the value to the nearest tenth.

a $\sqrt{6}$ b $\sqrt{30}$ c $\sqrt{29}$
d $\sqrt{40}$ e $\sqrt{59}$ f $\sqrt{70}$

6 **P** A logo contains 300 square pixels.
The logo has a total area of $600\,mm^2$.
a Estimate the side length of a pixel.
b Use a calculator to check your answer.

7 a Write down the value of 9^2 and 10^2.
b Estimate the value of 9.2^2 and 9.7^2. Round each estimate to the nearest whole number.
c Use a calculator to check your answers to part **b**.

8 Estimate to the nearest whole number.

a 4.1^2 b 5.2^2 c 2.6^2
d 9.6^2 e 3.4^2 f 10.3^2

9 a Estimate answers to these.
i $12.5 \times \sqrt{48.4} - 16.27$
ii $2.91 \times (12.4 - 2.87)^2$
iii $\dfrac{31.4^2}{18.5 + \sqrt{1.986}}$
iv $\dfrac{\sqrt{15.43 - 2.91}}{(3.4 - 2.091)^2}$
b Use your calculator to work out each answer.
Give your answers correct to 1 decimal place.

10 The sum of these two values is 9.

$\dfrac{14 + \sqrt{121}}{\sqrt{25}}$ $\dfrac{8^2 - \square}{\sqrt{9}}$

Work out the missing number.

11 **P** A cubic block has a side length of 12.3 cm.
Estimate the surface area of the block.

12 **P** The area of a square is $120\,cm^2$.
Estimate the perimeter of the square.

13 **P** Tiles measure 15 cm by 20 cm.
Each tile costs £1.09.
a Estimate the cost of tiles required for these rectangular spaces.
i 90 cm by 120 cm
ii 2 m by 4 m
iii 3.1 m by 2.6 m
b Use a calculator to work out each answer.

14 A spreadsheet is used to record the numbers of books on 10 shelves in the library.
The numbers are in cells A1 to J1.
The mean number is in cell K1.

	A	B	C	D	E	F	G	H	I	J	K
1	13	18	22	11	16	19	23	24	32	15	23.2

a Use estimates to show that the mean shown in cell K1 is wrong.
b Work out the correct mean to the nearest tenth.

1.3 HCF and LCM

1 a Copy and complete this factor tree for 54.

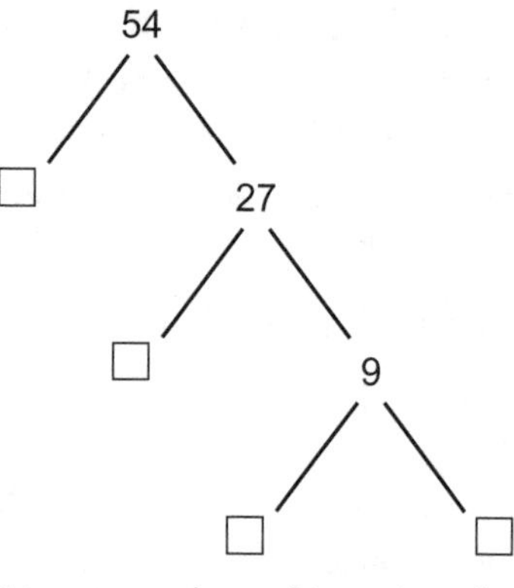

b Write 54 as a product of its prime factors.
$54 = \square \times \square \times \square \times \square = \square \times \square^{\square}$

Q1b hint Circle the prime factors in your factor tree.

2 Write 24 as a product of its prime factors.

3 Gail and Hermione are asked to find 90 as a product of its prime factors.
Gail begins by writing $90 = 9 \times 10$
Hermione begins by writing $90 = 3 \times 30$
a Work out a final answer for Gail.
b Work out a final answer for Hermione.
c Start the prime decomposition of 50 in two different ways: 2×25 and 5×10.

4 Write each number as a product of its prime factors in index form.

a 35 b 20 c 33
d 48 e 56 f 100

5 180 can be written as a product of its prime factors in the form $2^m \times n^2 \times p$.
Work out m, n and p.

6 Find the HCF and LCM of

a 12 and 20 b 15 and 40
c 18 and 24 d 7 and 12
e 30 and 42 f 9 and 60

Example

7 **P** Two lighthouses flash their lights at different intervals. One flashes every 6 seconds, the other every 9 seconds.
They have both just flashed together.
How long before they next flash together?

8 **P** Miranda needs 144 paper cups for a party. The cups come in packs of 15, 20 or 24.
Which single-size packs should she buy to ensure she has exactly the right number?

9 **P** The HCF of two numbers is 3.
Write down three possible pairs of numbers.

10 **P** The LCM of two numbers is 15.
One of the numbers is 15.

a Write down all the possibilities for the other number.
b Describe the set of numbers you have created.

11 $54 = 2 \times 3^3$ and $36 = 2^2 \times 3^2$
Write down, as a product of its prime factors,

a the HCF of 54 and 36
b the LCM of 54 and 36.

12 **Exam-style question**

Given that $A = 2^3 \times 5^2 \times 7$ and
$B = 2 \times 5^3 \times 7^2$
write down, as a product of its prime factors,

a the HCF of A and B
b the LCM of A and B. **(2 marks)**

13 Write 96 as a product of its prime factors.

14 **P** The prime factor decomposition of 3000 is $2^3 \times 3 \times 5^3$
Write down the prime factor decomposition of

a 8 b 25 c 24 d 75

15 a Daryl says the prime factors of 4 appear in the prime factor decomposition of 3000, so 3000 is divisible by 4.
Is 3000 divisible by 12, 75 or 21?
b Use prime factors to show that 945 is divisible by 27.
c Is 945 divisible by 25?
Explain your answer.
d Is 945 divisible by 35?
Explain your answer.

16 In prime factor form, $360 = 2^3 \times 3^2 \times 5$ and $2700 = 2^2 \times 3^3 \times 5^2$

a What is the HCF of 360 and 2700?
Give your answer in prime factor form.
b What is the LCM of 360 and 2700?
Give your answer in prime factor form.
c Which of these are factors of 180 and 2700?
i $2 \times 2 \times 3$ ii 25
iii 9 iv $2^2 \times 3 \times 5$
d Which of these are multiples of 180 and 2700?
i $2^3 \times 3^4 \times 5^2$
ii $2 \times 3^2 \times 5^2$
iii $2^3 \times 3^2 \times 5^3$

1.4 Calculating with powers (indices)

1 Work out

a $\sqrt[3]{8}$ b $\sqrt[3]{-8}$
c $\sqrt[3]{125}$ d $\sqrt[3]{-216}$

2 Work out these.
Use a calculator to check your answers.

a $\sqrt{5^2 + 12^2}$ b $\sqrt[3]{2 \times 5^2 + 2 \times 7}$
c $\sqrt[3]{8} - 3 + \sqrt{25}$ d $\sqrt[3]{-27} - \sqrt{25} + 4^2$
e $\sqrt{27 + \sqrt{8 + 1} \times 18}$ f $\frac{\sqrt[3]{-8} \times \sqrt{4}}{(-2)^2}$
g $\frac{\sqrt{100}}{\sqrt[3]{-1}} \times \frac{-\sqrt{16}}{\sqrt[3]{-8}}$ h $\frac{\sqrt[3]{-27} \times 5^2}{\sqrt[3]{-125}}$

Q2e hint The square root applies to the whole calculation. Work out the calculation inside the square root first.

3 Work out

a $[(18 - 2^2) \times 2)]^3$
b $54 - [21 \times 3 + (15 \times 9^2)]$
c $[73 - (3 \times 4^2) + 5] \div \sqrt{16}$

4 Work out

a $\sqrt[4]{256}$ b $\sqrt[4]{625}$

c $\sqrt[4]{100\,000\,000}$ d $\sqrt[5]{0.00001}$

5 a Work out

i $5^3 \times 5^4$ ii 5^7

iii $5^6 \times 5^3$ iv 5^9

b How can you work out the answers to part **a** by using the indices of the powers you are multiplying?

c Check your rule works for

i $5^3 \times 5^7$ ii $5^6 \times 5$

iii $5^{-3} \times 5^{-1}$

6 Write each product as a single power.

a $6^2 \times 6^8$ b $4^{-3} \times 4^5$ c $9^{-7} \times 9^3$

7 Find the value of n.

a $3^n \times 3^4 = 3^6$

b $9^4 \times 9^n = 9^2$

c $7^{-2} \times 7^n = 2^{-7}$

8 Write these calculations as a single power. Give your answers in index form.

a $8 \times 2^4 = 2^{\square} \times 2^4 = 2^{\square}$

b 125×5^3

c 27×81

d 2×32

e $9 \times 9 \times 9$

f $10 \times 1000 \times 100\,000$

9 **R** a i Work out $\dfrac{3 \times 3 \times 3 \times 3 \times 3 \times 3 \times 3}{3 \times 3 \times 3}$ by cancelling.
Write your answer as a power of 3.

ii Copy and complete. $3^7 \div 3^3 = 3^{\square}$

b Copy and complete.

$5^6 \div 5^2 = \dfrac{5 \times 5 \times 5 \times 5 \times 5 \times 5}{5 \times 5} = 5^{\square}$

c Work out $8^9 \div 8^3$

10 Work out

a $4^7 \div 4^3$ b $5^{-3} \div 5^2$ c $7^{-1} \div 7^{-4}$

11 Find the value of t.

a $7^5 \div 7^2 = 7^t$

b $5^9 \div 5^t = 5^3$

c $4^t \div 4^4 = 4^8$

12 **P** a Ying multiplies three powers of 8 together.

$8^{\square} \times 8^{\square} \times 8^{\square} = 8^{15}$

What could the three powers be when

i all three powers are different

ii all three powers are the same?

b Keiko divides two powers of 7.

$7^{\square} \div 7^{\square} = 7^5$

What could the two powers be when

i both numbers are greater than 7^{20}

ii one power is double the other power?

13 Work out these.
Write each answer as a single power.

a $3^5 \times 3^8 \div 3^2$ b $9^4 \div 9^{-2} \times 9^7$

c $\dfrac{3^2 \times 3^7}{3^{-4}}$ d $\dfrac{7^{-2} \times 7^4}{7^{-3}}$

14 The number of pixels on a computer screen is 2^{19}. A company logo uses 2^{12} pixels.

a How many logos can be displayed on the screen?

b Another company logo needs 2^{15} pixels to display all its data.
What fraction of the screen is blank?

15 Copy and complete.

a $(5^3)^6 = 5^3 \times \square^{\square} \times \square^{\square} \times \square^{\square} \times \square^{\square} \times \square^{\square} = 5^{\square}$

b $(2^2)^4 = \square^{\square} \times \square^{\square} \times \square^{\square} \times \square^{\square} = 2^{\square}$

c $(9^4)^7 = \square^{\square} \times \square^{\square} \times \square^{\square} \times \square^{\square} \times \square^{\square} \times \square^{\square} \times \square^{\square} = 9^{\square}$

16 Write as a single power

a $(5^2)^4$ b $(9^3)^4$ c $(4^{-1})^5$ d $(7^{-2})^{-3}$

17 **P** Write each calculation as a single power.

a $9 \times 27 \times 27$ b $\dfrac{4^8}{16}$ c $\dfrac{8 \times 4^2}{2^{-3}}$

1.5 Zero, negative and fractional indices

1 Work out the value of n.

a $18 = 2 \times 3^n$ b $2^n \times 2^n = 2^{10}$

c $7^{2n} \div 7^n = 7^4$ d $\frac{1}{3} \times 3^n = 3^5$

2 a Use a calculator to work out

i 3^{-1} ii 2^{-1} iii 100^{-1} iv 4^{-1}

b Write your answers to part **a** as fractions.

c Use a calculator to work out

i 3^{-2} ii 5^{-2} iii 100^{-2} iv 6^{-2}

d Write your answers to part **c** as fractions.

e Work out

i $\left(\dfrac{1}{4}\right)^{-1}$ ii $\left(\dfrac{2}{3}\right)^{-1}$

3 a Match the equivalent cards.

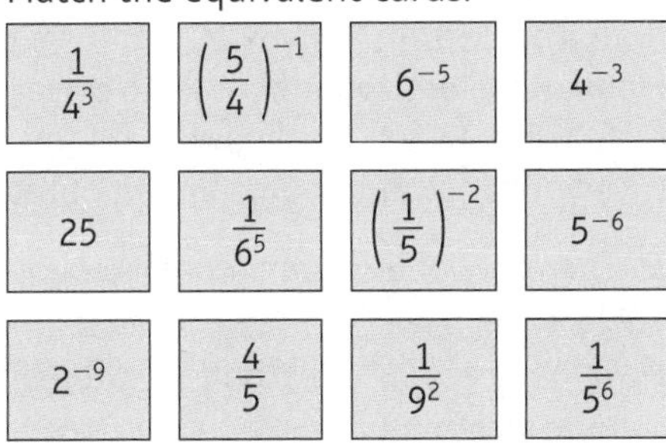

$\frac{1}{4^3}$	$\left(\frac{5}{4}\right)^{-1}$	6^{-5}	4^{-3}
25	$\frac{1}{6^5}$	$\left(\frac{1}{5}\right)^{-2}$	5^{-6}
2^{-9}	$\frac{4}{5}$	$\frac{1}{9^2}$	$\frac{1}{5^6}$

b Write a matching card for the two cards that are left over.

c Copy and complete.

$\left(\frac{3}{5}\right)^{-1} = \frac{\square}{\square}$ so $\left(\frac{a}{b}\right)^{-1} = \frac{\square}{\square}$

4 Work out these.
Write each answer as a single power.

a $4^{-5} \times 4^4 \div 4^2$ b $\frac{7^{-6} \times 7^7}{7^3}$ c $\frac{3^{-5} \times 3^{-2}}{3^{-4}}$

5 **P** a Copy and complete. $5^3 \div 5^3 = 5^{\square}$

b Write down 5^3 as a whole number.

c $5^3 \div 5^3 = 125 \div \square = \square$

d Copy and complete using parts **a** and **c**.
$5^3 \div 5^3 = 5^{\square} = \square$

e Repeat parts **a** and **b** for $2^5 \div 2^5$.

f Write down a rule for a^0, where a is any number.

6 Work out

a 5^{-1} b 3^{-2} c 10^{-5}

d $\left(\frac{1}{2}\right)^{-2}$ e $\left(\frac{4}{5}\right)^{-1}$ f $\left(2\frac{1}{2}\right)^{-1}$

g $\left(\frac{4}{5}\right)^{-2}$ h $\left(\frac{1}{10}\right)^{-3}$ i $(0.3)^{-1}$

j $(0.25)^{-3}$ k $(2^{-1})^0$ l $(8 - 4)^{-1}$

7 a Use a calculator to work out

i $4^{\frac{1}{2}}$ ii $100^{\frac{1}{2}}$ iii $144^{\frac{1}{2}}$ iv $\left(\frac{1}{4}\right)^{\frac{1}{2}}$

b Copy and complete.
$a^{\frac{1}{2}}$ is the same as the __________ __________ of a.

c Work out

i $8^{\frac{1}{3}}$ ii $64^{\frac{1}{3}}$ iii $\left(\frac{1}{27}\right)^{\frac{1}{3}}$ iv $-8^{\frac{1}{3}}$

d Copy and complete.
$a^{\frac{1}{3}}$ is the same as the __________ __________ of a.

e Copy and complete.

i $81 = 3^{\square}$ so $81^{\frac{1}{4}} = \square$

ii $1024 = \square^5$ so $1024^{\frac{1}{5}} = \square$

8 Evaluate

a $100^{\frac{1}{2}}$ b $8^{\frac{1}{3}}$ c $\left(\frac{4}{9}\right)^{\frac{1}{2}}$ d $\left(\frac{25}{49}\right)^{\frac{1}{2}}$

e $-27^{\frac{1}{3}}$ f $-\left(\frac{1}{27}\right)^{\frac{1}{3}}$ g $-\left(\frac{4}{9}\right)^{\frac{1}{2}}$ h $\left(\frac{1}{1000}\right)^{\frac{1}{3}}$

9 Work out

a $36^{-\frac{1}{2}}$ b $9^{-\frac{1}{2}}$ c $\left(\frac{16}{49}\right)^{-\frac{1}{2}}$

10 Work out

Example

a $125^{\frac{2}{3}}$

b $100\,000^{\frac{2}{5}}$

c $9^{\frac{3}{2}}$ d $\left(\frac{16}{25}\right)^{\frac{3}{2}}$

e $16^{-\frac{3}{2}}$ f $1000^{-\frac{2}{3}}$

11 Work out

a $25^{\frac{1}{2}} \times 8^{\frac{2}{3}}$ b $\left(\frac{16}{25}\right)^{-\frac{1}{2}} \times \left(\frac{1}{8}\right)^{\frac{1}{3}}$

c $\left(\frac{1}{10\,000}\right)^{-\frac{1}{4}} \times \left(\frac{1}{100}\right)^{\frac{5}{2}}$

12 Find the value of n.

a $100 = 10^n$ b $\sqrt[n]{1000} = 10$

c $\frac{1}{8} = 2^n$ d $\sqrt[3]{\frac{27}{64}} = \left(\frac{64}{27}\right)^n$

e $\sqrt[3]{2^5} = 2^n$ f $(\sqrt{25^3})^4 = 5^n$

13 **P / R** Dan says that $8^{-\frac{1}{3}} \times 25^{\frac{1}{2}} = 10$

a Show that Dan is wrong.

b What mistake did he make?

14 **P / R** Match the expressions with indices to their values.

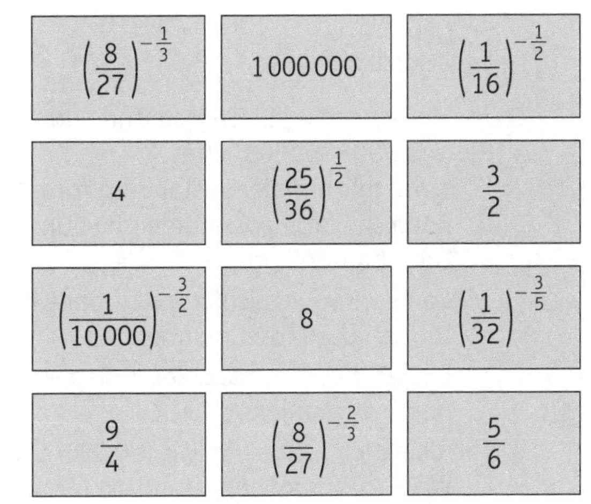

$\left(\frac{8}{27}\right)^{-\frac{1}{3}}$	$1\,000\,000$	$\left(\frac{1}{16}\right)^{-\frac{1}{2}}$
4	$\left(\frac{25}{36}\right)^{\frac{1}{2}}$	$\frac{3}{2}$
$\left(\frac{1}{10\,000}\right)^{-\frac{3}{2}}$	8	$\left(\frac{1}{32}\right)^{-\frac{3}{5}}$
$\frac{9}{4}$	$\left(\frac{8}{27}\right)^{-\frac{2}{3}}$	$\frac{5}{6}$

1.6 Powers of 10 and standard form

1 Write these prefixes in size order, starting with the smallest.
pico, milli, nano, tera, kilo, deci, micro, giga, centi, mega

2 Convert

a 3 Tg into grams b 8 μm into metres

c 3.5 Gg into kg d 9000 ps into seconds.

3 Write these measurements in grams.

a 1 grain of rice weighs about 2 mg.

b An elephant weighs about 5.5 Mg.

c A bacterium weighs about 1.2 pg.

4 a Copy and complete. $39\,000\,000 = 3.9 \times \square$

b $10\,000\,000 = 10^{\square}$

c $39\,000\,000 = 3.9 \times 10^{\square}$

5 Which of these numbers are in standard form?

A 5.4×10^3 B 91×10^2

C 0.7×10^{-4} D 9.2×10^{-13}

E 5.2 nm F 8.6×10

6 Write these numbers in standard form.

a 540 000 b 1 903 000

c 209 000 d 0.000 754

e 0.9015 f 0.0007

Q6a hint Write the number between 1 and 10 first. Multiply by a power of 10 to give 540 000.

7 Write these as ordinary numbers.

a 5×10^2 b 7.1×10^5

c 8.04×10^7 d 9.25×10^{-3}

e 8.04×10^{-5} f 3.001×10^{-2}

8 **R** a The average distance from the Sun to Pluto is 5 906 000 000 000 m.

i Write this number in standard form.

ii Enter the ordinary number in your calculator and press the = key.

Compare your calculator number with the standard form number.

Explain how your calculator displays a number in standard form.

b The diameter of a hair follicle is 0.000 69 m.

i Write this number in standard form.

ii Enter the ordinary number in your calculator and press the = key.

Compare your calculator number with the standard form number.

9 Work out these.
Use a calculator to check your answers.

a $(2 \times 10^4) \times (3 \times 10^8)$

b $(5 \times 10^2) \times (7 \times 10^3)$

c $(9 \times 10^{-3}) \times (4 \times 10^5)$

d $(8 \times 10^3) \div (2 \times 10^2)$

e $(12 \times 10^5) \div (3 \times 10^{-2})$

f $(3 \times 10^{-2}) \div (3 \times 10^{-4})$

g $(2 \times 10^5)^2$ h $(3 \times 10^{-2})^4$

Example

10 **P** A space shuttle travels at 2.8×10^4 km per hour. The Earth is 3.844×10^5 km from the Moon. How long would it take to travel the distance between Earth and the Moon? Give your answer in hours and minutes.

11 **P** The mass of a carbon atom is 1.994×10^{-23} g.
How many atoms of carbon are there in a sample weighing 20 g?
Give your answer in standard form to 2 d.p.

12 a Write these numbers as ordinary numbers.

i 5×10^3 ii 7×10^5

b Work out $(5 \times 10^3) + (7 \times 10^5)$, giving your answer in standard form.

13 Work out these.
Give your answers in standard form.

a $4.8 \times 10^3 + 2.1 \times 10^2$

b $4.2 \times 10^4 - 3.5 \times 10^3$

c $9.5 \times 10^{-2} + 6.8 \times 10^{-1}$

d $1.2 \times 10^3 - 9 \times 10^{-2}$

14 **Exam-style question**

$(7 \times 10^x) + (8 \times 10^y) + (9 \times 10^z) = 9.007\,08$

Write down a possible set of values for x, y and z. **(3 marks)**

Exam hint
Don't just write down the possible values – give your working to show how you worked out the values.

1.7 Surds

1 Write to 2 decimal places.

a $\sqrt{6}$ b $\sqrt{11}$ c $\sqrt{28}$ d $\sqrt{37}$

2 a Work out

i $\sqrt{2} \times \sqrt{5}$ ii $\sqrt{10}$

b Work out

i $\sqrt{3}\sqrt{7}$ ii $\sqrt{21}$

c What do you notice about your answers to parts **a** and **b**?

d Fill in the missing numbers.

i $\sqrt{2} \times \sqrt{3} = \sqrt{\square}$

ii $\sqrt{2} \times \sqrt{\square} = \sqrt{12}$

iii $\sqrt{\square}\sqrt{7} = \sqrt{14}$

3 Find the value of the integer k to simplify these surds.

a $\sqrt{72} = \sqrt{\square}\sqrt{\square}\,\sqrt{2} = k\sqrt{2}$

b $\sqrt{125} = \sqrt{\square}\sqrt{5} = k\sqrt{5}$

c $\sqrt{48} = k\sqrt{3}$

d $\sqrt{63} = k\sqrt{7}$

4 Simplify these surds.

a $\sqrt{18}$ b $\sqrt{75}$ c $\sqrt{98}$

d $\sqrt{200}$ e $\sqrt{45}$ f $\sqrt{20}$

> **Q4 hint** Find a factor that is also a square number.

5 Use a calculator to work out $\sqrt{128}$

a as a simplified surd

b as a decimal.

6 a A surd simplifies to $4\sqrt{3}$. What could the original surd be?

b How did you find the surd?

7 Simplify

a $\sqrt{\frac{5}{4}} = \frac{\sqrt{5}}{\sqrt{4}} =$ b $\sqrt{\frac{17}{25}}$

c $\sqrt{\frac{10}{64}} = \frac{\sqrt{10}}{\sqrt{64}} =$ d $\sqrt{\frac{6}{16}}$

8 Copy and complete the table using the numbers below.

Rational	Irrational

$\sqrt{15}$ $\frac{5}{9}$ $0.\dot{3}$ -12 $\sqrt{0.01}$

$\sqrt{\frac{2}{25}}$ 0.7 $-\sqrt{100}$ $\sqrt[3]{9}$

9 Solve the equation $x^2 + 20 = 70$, giving your answer as a surd in its simplest form.

10 Solve these equations, giving your answer as a surd in its simplest form.

a $3x^2 = 54$ b $\frac{1}{3}x^2 = 225$

c $x^2 - 5 = 75$ d $2x^2 + 2 = 42$

11 The area of a square is $75\,\text{cm}^2$.
Find the length of one side of the square.
Give your answer as a surd in its simplest form.

12 a Work out

i $3\sqrt{5} \times 2\sqrt{3}$ ii $2\sqrt{7} \times 3\sqrt{10}$

iii $2\sqrt{8} \times 5\sqrt{4}$ iv $6\sqrt{2} \times 7\sqrt{11}$

b Use a calculator to check your answers to parts **i** to **iv**.

13 Rationalise the denominators. Simplify your answers if possible.

a $\frac{1}{\sqrt{2}}$ b $\frac{1}{\sqrt{8}}$ c $\frac{1}{\sqrt{12}}$

d $\frac{1}{\sqrt{30}}$ e $\frac{3}{\sqrt{18}}$ f $\frac{4}{\sqrt{13}}$

g $\frac{12}{\sqrt{16}}$ h $\frac{5}{\sqrt{5}}$

14 **R / P** Henry types $\frac{3}{\sqrt{3}}$ into his calculator.
His display shows $\sqrt{3}$.

a Show that $\frac{3}{\sqrt{3}} = \sqrt{3}$

b Use your calculator to check your answers from **Q12**.

15 The area of a rectangle is $33\,\text{cm}^2$. The length of one side is $\sqrt{3}\,\text{cm}$.
Work out the length of the other side. Give your answer as a surd in its simplest form.

16 Work out the area of these shapes. Give your answer as a surd in its simplest form.

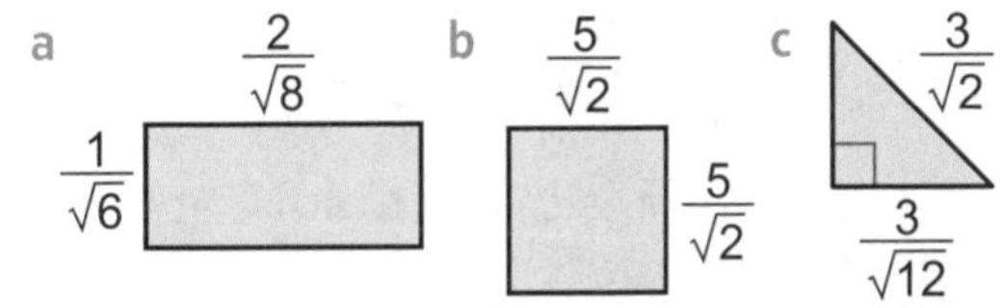

1 Problem-solving

Solve problems using this strategy where appropriate:

- **Use pictures or lists.**

1 **R** Kevin cuts a 90 cm length of wood into thirds by estimating the places to cut.
One piece is exactly a third of the total length. The difference in length between the other two pieces is 5 cm. How long is each piece?

> **Q1 hint** Draw a picture of the three lengths.

2 **R** 126 314 tickets have been sold to a concert in the park. On average, each ticket holder throws away 942 g of rubbish. The rubbish will be taken away in trucks that must not carry more than 575 kg each.

a Estimate the number of trucks that will be needed.

b Is this an under-estimate or an over-estimate? How can you tell?

3 **R** Here are some clues to fill in these boxes.

□ – □ = 6

- Both numbers have two digits.
- They round to the same number when rounded to 1 significant figure.
- They round to different numbers when rounded to the nearest 100.

a Write a pair of numbers that fit the clues.

b Write the pair of numbers that fit which have the lowest sum.

4 **R** At a charity quiz night, each team can have 3, 4 or 5 people. 21 people take part.

a How many different possible arrangements of teams are there?

Each team pays £8. The total raised is £40.

b How many people could there be in each team?

c What is the maximum amount of money that could be raised from this number of people?

5 Rachel estimates that $\sqrt[3]{\frac{2(9.8)^2 + 7(2.14)^3}{1.98^2}}$ is about 8.

a Without using a calculator, check if she is correct.

b Now use a calculator to arrive at the correct answer.

c Comment on both your estimate and Rachel's.

6 The large gear has 24 teeth, the medium gears both have 20 teeth and the smallest gear has 16 teeth.

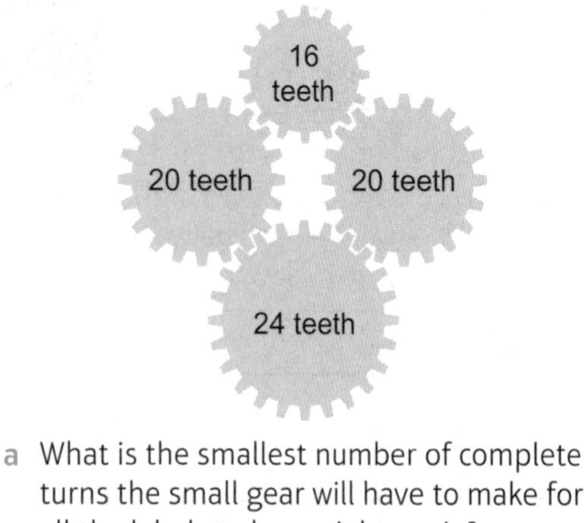

a What is the smallest number of complete turns the small gear will have to make for all the labels to be upright again?

b Every second, the small gear makes the number of turns worked out in part **a**. How many turns will the large gear make

i every second ii every minute?

7 **R** Fill in the boxes to make all five expressions equivalent.

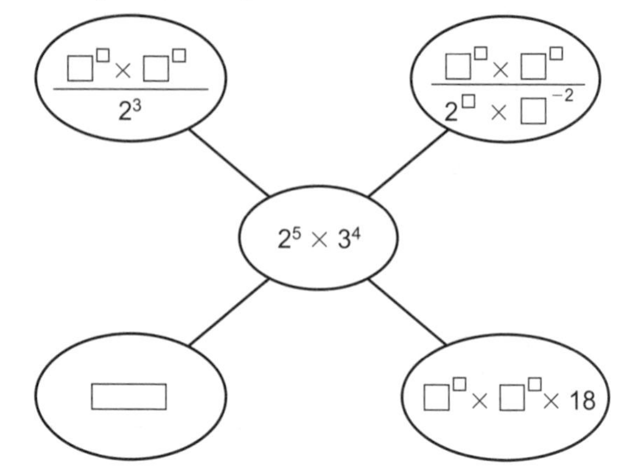

8 **R** On average, a grain of very coarse sand weighs 11 mg and has a volume of 0.004 19 cm³.

a 1 litre = 1000 cm³. How many cubic centimetres are in a 12 litre bucket?

b How many grains of very coarse sand will fill a 12 litre bucket? Give your answer in standard form to 3 significant figures.

The bucket will break if its contents weigh more than 15 kg.

c Can the bucket be lifted when it is half full without breaking it? How can you tell?

9 **Exam-style question**

One sheet of paper is 9×10^{-3} cm thick.

Mark wants to put 500 sheets of paper into the paper tray of his printer.

The paper tray is 4 cm deep.

Is the paper tray deep enough for 500 sheets of paper?

You must explain your answer. **(3 marks)**

June 2013, Q15, 1MA0/1H

10 **R** This square has been split into four sections. Section D has an area of 16.

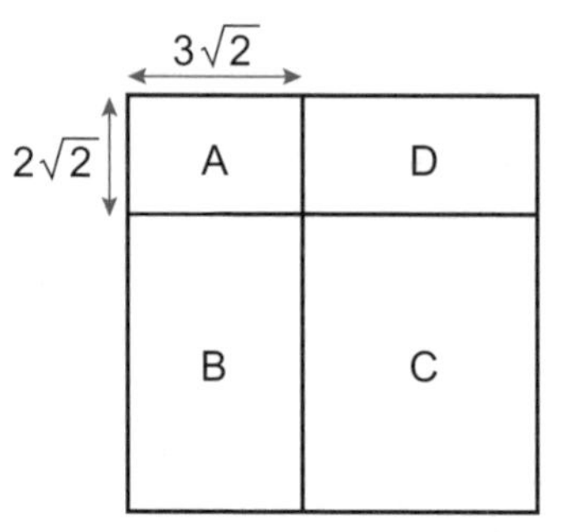

a What is the area of the square?

b What is the ratio of area A : area C?

2 ALGEBRA

2.1 Algebraic indices

1 Simplify

a $y^4 \times y^2$ b $x^3 \times x^5$ c $b^2 \times b$

d $y^3 \times y^3 \times y^3$ e $m^{\frac{1}{2}} \times m^6$

2 Simplify

a $3b^2 \times 5b^5$ b $2a \times 3a^2$

c $5m^3 \times 10m^4$ d $t^6 \times 2t^2$

e $4s^3t^2 \times 2st^4$ f $3a^3b^2 \times 2ab^3 \times 4ab$

3 Simplify

a $x^3 \div x^2$ b $a^5 \div a^3$ c $\frac{a^{10}}{a^4}$

d $\frac{y^3}{y}$ e $\frac{r^6}{r^5}$ f $\frac{p^7}{p \times p^3}$

4 Simplify

a $\frac{12g^4}{2g^3}$ b $\frac{9p^3}{3p}$

c $18x^4 \div 6x^2$ d $16t^{10} \div 8t^5$

5 Simplify

a $(x^5)^2$ b $(y^3)^3$ c $(t^4)^5$ d $(r^3)^7$

6 Simplify

a $(3t^2)^3$ b $(4m^3)^2$ c $\left(\frac{r^3}{2}\right)^3$

7 The term in each brick in the wall is found by multiplying together the two terms in the bricks underneath.

Copy and complete the wall.

$24x^8y$

[ ] $2x^3y$

[ ] $2x^2$ [ ]

[ ] $2x$ [ ] [ ]

Q7 hint To find the term in the yellow brick you must work out $24x^8y \div 2x^3y$

8 Simplify

a $(3x^2y^3)^2$ b $(2x^3y^5)^3$ c $(5x^2y)^4$ d $\left(\frac{4x^3y^6}{2xy^5}\right)^2$

9 **R** Copy and complete

a $x^4 \div x^4 = x^{\square - \square} = x^{\square}$

$x^4 \div x^4 = \frac{x^4}{x^4} = \square$

Therefore $x^{\square} = \square$

b $x^5 \div x^6 = x^{\square - \square} = x^{\square}$

$x^5 \div x^6 = \frac{x \times x \times x \times x \times x}{x \times x \times x \times x \times x \times x} = \frac{\square}{\square}$

Therefore $x^{\square} = \frac{\square}{\square}$

c $x^2 \div x^5 = x^{\square - \square} = x^{\square}$

$x^2 \div x^5 = \frac{x \times x}{x \times x \times x \times x \times x} = \frac{\square}{\square}$

Therefore $x^{\square} = \frac{\square}{\square}$

10 Simplify

a a^{-2} b m^{-4} c c^{-1} d r^0

11 **Exam-style question**

a Simplify $12x^3y^{-2} \div 3x^2y^4$ **(3 marks)**

b $a^b \times a^3 = a$

Work out b. **(1 mark)**

12 Simplify

a $(y^{-3})^{-2}$ b $(a^{-4})^{-3}$ c $(m^1)^{-1}$ d $(w^{-4})^0$

13 Simplify

a $(a^3b^5)^2$ b $(x^3y^3)^{-3}$

c $(3mn^{-2})^{-1}$ d $\left(\frac{5b^6}{2a^3}\right)^{-2}$

14 Simplify

a $\sqrt{y^{10}}$ b $\sqrt{4x^4}$ c $\sqrt{9a^6}$ d $\sqrt{16x^8y^2}$

15 **R** Copy and complete

a $a^{\frac{1}{4}} \times a^{\frac{1}{4}} \times a^{\frac{1}{4}} \times a^{\frac{1}{4}} = a^{\square} = \square$

$\sqrt[4]{a} \times \sqrt[4]{a} \times \sqrt[4]{a} \times \sqrt[4]{a} = \square$

Therefore $a^{\frac{1}{4}} = \square$

b $a^{\frac{1}{5}} \times a^{\frac{1}{5}} \times a^{\frac{1}{5}} \times a^{\frac{1}{5}} \times a^{\frac{1}{5}} = a^{\square} = \square$

$\sqrt[5]{a} \times \sqrt[5]{a} \times \sqrt[5]{a} \times \sqrt[5]{a} \times \sqrt[5]{a} = \square$

Therefore $a^{\frac{1}{5}} = \square$

16 Simplify

a $(2x^2q^3)^{-2}$ b $(16b^8)^{\frac{1}{2}}$

c $(27x^3y^9)^{-\frac{1}{3}}$ d $(16x^{12}y^{16})^{\frac{1}{4}}$

2.2 Expanding and factorising

1 **R** a Write down an expression containing brackets for the area of the rectangle.

$x + 3$

7

b Copy and complete this diagram to show the areas of the two small rectangles.

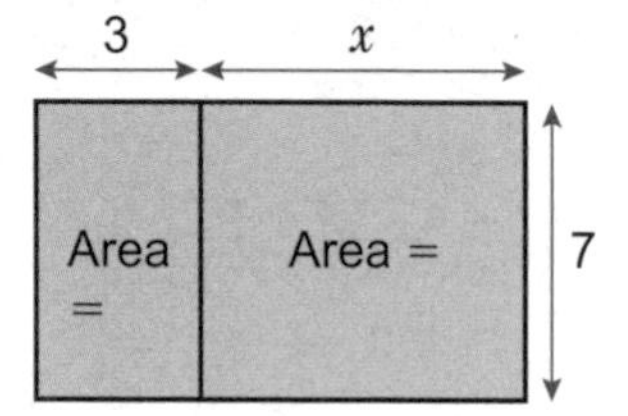

c What do you notice about your answers to parts **a** and **b**?

2 **R** State whether each relation is an equation or an identity.
Rewrite the identities using ≡.

a $x + 1 = 1 + x$ b $xy = yx$

c $4x + 1 = 5x$ d $\frac{2x}{4} = 0.5x$

3 **R** By drawing rectangles show that

a $2x(x + 1) = 2x^2 + 2x$

b $a(2b + c) = 2ab + ac$

c $2x(5 + 3x) = 10x + 6x^2$

4 a Expand

i $2a(b + 3)$ ii $3b(10 + 3a)$

b Use your answers to part **a** to expand and simplify

$2a(b + 3) + 3b(10 + 3a)$

5 Expand and simplify

a $3(x + 1) + 4x$ b $5y + 3(y + 2)$

c $9(2x + 1) + 5x$ d $2(x + 5) + 4(x + 7)$

e $4b + 3(b + 5) + 2b$ f $4(5x + 2) + 5(2 + 5x)$

6 Expand and simplify

a $2(x - 4)$ b $3(x - 5) + 8x$

c $x(2x - 4)$ d $2x(x + 1) - x(x + 3)$

e $5x + 2x(10 - x)$ f $5x(2 + 5x) - 2(5x + 1)$

7 Find the HCF of

a $2a$ and $4ab$ b $7ab$ and $8b$

c $6ab$ and $3a$ d $15a^2b^2$ and $5a$

8 Factorise completely

a $4x + 6$ b $2xy + 8y$

c $7xy + 2x$ d $3ab + 6ac$

e $xyz + 2xy$ f $a^2 + 3a^3$

g $10ab^2 + 5ac$ h $4x^2y^2 + 6xz^2$

i $12n^3k - 8n^2k^2$ j $15tm^2 - 12mq^3$

9 a What is the HCF of $2(a + 3b)^3$ and $12(a + 3b)$?

b Copy and complete

$2(a + 3b)^2 - 12(a + 3b)$

$= \square(a + 3b)[(a + 3b) - \square]$

$= \square(a + 3b)(a + 3b - \square)$

10 Factorise completely

a $12(a + 3)^2 + 4(a + 3)$

b $6(m - 1)^2 - 12(m - 1)$

c $2(y + 7)^2 - 4(y + 7)$

d $(x + 2y)^2 - 3(x + 2y)$

e $7(a - 2) + 14a(a - 2)$

f $3(x - y)^2 - 9(x - y)$

11 **R** Show algebraically that the product of five consecutive numbers must be a multiple of 120.

Example

12 **Exam-style question**

a Expand $2a(10b - 3a)$ **(1 mark)**

b Factorise completely $12x^2y + 18xy^2$ **(1 mark)**

c Simplify $\sqrt{16x^4y^2}$ **(2 marks)**

2.3 Equations

1 a Copy and complete to begin to solve the equation.

$4x - 3 = 2x + 7$

$4x - 3 - \square = 2x + 7 - \square$

$\square x - 3 = \square$

b Solve the equation.

2 Solve

a $4x + 2 = 5x$ b $9x - 5 = 7x + 1$

c $x - 4 = 2x - 9$ d $10x + 3 = 12x - 5$

3 a Expand

i $4(x - 2)$ ii $2(3x - 2)$

b Use your answers to part **a** to solve $4(x - 2) = 2(3x - 2)$

4 a Expand and simplify $7(x - 2) - 3(x - 3)$

b Use your answer to part **a** to solve $7(x - 2) - 3(x - 3) = 11$

5 Solve these equations.

a $3(2x - 1) + 5(x + 2) = 40$

b $3(x + 1) - (2x - 5) = 12$

6 Solve

a $3(2x + 1) + 2(x - 7) = 20$

b $9(2x - 4) = 3(4x + 1)$

c $2(3x - 4) = 5(10 - 2x)$

d $10x - 3(x + 7) = 12$

e $9(2x - 1) - 2(3x + 4) = 15$

f $6(10 - 2x) = 5(2 - 3x)$

7 Simplify these expressions by cancelling.

a $\frac{12x}{6}$ b $\frac{7y}{21}$ c $\frac{3a}{12}$ d $\frac{10m}{2}$

8 a Copy and complete to begin to solve the equation.

$$\frac{3x-5}{2} = 8$$

$$\frac{3x-5}{2} \times \square = 8 \times \square$$

$$3x - 5 = \square$$

b Solve the equation.

9 a Copy and complete to begin to solve the equation.

$$\frac{5}{x+1} = 2$$

$$\frac{5}{x+1} \times (\square) = 2 \times (\square)$$

$$5 = \square x + \square$$

b Solve the equation.

10 a By multiplying both sides of the equation $\frac{3x+1}{2} = \frac{x-1}{4}$ by 4, and cancelling, show that $2(3x + 1) = x - 1$.
Then solve the equation.

b By multiplying both sides of the equation $\frac{x}{2} - \frac{x}{3} = \frac{5}{6}$ by 6, and cancelling, show that $3x - 2x = 5$.
Then solve the equation.

11 Solve these equations.

a $\frac{x+2}{3} = \frac{3x-2}{2}$ b $\frac{a}{3} - \frac{a}{6} = 5$

c $\frac{y+3}{2} + \frac{y+1}{3} = \frac{1}{6}$ d $\frac{3}{2x-1} = 5$

e $\frac{x+1}{3} + \frac{x-1}{5} + \frac{x}{15} = 2$

12 **P** Find the size of the smallest angle in the kite.

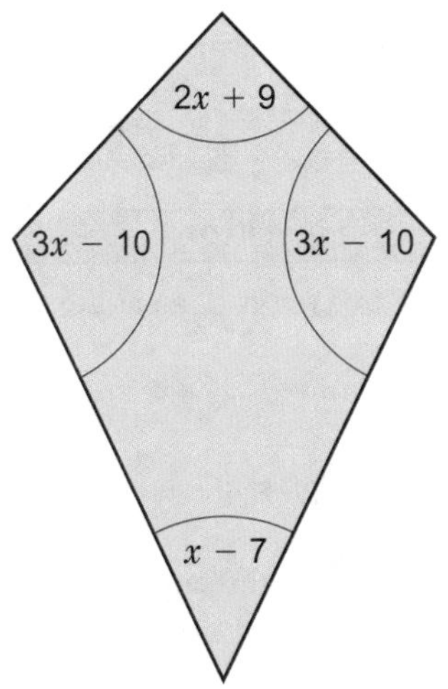

13 **R** A scientist mixes the *same mass* of two fluids with different densities.
The resulting mixture has a density of $9\,\text{g/cm}^3$.
He uses $40\,\text{cm}^3$ of the first fluid and $5\,\text{cm}^3$ of the second fluid.

a Assume the mass of each fluid is x kg. Write down an expression for the density of each fluid.

b Write down an expression for the density of the mixture.

c Form and solve an equation to find the value of x.

Q13a hint Density = $\frac{\text{mass}}{\text{volume}}$

2.4 Formulae

1 Write whether each of these is an expression, an equation, an identity or a formula.

a $y = 2x + 5$ b $2x - 3 = 4x + 1$

c $2x + 2y = 2(x + y)$ d $3xyz$

e $A = \pi r^2$ f $s = u + at$

g m^3 h $3(x + 1) - 4 = 3x - 1$

i $V = \pi r^2 h$ j $s = \frac{d}{t}$

k $(ab^3)^2 = a^2b^6$ l $\frac{2x-3}{5} = 7x$

2 Use the formula $y = x^2 - x^3$ to work out the value of y when

a $x = 3$ b $x = -2$

3 Use the formula $c = (a + 1)^3 - b$ to work out the value of c when

a $a = -1$ and $b = 75$ b $a = -4$ and $b = -50$

4 **R** The cost of posting a package is a fixed cost of £5 plus £2 per kg over 10 kg.

a Work out the total cost for a 14 kg package.

b Write a formula for the total cost (C) for a package of weight p kg.

5 **P** a Write a formula, in terms of a and b, for the perimeter, P, of the rectangle.

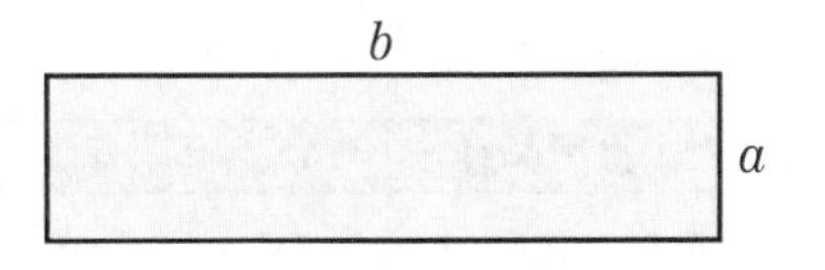

b Use the formula to work out the value of

i P when $a = 4$ and $b = 7$

ii a when $P = 30$ and $h = 7$

6 A credit card company charges e% monthly interest on all debts.
After n months the debt (D) for an initial loan of £R is given by the formula

$$D = R\left(\frac{100 + e}{100}\right)^n$$

Mr Bloomer borrows £500 at a monthly interest rate of 12%.
How much will he owe after 6 months?
Give your answer to the nearest penny.

7 **P** The final velocity of an object can be calculated using the formula

$$v^2 = u^2 + 2as$$

where v = final velocity, u = initial velocity, a = acceleration and s = distance.
A ball is released from the top of the Empire State Building, at a height of 380 m.
It accelerates at 9.8 m/s².
Work out the final velocity of the ball when it hits the ground if

a the ball is dropped (assume that the initial velocity is 0 m/s)

b the ball is thrown vertically down with an initial velocity of 15 m/s.

8 Change the subject of each formula to the letter given in the brackets.

Example

a $y = mx + c$ [x]

b $E = mc^2$ [m]

c $\frac{2b}{a} = c$ [a]

d $C = 2\pi r$ [r]

e $A = \frac{1}{2}bh$ [b]

f $v^2 = u^2 + 2as$ [s]

9 The formula $K = \frac{8M}{5}$ can be used to convert between miles (M) and kilometres (K).

a Convert 38 miles into kilometres.

b Make M the subject of the formula.

c Convert 108 km into miles.

10 The formula for the surface area of a sphere is $A = 4\pi r^2$

a Make r the subject of the formula.

b Work out the radius of a circle with surface area 100 cm². Give your answer to 1 d.p.

11 **Exam-style question**

a Make t the subject of the formula
$2s = 4(3rt + 2)$ **(2 marks)**

b Find t when $r = 2$ and $s = 52$. **(3 marks)**

12 **R** The formula for converting between temperatures in Fahrenheit (F) and Celsius (C) is

$$F = \frac{9C}{5} + 32$$

It is recommended that paint is stored between 59 °F and 80 °F.
Convert this range of temperatures to degrees Celsius.

2.5 Linear sequences

1 Work out the 1st, 2nd, 3rd, 10th and 100th terms of the sequence with nth term

a $u_n = 2x - 5$ b $u_n = 24 - 3n$ c $u_n = -5$

2 For each arithmetic sequence, work out the common difference and hence find the 3rd term.

a 0.21, 0.23, …

b $\frac{1}{3}$, 1, …

c 12, −1, …

d 0.101, 0.202, …

3 **R** Write down the common difference for each sequence.

Example

a nth term $4n - 5$

b $u_n = -2n - 3$

c nth term $10 - 0.5n$

d $u_n = 0.5 - 0.2n$

4 Write down, in terms of n, expressions for the nth term of these arithmetic sequences.

a 2, 5, 8, 11, …

b 19, 25, 31, 37, …

c 18, 15, 12, 9, …

d 12, 24, 36, …

e 9, 9.5, 10, 10.5, …

5 **R** Which of the values in the cloud are terms in the sequence $u_n = 3n + 2$?

6 **Exam-style question**

The first three terms of an arithmetic sequence are

$\frac{2}{3}$, $1\frac{1}{3}$, 2

a Write an expression in terms of n for the nth term. **(2 marks)**

b Is 15 a term in the sequence? Explain your answer. **(2 marks)**

7 **R** The nth term of the sequence 2, 15, 28, 41, … is $13n - 11$.

a Solve $13n - 11 = 200$

b Use your answer to part **a** to find how many terms in the sequence are smaller than 200.

8 **R** a Find the first term in the arithmetic sequence 24, 31, 38, 45, … which is greater than 300.

b Find the first term in the arithmetic sequence 400, 330, 260, … which is less than zero.

9 A baby girl gains 0.5 lb in weight per week.

a If the baby weighs 8.2 lb at age 2 weeks, how much does she weigh when she is age

i 3 weeks

ii 4 weeks

iii 5 weeks?

b How long will it be before she weighs more than 2 stone?

Q9b hint 1 stone = 14 lb

10 Jen opens a new email account.
In the first week she receives 12 junk emails.
Each week after that she receives 5 more junk emails than the previous week.
How long before she receives over 200 junk emails a week?

11 **R** The nth term of an arithmetic sequence is $u_n = 3n - 2$

a Write down the values of the first four terms, u_1, u_2, u_3, u_4.

b Write down the value of the common difference, d.

c By substituting $n = 0$, work out the value of the zero term, u_0.

12 a Find the outputs when the terms in each of these arithmetic sequences are used as inputs to the function machine.

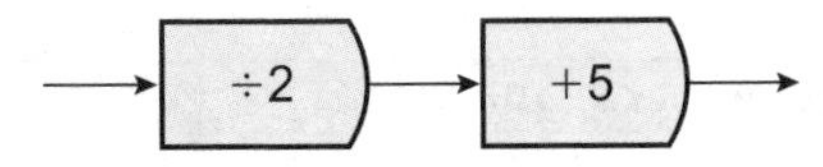

i 1, 5, 9, 13, 17, …

ii 5, 10, 15, 20, 25, …

b Compare the common difference for each input sequence with the common difference for the output sequence. How are these related to the operations used in the function machine?

13 **R** When 5 is input into this function machine, the output is 23.
When 1 is input into the function machine, the output is 7.

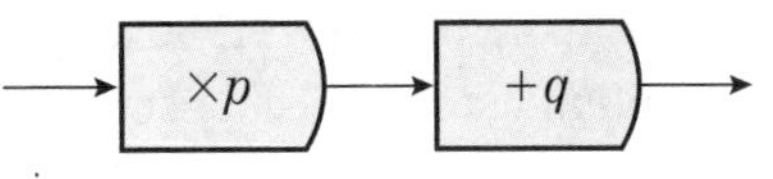

a Work out the difference between the two inputs.

b Work out the difference between the two outputs.

c Use your answers to parts **a** and **b** to find the value of p in the function machine.

d Work out the value of q.

14 **R** Find the values of p and q in this function machine when the inputs −1 and 12 produce outputs of −10 and 42 respectively.

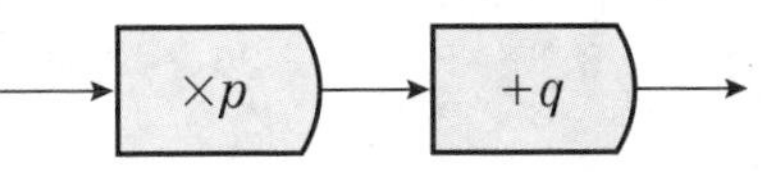

2.6 Non-linear sequences

1 Find the next three terms in each of these Fibonacci-like sequences.

a 1, 7, □, □, □, …

b −1, 1, □, □, □, …

c 2, 4, □, □, □, …

2 Write down the first four terms of each sequence.

a $u_n = \frac{n}{n^2}$ b $u_n = 1^n$ c $u_n = 0.1^n$

3 Write down the first five terms of these geometric sequences.

a first term = $2\sqrt{2}$;
term-to-term rule is 'multiply by $\sqrt{2}$'

b first term = $\sqrt{3}$;
term-to-term rule is 'multiply by $\frac{1}{\sqrt{3}}$'

4 **P** A paperboy is offered a payment of 1p for the first day, 2p for the second, 4p for the third and so on.
How long before he is paid over £100 for the paper round?

5 The surface area of bacteria in a Petri dish increases by 10% each day.

a If the area of bacteria is 2 cm² on day 1, what will the area be

i on day 2 ii on day 3?

b How long before the area is over 10 cm²?

6 A job has a starting salary of £22 000.
The company offers two different pay options.
Option 1: An increase of 5% of the current salary each year.
Option 2: An increase of £1500 per year.
Which option would you choose? Explain why.

7 **R** **a** Write down the first six terms of the sequence $u_n = n^2$.

b Work out a formula for the nth term of each sequence.

i 3, 6, 11, 18, 27, …

ii −1, 2, 7, 14, 23, 34, …

iii 2, 8, 18, 31, 50, 72, …

8 Copy and complete this diagram to work out the next term in the sequence 3, 12, 25, 42, …

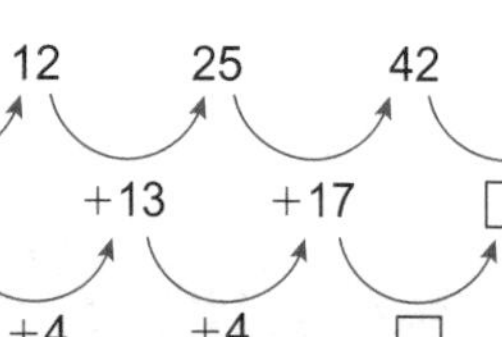

sequence 3 12 25 42 □

1st differences +9 +13 +17 □

2nd differences +4 +4 □

9 Work out the next term of each sequence.

a 2, 3, 6, 11, 18, …

b 5, 10, 20, 35, 55, …

c 0, 2, 6, 12, 20, …

10 **a** Copy and complete to work out the first and second differences for the sequence $u_n = n^2 - 1$

sequence 0 3 8 15 24

1st differences +3 +5 +7 □

2nd differences +2 □ □

b Copy and complete for the sequence $v_n = 2n^2 + n - 5$

sequence −2 5 16 31 50

1st differences +7 +11 □ □

2nd differences +4 □ □

11 **R** Find a formula for the nth term of each of these quadratic sequences.

a 5, 14, 29, 50, 77, …

b 1, 7, 17, 31, 49, 71, …

c 12, 27, 52, 87, 132, …

Example

12 **R** The triangular numbers can be found by adding increasing integers.

Term	Triangular number
1	1
2	1 + 2 = 3
3	3 + 3 = 6
4	6 + 4 = 10
5	10 + 5 = 15

a Copy and complete the table for the triangular numbers.

Term	1	2	3	4	5	6	7	8	9
Triangular number	1	3							

b Work out a formula for the nth triangular number.

13 The sequence 4, 13, 26, 43, 64 … has nth term in the form $u_n = an^2 + bn + c$

a Find the second differences and show that $a = 2$.

b Subtract the sequence $2n^2$ from the given sequence.

	4	13	26	43	64
−	2	8	18	32	50
	2	□	□	□	14

c Find the nth term of this linear sequence.

d Write the nth term of 4, 13, 26, 43, 64 …
$2n^2 + \square n - \square$

14 Find the nth term of each sequence.

a 1, 6, 13, 22, 33, …

b −2, 5, 16, 31, 50, …

c 7, 14, 27, 46, 71, …

d −1.5, 1, 4.5, 9, 14.5, …

Q14 hint Use the method in **Q13**.

15 The nth term of a sequence is $u_n = 2^n$.
Show that the product of the 5th, 6th and 7th terms is 2^{18}.

16 **Exam-style question**

a Write down the first four terms in the sequence with nth term $u_n = \frac{1}{2^n}$ **(2 marks)**

b State the term-to-term rule. **(1 mark)**

c Use algebra to show that the product of any two terms in the sequence is also a term in the sequence. **(2 marks)**

2.7 More expanding and factorising

1 Expand and simplify

Example

a $(x + 3)(x + 4)$
b $(x + 3)(x - 6)$
c $(x - 2)(x + 4)$
d $(x - 2)(x - 6)$

2 **P** Find the missing terms in these quadratic expansions.

a $(x + \square)(x + 5) = x^2 + \square x + 15$
b $(x - \square)(x - 3) = x^2 - 8x + \square$

3 Expand and simplify

a $(x + 1)^2$ b $(x - 1)^2$
c $(x + 7)^2$ d $(x - 5)^2$

4 a Copy and complete to evaluate $35^2 - 15^2$ without a calculator.

$(35 - 15)(35 + 15) = 20 \times \square = \square$

b Without using a calculator work out

i $55^2 - 35^2$
ii $1.5^2 - 0.5^2$

5 Expand and simplify

a $(x + 1)(x - 1)$ b $(x + 3)(x - 3)$

6 Factorise

a $x^2 - 16$ b $y^2 - 121$ c $t^2 - 64$

7 Factorise

Example

a $x^2 + 5x + 6$ b $x^2 + 5x + 4$
c $x^2 + 6x + 5$ d $x^2 - x - 6$
e $x^2 + 2x - 8$ f $x^2 - 2x - 3$
g $x^2 + x - 20$ h $x^2 + 9x + 14$
i $x^2 - 3x - 18$ j $x^2 + 2x - 15$
k $x^2 + 7x - 30$ l $x^2 + 20x + 100$

8 **P** Charlie and James are given a mystery number.

Charlie squares the number.

a Write down an algebraic expression to represent Charlie's number.

James adds 2 to the number and then squares then result.

b Write down an algebraic expression to represent James's number.

James's number is 32 larger than Charlie's.

c What was the original number?

Q8a hint Let x represent the mystery number.

9 **P / R** The rectangle and triangle shown have the same area. Find x.

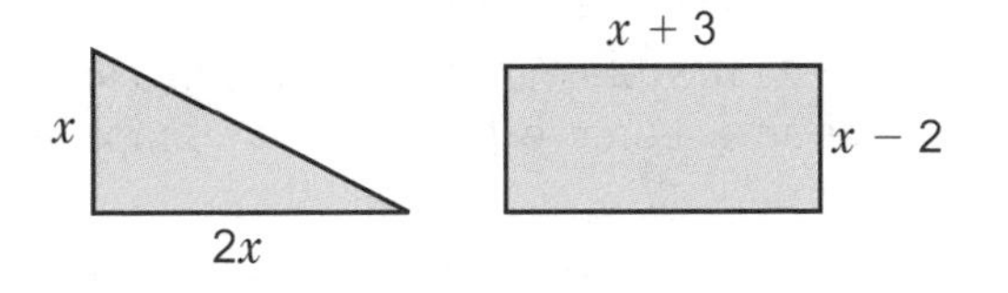

10 Copy and complete these factorisations.

a $9x^2 - 100 = (3x)^2 - \square^2$
$= (3x - \square)(3x + \square)$

b $4y^2 - 25 = (\square y)^2 - \square^2$
$= (\square y - \square)(\square y + \square)$

11 Factorise

a $100m^2 - 1$ b $16c^2 - 36$ c $x^2 - y^2$

12 **Exam-style question**

a Factorise $b^2 - 5b + 6$ **(2 marks)**
b Expand $(5r - 2s)^2$ **(2 marks)**

2 Problem-solving

Solve problems using these strategies where appropriate:

- **Use pictures or lists**
- **Use smaller numbers.**

Example

1 **R** Tom finds the original plan of his house and sees that a rectangular extension has been built on the side. The rectangular area of the ground floor on the original plan is 10 m long × 8 m wide. The extension is the full width of the house but Tom does not know the length.

a What is the area of the whole ground floor if the unknown length is x?
Expand your answer.

Tom wants to widen his house by 3 m.

b What would be the new total area of the ground floor?
Expand and simplify your answer.

After his extension is built, Tom decides to lay flooring on the whole of the ground floor but not in the kitchen.
He knows that the kitchen is 4 m wide but he does not know the length y.

c What is the total area of floor to be covered?

Q1 hint Draw a picture.

2 There are x cakes in a packet, x packets in a box, x boxes in a case, y cases in a single load and six loads to fill a truck.

a Write the total number of cakes in a load as an algebraic expression in its simplest form.

Three trucks deliver equal numbers of cases to x^2 stores.

b How many cakes are delivered to each store? Write the total as a simplified algebraic expression.

3 The surface area s of a cuboid box is $6200\,\text{cm}^2$.
The length is 30 cm and the height is 20 cm.
What is the width?
Show your working.

4 **Exam-style question**

Here are the first five terms of an arithmetic sequence:
5, 8, 11, 14, 17

a What is the next number in this sequence? **(1 mark)**

b Find an expression, in terms of n, for the nth term of this sequence. **(1 mark)**

c Johann says that 52 is in this sequence.
Is Johann correct?
Explain your answer. **(2 marks)**

5 The term-to-term rule of a sequence is 'multiply by 3'. The second term is 24.
Write the first five terms of the sequence.

6 Find the missing terms in these quadratic expressions.

a $(x + 5)(x + \square) = x^2 + \square x + 20$

b $(x + 8)(x - \square) = x^2 + \square x - 24$

c $(x - \square)(x - 4) = x^2 + \square x + 8$

7 **R** Oliver is designing a rug with a Fibonacci-like sequence. He starts with two squares of side length 2 cm and then adds a square of side length 4 cm along the side of the first two squares.

a What are the sizes of the next three squares of the rug?

b Oliver wants to make the rug with 10 squares.
What is the size of the tenth square?

c Oliver has arranged the 10 squares in a spiral.
What is the size of the whole rug?

8 **R** Joshua and Charlie both expand and simplify $(x + 2)(2x - 3)$.
Joshua says that the answer is $2x^2 + x - 6$
Charlie says that the answer is $2x^2 + 4x + 6$
Who is correct?
Give reasons for your answer.

9 Look at these simultaneous equations:
$2x + 2y = 8$ and $x - 2y = 1$.
Solve to find the values for x and y.

10 Suki measured the distances a group of animals travelled in one hour to find out their average speed. She measured the distances in millimetres.

Animal	Distance (mm)
snail	540
worm	12
slug	22
ladybird	980
spider	110

a Convert each of the distances into kilometres.
1 millimetre = 0.000 001 kilometres

b Write each of your answers to part **a** in standard form.

3 INTERPRETING AND REPRESENTING DATA

3.1 Statistical diagrams 1

1 The pie charts show the sizes of jeans sold in a shop on a Friday and Saturday.
60 pairs of jeans were sold on Friday and 120 pairs on Saturday.

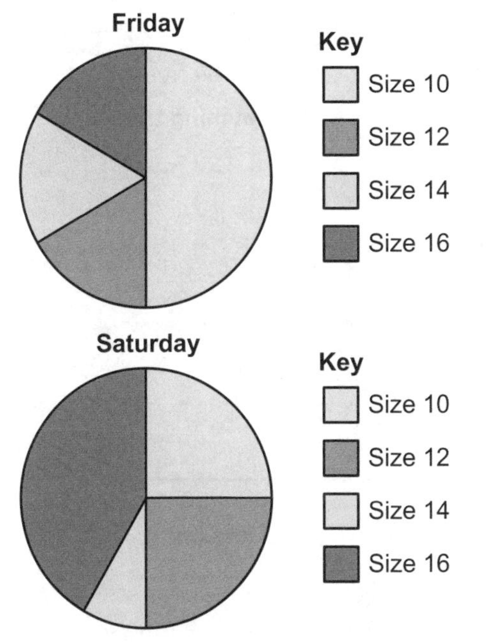

a How many pairs of size 10 jeans were sold on Saturday?
b Which pie chart has the larger sector for size 14 jeans?
c Show that the same number of pairs of size 14 jeans were sold on both days.

2 **R** The stem and leaf diagram shows the ages of customers in a restaurant.

Stem	Leaf
1	2 5 7
2	4 9
3	8 8 9 9
4	0 2

Key
2 | 4 means 24 years

a How many people are in the restaurant?
b What is the age of the oldest person in the restaurant?
c What is the range?
d What is the median?
e Calculate the mean age of the people in the restaurant.

> **Q2d hint** The median is the $\frac{n+1}{2}$th value, where n is the total number of values.

3 **P** The ages of participants in two different exercise classes were recorded.
The back-to-back stem and leaf diagram shows the results.
Compare the distribution of ages in the two classes.

Yoga		Body pump
	1	8 9
8 8 7	2	3 4 4 6 6 9
8 3 2 2 0	3	2 4 4 6
8 7 5 0 0	4	1 9
5 1	5	0

Key Yoga 7 | 2 represents 27 years   Body pump 2 | 3 represents 23 years

4 **P** The masses (in g, to the nearest gram) of chicks were measured at two different farms.
Farm A: 23, 24, 24, 24, 25, 32, 33, 33, 34, 34, 35, 36, 37, 39, 40, 41, 41, 42, 42, 42
Farm B: 29, 36, 36, 37, 38, 40, 44, 44, 45, 46, 46, 46, 49, 51, 52, 54, 54, 54, 55, 56

a Draw a back-to-back stem and leaf diagram for this data.
b Use the shape of your diagram to compare the distribution of the masses of chicks at the two different farms.

5 **Exam-style question**

The numbers of digital songs downloaded by 20 customers in a month were recorded.

0	9
1	2 3 3 5
2	0 6 8 8 8 9
3	1 2 3 3 4
4	0 5 5
5	3

Key
1 | 2 represents 12 downloads

a Write down the number of customers who downloaded 45 songs.
b Work out the range.
c What is the modal number of downloads?
d Work out the median. **(4 marks)**

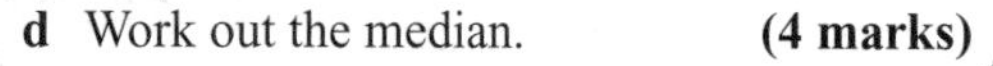

6 The table shows the prices, P, of 190 houses.

Price, P (£000s)	Frequency
$0 < P \leqslant 100$	0
$100 < P \leqslant 200$	12
$200 < P \leqslant 300$	19
$300 < P \leqslant 400$	27
$400 < P \leqslant 500$	32
$500 < P \leqslant 600$	89
$600 < P \leqslant 700$	11

a Copy and complete the frequency diagram.

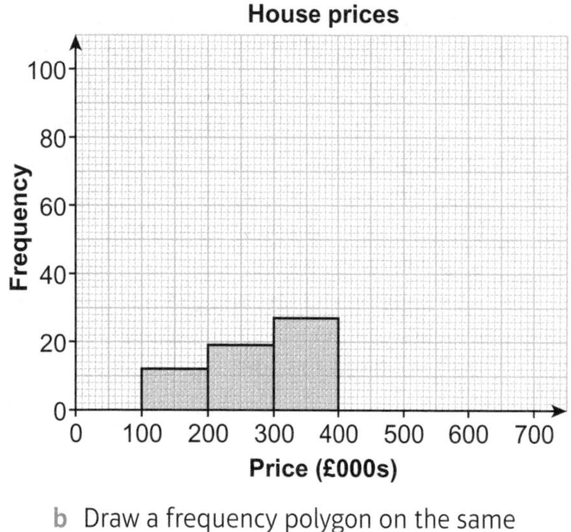

b Draw a frequency polygon on the same diagram.

7 The frequency table shows the lengths of time, t (in minutes), some people spend watching TV each day.

Time, t (min)	Frequency
$0 < t \leqslant 30$	35
$30 < t \leqslant 60$	51
$60 < t \leqslant 90$	78
$90 < t \leqslant 120$	28
$120 < t \leqslant 150$	8

a How many people were surveyed?
b What percentage watched TV for more than 1 hour?
c Estimate the range.
d Copy and complete the frequency polygon.

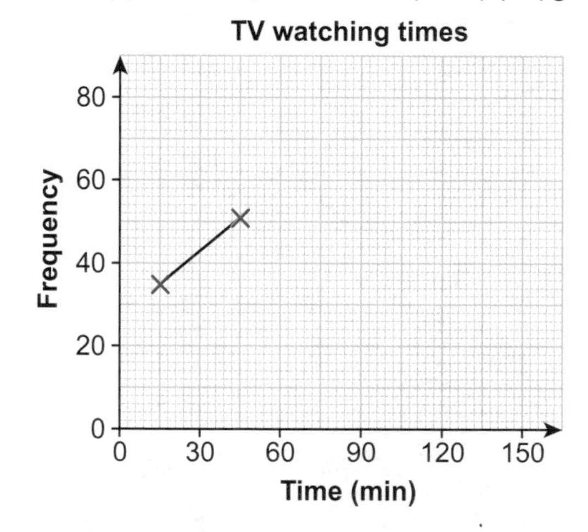

8 **R** The heights of 30 seedlings at two different nurseries were measured.
The results are displayed on the frequency polygons.

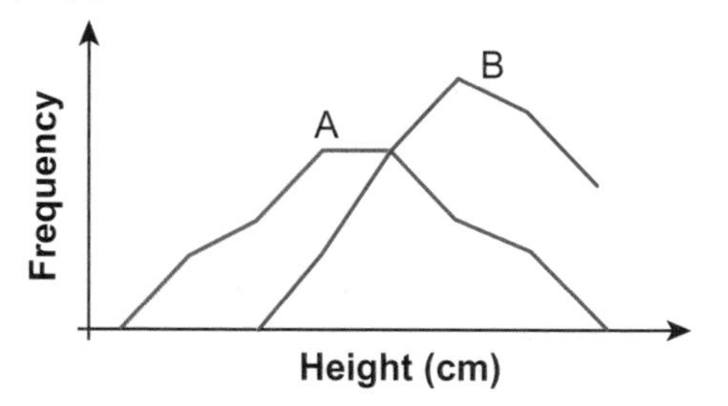

a Which data set has the greater range?
b Would you expect the median of data set A to be greater than, less than or about the same as the median of data set B?
c Which data set do you think shows seedlings that were planted earlier?

3.2 Time series

1 **R** A leisure centre recorded the number of swimmers in the pool at 2-hour intervals during a 12-hour period.

Time	0900	1100	1300	1500	1700	1900	2100
Number of swimmers	35	9	12	2	35	41	28

a How many swimmers were in the pool at 1300?

b What is the smallest number of swimmers in the pool at any one time?

c Work out the average number of swimmers in the pool. Give your answer to the nearest whole number.

d Represent this time series on a line graph. Comment on the variation in the number of swimmers.

2 Mr Jayshuk catches the train to work. He records the number of minutes the train is late over a 2-week period.

Day	1	2	3	4	5
Number of minutes late	2	5	3	8	2

Day	6	7	8	9	10
Number of minutes late	4	5	7	8	6

Draw a time series graph for this data. Comment on how the lateness of the train varies over the 2-week period.

3 **R** The time series graph shows the viewing figures for two TV programmes, A and B.

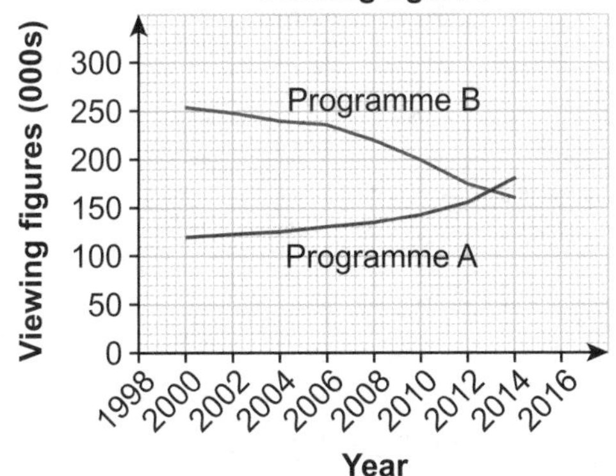

a What were the viewing figures for Programme A in 2004?

b The director of Programme A says that the viewing figures for Programme A have risen more than the figures for programme B have fallen. Is she correct?

c She also thinks that the number of viewers of programme A is rising at an increasing rate. Is she correct?

d Predict the viewing figures for both programmes in 2016.

4 **P** The tables show the numbers of hits (in thousands) on a tourist information website during each quarter over the last 3 years.

2012

Q1	Q2	Q3	Q4
12	54	35	48

2013

Q1	Q2	Q3	Q4
9	57	40	46

2014

Q1	Q2	Q3	Q4
12	61	40	49

a How many hits were there in the first quarter of 2014?

b In which quarter were there the most hits?

c Draw a time series graph for this data.

d Describe the variation in numbers of hits during the 3-year period and comment on the overall trend.

Q4a hint The numbers of hits are in thousands.

5 **P** The tables show the population (in millions) of the UK over a 10-year period.

Year	2003	2004	2005	2006	2007
Population (in millions)	59.4	59.7	60.1	60.4	?

Year	2008	2009	2010	2011	2012
Population (in millions)	61.2	61.6	62	63.3	63.7

The data for 2007 has been lost.

a Draw a line graph for this time series.

b Describe the overall trend.

c Estimate what the population might have been in 2007.

d Use your graph to predict the population in 2013.

6 **Exam-style question**

In a grandfather clock a pendulum swings from side to side.

The pendulum starts in a vertical position and begins to swing.

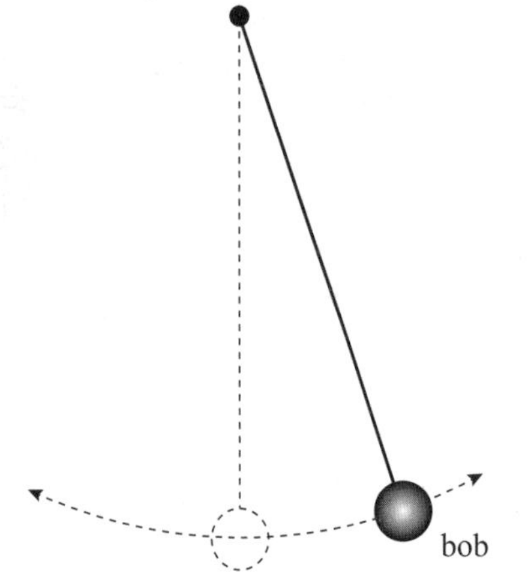

The tables show the horizontal distance of the bob of the pendulum from the vertical during 1 second.

Time (s)	0	0.1	0.2	0.3	0.4	0.5
Distance (cm)	0	1	1.8	2.4	2.8	3

Time (s)	0.6	0.7	0.8	0.9	1.0
Distance (cm)	2.8	2.4	1.8	1	0

a Draw a time series graph to show the data.

b At what time will the bob be 3 cm from the vertical again?

c Predict the distance of the bob from the vertical 2.5 seconds after it starts to swing.

d For how long is the bob more than 1.8 cm from the vertical during 1 second? **(7 marks)**

3.3 Scatter graphs

1 A hospital recorded the heights and weights of 8 patients.
The measurements are shown in the table.

Patient	A	B	C	D	E	F	G	H
Height (cm)	160	145	171	165	167	162	155	149
Weight (kg)	60	53	66	62	65	62	59	57

a Copy and complete the scatter graph.
Patient A was 160 cm tall and weighed 60 kg so draw a cross at (160, 60).
For patient B, draw a cross at (145, 53).
Complete the scatter graph with crosses for all the patients.

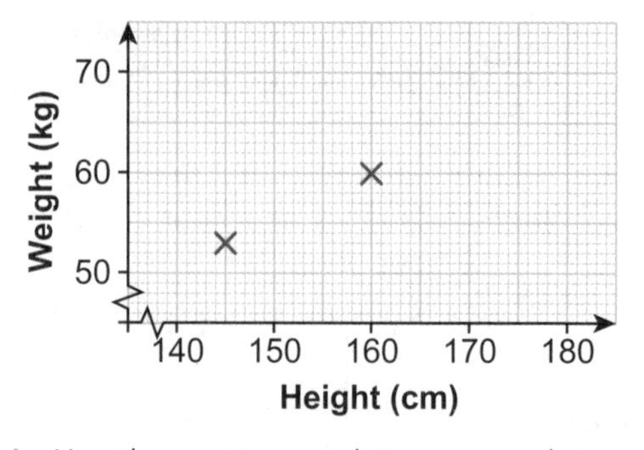

b Use the scatter graph to copy and complete the sentence.
In general, patients who are taller weigh .... and patients who are shorter weigh ....

2 The height and age of a number of trees are recorded, together with the average annual rainfall in the area each tree is growing.
These scatter diagrams are plotted from the data.

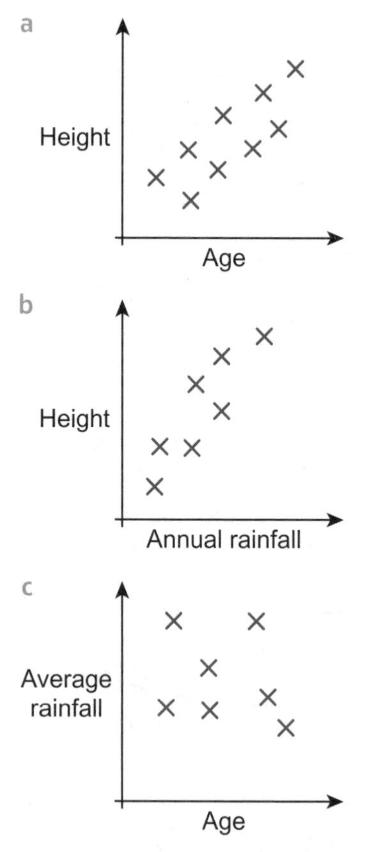

For each graph state whether there is positive, negative or no correlation and describe in words what this means.

Q2a hint To describe what the correlation means in words, you could say, 'As age increases, height ...'.

3 A farmer records the ages of some chickens in years and the number of eggs they each produce in a week.

Age (years)	4	6	1	2	8	5	7	3
Number of eggs	6	4	6	6	1	4	3	5

a Draw a scatter graph for this data.

b Describe any relationship between these two variables.

4 A newsagent records how many of a particular type of chocolate bar are sold per day as she gradually reduces the price. The table shows her results.

Price (p)	45	44	43	42	41	40	39	38
Number sold	12	13	15	14	8	12	4	12

Draw a scatter graph and describe any relationship between the price of the chocolate bars and the number of bars sold. Suggest a possible reason for this relationship.

5 **R** The manager of a coffee shop records the daily temperature and the number of hot chocolates sold in one week.

Day	Mon	Tues	Wed	Thurs	Fri	Sat
Temperature (°C)	9	8	5	4	7	6
Number of hot chocolates sold	5	6	5	7	9	15

a Draw a scatter graph and comment on any relationship between the two variables.

b The manager says that the colder it is the more hot chocolates are sold. Does the scatter graph provide statistical evidence to support the manager's view?

6 What sort of correlation would you expect to find between

a a student's mark in a maths test and in a science test

b the speed of a train and the time taken to travel 100 miles

c the number of magazines a student buys and their weight?

7 **R** In a physics experiment the weight attached to a spring (in newtons, N) and the length of the spring's extension are recorded.

Weight (N)	1	2	3	4	5	6	7	8
Length of spring (cm)	1.2	2.4	3.6	4.8	6.0	7.2	7.4	7.5

a Plot these points on a scatter graph.

b State the type of correlation between weight and length of spring extension for weights between 1 N and 6 N.

c Describe in words what happens to the spring when the weight exceeds 6 N.

8 **Exam-style question**

Some students sit two English tests.
Their results are shown on the scatter graph.

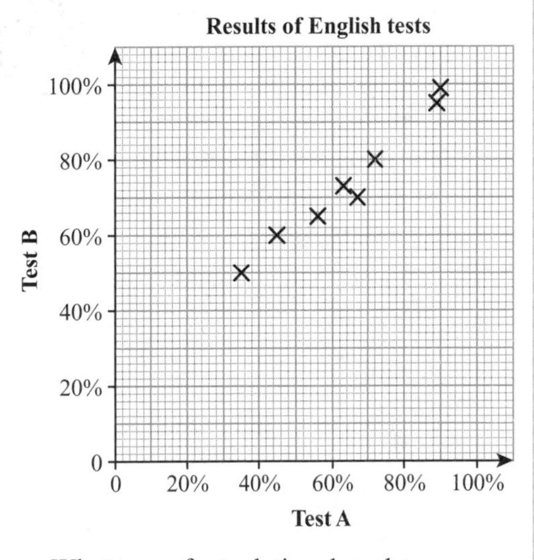

a What type of correlation does the scatter graph show? Interpret your answer. **(1 mark)**

b What was the highest mark in Test B? **(1 mark)**

c What was the range of marks for Test A? **(2 marks)**

3.4 Line of best fit

1 Which line, A, B or C, is the best line of best fit for the data points on the scatter graph?

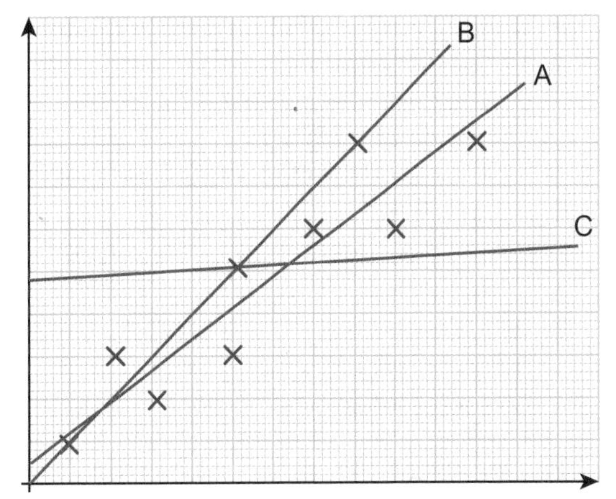

2 The table shows the percentage marks of 8 students in two maths tests.

Test A (%)	73	45	84	91	63	53	67	76
Test B (%)	64	39	74	80	55	50	60	66

a Draw a scatter graph for this data.

b Draw a line of best fit on your graph.

c Use your line of best fit to estimate the score in test B of a student who scores 80% in test A.

d Estimate the score in test A of a student who scores 50% in test B.

3 A lorry driver records the distance travelled and the number of litres of fuel used on each of 6 journeys.

Distance (miles)	105	124	78	125	102	91
Fuel (litres)	158	195	110	185	160	140

a Draw a line of best fit on a scatter graph and use the line to estimate the number of litres of fuel used for a 110-mile journey.

The fuel tank on the lorry holds 1500 litres.

b Estimate how far the lorry can go between refuellings.

4 **R** The table shows the outside temperature and the daily heating costs of a house.

Temperature (°C)	10	6	7	3	0	4
Heating cost (£)	12	16	14	20	24	18

a Draw a line of best fit on a scatter graph and use it to estimate

i the daily heating cost if the outside temperature is 8 °C

ii the outside temperature if the heating cost is £15

iii the daily heating cost if the outside temperature is 15 °C.

b Which of these estimates do you think is the least reliable?
Give a reason for your answer.

5 **R** Two surveys are carried out on crop growth.
The results are shown on the scatter graphs.

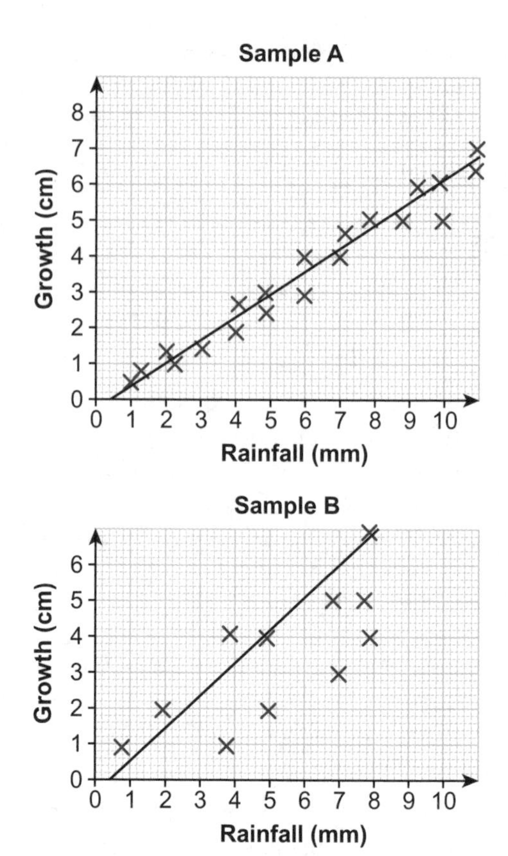

a Use the given lines of best fit to work out two estimates for the growth of the crops if the rainfall is 4.5 mm.

b Which of the estimates is likely to be more reliable? Give two reasons for your answer.

6 **R** The heights (to the nearest metre) and trunk circumferences of 6 trees are recorded. The table shows the height, h (in metres) and circumference, C (in cm) of the trees.

h (m)	2	3	4	5	6	7
C (cm)	30	82	93	101	112	120

a Draw a scatter graph for this data.

b Why is the first point classified as an outlier? Suggest a possible reason for this.

c Draw a line of best fit passing close to the remaining five points.

d Use the line to estimate the circumference of a tree which is 3.5 m in height.

e Estimate the height of a tree with circumference 125 cm.

7 The table shows the age, x (in years) and shoe size, y, of a sample of 9 boys.

Age, x	7	8	9	10	11	12	13	14	15
Shoe size, y	1	3	5	6	7	8	8.5	8.5	9

a Draw a scatter graph of this data.

b Assuming that shoe size can be modelled using a line of best fit
   i estimate the shoe size of a 16-year-old
   ii estimate the shoe size of an 11-year-old.

c Which of the answers in part **b** is likely to be the more reliable?

d By drawing a smooth curve close to the data points, make new estimates of the shoe sizes in part **b**.

e Which of the two models is the more accurate? Give a reason for your answer.

8 **Exam-style question**

The tables show the distance, d (in km), cars with different engine sizes, E, can travel on one litre of petrol.

Engine size, E (litres)	1	1.4	1.6	2
Distance, d (km)	16	14.4	13.5	12.2

Engine size, E (litres)	2	3	3.5	4
Distance, d (km)	11.8	9.4	8.4	7.1

a Plot the points on a scatter graph. **(2 marks)**

b Describe the relationship between engine size and the distance a car can travel on one litre of petrol. **(1 mark)**

Another car has an engine size of 2.5 litres.

c Estimate the distance this car can travel on one litre of petrol. **(2 marks)**

Exam hint
Always draw lines on your diagram for any readings from your graph. If you get the answer wrong, you may still get marks for using the correct method.

3.5 Averages and range

1 The annual numbers of burglaries reported in a town over the past 5 years are
45, 33, 47, 47, 93

a Work out the mean, median and mode of the number of burglaries.

b An insurance company bases how much it charges on the average number of burglaries.
Which of the averages would be the most appropriate? Give reasons for your answer.

2 **R** The sizes of jeans sold in a shop one day are
8, 8, 10, 10, 12, 12, 12, 12, 12, 12, 14, 14, 14, 16, 16, 18, 20, 20, 22, 22

a Work out the mean, median and mode of the sizes.

b The shop owner wants to order some more jeans but can only order one size.
Which size should he order?
Give reasons for your answer.

3 The numbers of passengers using a train service one week are recorded in the table.

Day	Number of passengers
Monday	230
Tuesday	180
Wednesday	170
Thursday	180
Friday	210

a Work out the mean, median and mode of the number of passengers.

b The train company wishes to work out the average daily profit.
Which average should be used to calculate an accurate figure?
Give reasons for your answer.

4 **R** State whether it is best to use the mean, median or mode for these data sets.
Give reasons for your answers.

a Colour of tablet case:
red, blue, green, orange, blue

b Number of customers in a shop:
12, 12, 13, 17, 19

5 Identify the outliers of the data sets and find the range of each.

a The heights of players in a netball team:
155 cm, 145 cm, 160 cm, 21 cm, 148 cm, 150 cm, 163 cm

b The monthly fuel bills of a home:
£130, £125, £143, £192, £135, £33

Q5 hint Think about whether you should include the outlier in your calculations.
Q5a hint Is it possible for someone in the team to be 21 cm tall?

6 Identify the outliers of the data sets. Calculate a sensible value of the range. Give a reason why the outlier has been included or excluded in your calculation.

a The average daily temperatures (°C) during a week in June in the UK:
19, 21, 14, 2, 18, 17, 16

b The heights of 10 sunflowers:
1.2 m, 1.8 m, 0.9 m, 1.12 m, 1.0 m, 0.2 m, 1.3 m, 1.45 m, 1.66 m, 1.44 m

7 The grouped frequency table shows the distance, D, students in a class travel to school.

Distance, D (miles)	Frequency, f	Midpoint, x	xf
$0 \leqslant D < 2$	12	1	$1 \times 12 = 12$
$2 \leqslant D < 4$	11		
$4 \leqslant D < 6$	7		
$6 \leqslant D < 8$	3		
$8 \leqslant D < 10$	2		
Total			

a Copy and complete the table to estimate the mean distance the students travel to school.

b Another student joins the class.
If they travel 9 miles to school, will the mean increase, decrease or stay the same?

8 **R** The times, T (in minutes), a group of students spent watching TV one night are shown in the table.

Time, T (min)	Frequency, f
$0 \leqslant T < 20$	2
$20 \leqslant T < 40$	19
$40 \leqslant T < 60$	14
$60 \leqslant T < 80$	2

a How many students spent less than 20 minutes watching TV?

b How many students altogether spent less than 40 minutes watching TV?

c How many students altogether spent less than 60 minutes watching TV?

d State the modal class.

e Explain why the median is the 19th value.

f Use your answers to parts **a** to **d** to work out which class interval contains the median.

9 **P** The table shows the times taken, t (in seconds), for two different groups of athletes to run a 100 m race.

Time, t (s)	Group A	Group B
$9.5 \leqslant t < 10$	3	6
$10 \leqslant t < 10.5$	5	9
$10.5 \leqslant t < 11$	10	3
$11 \leqslant t < 11.5$	2	2

a How many athletes were in each group?

b Explain why the median time for each group will be between the 10th and 11th values.

c In which class interval is Group A's median time?

d Work out which class interval contains Group B's median time.

e On average, which group has the shorter time?

f State the modal class for each group.

g At an athletics competition, both groups must compete against the current champions, Group C, who have a mean time of 10.4 s. Which group stands the better chance of beating them?
Explain your answer.

10 **Exam-style question**

Bob asked each of 40 friends how many minutes they took to get to work.
The results are shown in the frequency polygon.

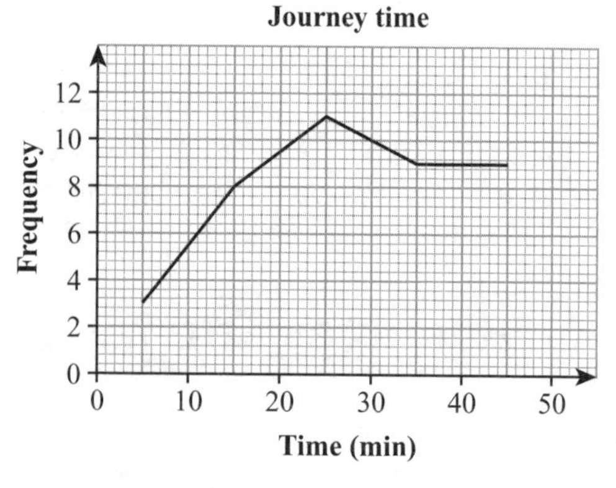

a How many people travelled for more than 20 minutes? **(1 mark)**

b If the average time spent travelling to work is more than 30 minutes, the company will consider allowing its employees to work from home.
Should the company consider allowing employees to work from home? **(4 marks)**

3.6 Statistical diagrams 2

1 A fixed menu at a café offers two main course options followed by two pudding choices.
One lunchtime there were 50 customers.
The table shows some information about their choices.

	Chicken	Vegetarian	Total
Cheese	15		23
Ice cream		5	
Total			50

Copy and complete the table.

2 **R** A school offers three language options and two humanities options at GCSE.
Students must choose one language and one humanities option.

	French	German	Mandarin	Total
History	57		18	126
Geography		12		
Total			35	200

a Copy and complete the table.

b What percentage of the students study Mandarin?

c Which humanities option was more popular?
Give reasons for your answer.

3 **R** A survey was carried out in a town to find out if more parking spaces were needed.
Of 60 men surveyed, 23 thought there were not enough spaces.
Of 60 women surveyed, 30% thought there were enough spaces.
The town has equal numbers of men and women.

a Display this information in a two-way table.

b The town council will provide more parking spaces if more than 60% of residents want more parking.
Will this happen?

4 The time series graph shows the increase in a supermarket's profits.

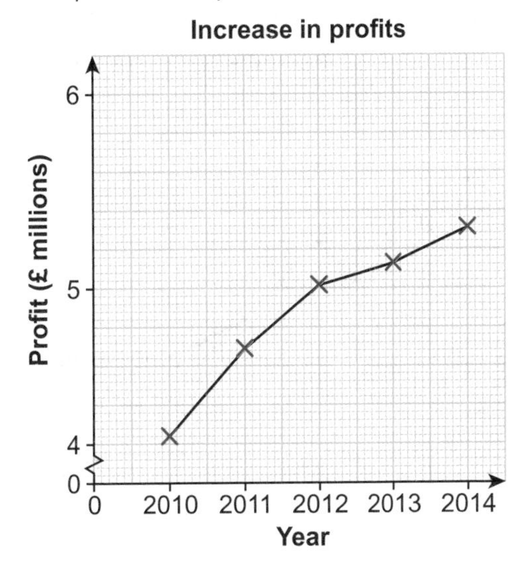

The supermarket chain says that this shows that the profit has increased by over 50% in the 4 years between 2010 and 2014.

a Explain why this graph is misleading.

b Draw a correct version.
Comment on what this shows about the actual increase in percentage profit.

Q4a hint Look closely at the vertical scale.

5 A soft drinks supplier produces a bar chart showing sales of their soft drinks and uses it to support the claim that sales of cola are more than twice those of the other drinks.
Give *two* reasons why the diagram is misleading.

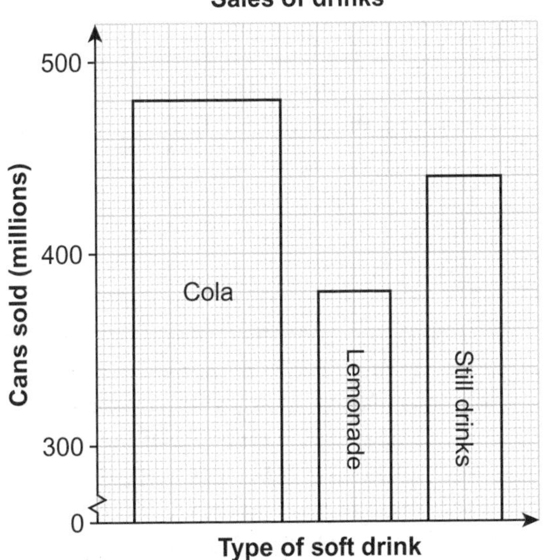

6 **R** A hospital patient's temperature is taken every hour. The time series graph is plotted from the measurements.

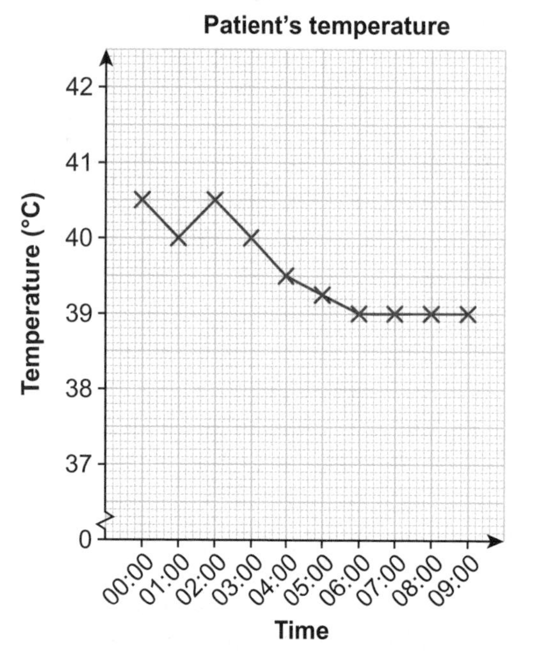

a What was the patient's temperature at 03:00?

b A temperature of over 39.4 °C is considered dangerous. At what time did the patient's temperature drop below this?

c Between which times was the drop in temperature the greatest?

7 **R** A librarian records the items borrowed by the over-50s and the under-50s, using four categories: modern fiction (MF), classic fiction (CF), non-fiction (NF) and other (O).

During one hour the items borrowed are as follows:

Over-50s: CF, MF, CF, NF, O, NF, NF, O, NF, NF, CF, MF, O, MF, CF, NF, O, NF, NF, O

Under-50s: MF, MF, NF, NF, O, O, O, MF, MF, MF, O, O, MF, MF, MF, CF, O, O, MF, O

a Explain why it is not possible to display this data as a frequency polygon.

b The librarian decides to display the data in either a pie chart or a bar chart. Which should she use if she is most interested in

i the proportions of items borrowed by over-50s and under-50s combined

ii comparing the numbers of the different items borrowed by each age group?

c The librarian is going to buy 400 more books for the library. How many modern fiction titles should she buy?

8 The times taken (in minutes) for employees of two companies to travel to work are recorded.

Company A: 35, 42, 39, 30, 23, 43, 47, 39, 38, 42

Company B: 18, 22, 19, 12, 14, 17, 32, 39, 42, 25

a Explain why you would display this data in a back-to-back stem and leaf diagram instead of in a dual bar chart.

b Draw a back-to-back stem and leaf diagram for this data.

c Find the median times to compare the travel times for the employees of the two companies.

9 **Exam-style question**

A company wants to compare the salaries of its employees.

The table shows their current salaries.

Salary, S, (£)	Frequency
$10\,000 \leqslant S < 12\,000$	3
$12\,000 \leqslant S < 14\,000$	15
$14\,000 \leqslant S < 16\,000$	19
$16\,000 \leqslant S < 18\,000$	22
$18\,000 \leqslant S < 20\,000$	11

a Choose *one* of the following statistical diagrams to display the data.

stem and leaf — scatter graph

frequency polygon — time series graph

b Give reasons for your choice. **(4 marks)**

Q9 hint Imagine trying to draw each diagram. You should also think about what the diagram is going to be used for.

3 Problem-solving

Solve problems using these strategies where appropriate:

- **Use pictures or lists**
- **Use smaller numbers.**

1 Sequences A and B are Fibonacci-like sequences. Work out the difference between the 6th term in sequence A and the 7th term in sequence B.

Sequence A: 4, 7, 11, 18, …

Sequence B: 5, 8, 13, 21, …

2 **R** Two teams are competing in a trampoline competition. Their scores are displayed in the back-to-back stem and leaf diagram. The coach needs to decide which team to take to the championships.

Team A	Stem	Team B
9 6 5	1	8
9 7 6 4 4	2	5 7 9
6 3 1	3	1 4 5 5 7
1	4	3 9 9

Key Team A 5 | 1 represents 15 marks   Team B 1 | 8 represents 18 marks

a What are the ranges for the two teams?

b What is the median score for each team?

c Use the range, the median and the shape of the diagram to compare the distribution of scores of both teams.
Which team should the coach take to the championships?

3 **R** Tina is working on a weather project. She has found the average high temperature and the average low temperature for four different cities around the world at three different times of the year.
The table shows her data. In it, the average high and average low temperatures are displayed together like this: 9 / 5.

City	Temperature (°C)		
	January	April	July
London	9 / 5	15 / 7	23 / 15
Chicago	−1 / −11	15 / 4	29 / 17
Sharm el-Sheikh	22 / 13	30 / 20	38 / 27
Sydney	26 / 19	23 / 15	17 / 8

a What type of statistical diagram would be a good way for Tina to display this data?

b Draw a statistical diagram to display the data.

c Which city has the widest range of temperatures throughout the year?

d London is in the Northern Hemisphere. By looking at your diagram, how can you tell which city is in the Southern Hemisphere?

4 Loughborough University and Birmingham University are about 43 miles apart.
Gayle and Luke both travelled from one university to the other for interviews.
Gayle went one way and got there in 57 minutes. Luke went the other way and travelled at an average speed of 50 mph.

a Use the formula $S = \frac{D}{T}$ to find how many minutes it took Luke to travel between the universities. Round your answer to the nearest whole number.

b Who travelled faster?

Marcie travelled between the universities on a different day.
She travelled at a speed of 72 km/h.

c Did Marcie take more or less time than Gayle?

5 **R** A museum increased its ticket prices every year for 8 years.
The museum manager wants to see if this has affected the number of tickets sold.
The table shows the average number of tickets sold per week at the different ticket prices.

Ticket price (£)	1.00	1.50	2.00	2.50	2.75	3.00	3.75	4.25
Number of tickets sold	250	246	262	235	228	236	182	135

a Draw a scatter graph to show the data in the table.

b Is there a relationship between the two variables? Explain your answer.

c The cost of running the museum increases each year. Is raising the price of tickets the best way to cover the increase in running costs? Explain your reasoning.

6 **R** The tables show the profits of a small publishing company over a 10-year period.

Year	2004	2005	2006	2007	2008	2009
Profit (£)	110	575	1500	3400	3620	4340

Year	2010	2011	2012	2013	2014
Profit (£)	6120	9435	12 350	18 500	23 900

a Use the tables to draw a time series graph of the profits made by the publishing company.

b What does the graph show you?

c Predict the profit for 2015.

d A large publishing company is looking to buy smaller companies. Would this be a good company to buy? Explain your reasoning.

7 **Exam-style question**

The frequency table gives information about the times it took some office workers to get to the office one day.

Time, t (minutes)	Frequency
$0 < t \leqslant 10$	4
$10 < t \leqslant 20$	8
$20 < t \leqslant 30$	14
$30 < t \leqslant 40$	16
$40 < t \leqslant 50$	6
$50 < t \leqslant 60$	2

a Copy the axes below and draw a frequency polygon for this information.

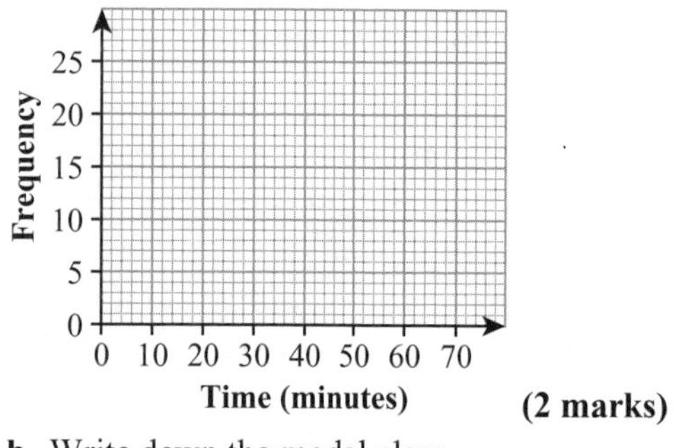

(2 marks)

b Write down the modal class interval. **(1 mark)**

One of the office workers is picked at random.

c Work out the probability that this worker took more than 40 minutes to get to the office. **(2 marks)**

November 2012, Q12, 1MA0/2H

8 **R** If you continue to write the arithmetic sequence 4, 11, 18, 25 …, will you eventually write 500?
Explain with reference to the nth term.

9 The frequency table shows the heights of sunflowers grown by a Brownie pack.

Height, h (cm)	Frequency
$150 \leqslant h < 180$	2
$180 \leqslant h < 210$	4
$210 \leqslant h < 240$	7
$240 \leqslant h < 270$	8
$270 \leqslant h < 300$	11
$300 \leqslant h < 330$	3
$330 \leqslant h < 360$	1

a How many sunflowers were grown?
b Which is the modal group?
c Estimate the range.
d Draw a frequency polygon for the data.
e Was the number of sunflowers 240 cm or higher greater or less than the number that were shorter than 240 cm?
Explain how you know from looking at the graph.

10 Stacey, Paula and Mark are looking at the same scatter graph. They each draw a line of best fit.

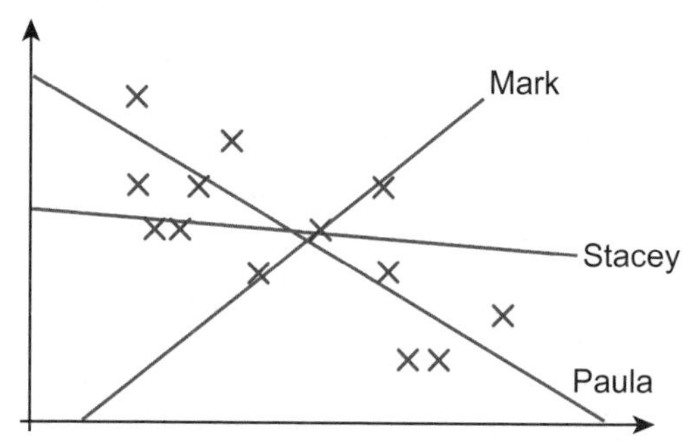

a Who has drawn the line of best fit?
b Is there a correlation between the two variables? If so, what type of correlation is it and what does it mean?

Q10a hint What are the features of a line of best fit?

4 FRACTIONS, RATIO AND PERCENTAGES

4.1 Fractions

1 Simon and Cathy each buy an identical pizza.
Simon eats $\frac{2}{5}$ of his pizza and Cathy eats $\frac{3}{8}$ of her pizza.
How much pizza is left?
Give your answer as a mixed number.

2 Find the reciprocal of each number.
a 5 b 0.125 c 2.25 d $\frac{5}{6}$
Use a calculator to check your answers.

3 Find the reciprocal of
a $\frac{1}{4}$ b $\frac{3}{5}$ c $\frac{11}{8}$ d $4\frac{5}{6}$

4 Giving your answer as a mixed number where appropriate, work out
a $2\frac{1}{5} \times \frac{3}{4}$ b $1\frac{3}{8} \times \frac{4}{5}$
c $3\frac{4}{7} \times 1\frac{2}{5}$ d $3\frac{7}{10} \times 2\frac{5}{6}$

5 Work out
a $3\frac{1}{4} \div 8$ b $10 \div \frac{5}{9}$

6 Giving your answer as a mixed number where appropriate, work out
a $1\frac{1}{8} \div \frac{2}{9}$ b $4\frac{3}{7} \div 1\frac{1}{3}$
c $2\frac{5}{6} \div \frac{3}{10}$ d $3\frac{5}{6} \div 5\frac{1}{12}$

7 Samir says, 'Multiplying a fraction by its reciprocal always results in an answer of 1.'
Is he correct?
Show some working to explain your answer.

8 Copy and complete the calculation.

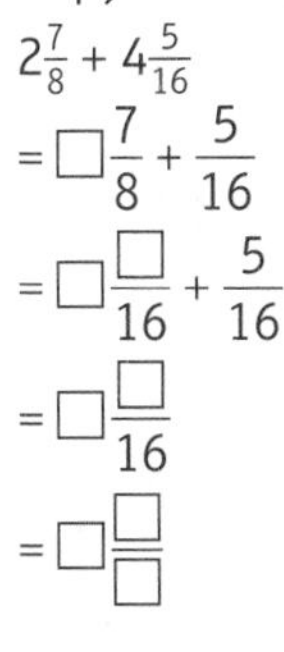

9 Work out
a $2\frac{7}{12} + 1\frac{1}{6}$ b $3\frac{2}{3} + 4\frac{11}{12}$
c $3\frac{2}{7} + 2\frac{4}{5}$ d $2\frac{5}{8} + 3\frac{5}{6}$

10 **Exam-style question**

Sally has ordered a wedding cake.
She has asked that the cake be between $8\frac{1}{2}$ inches and 9 inches high.
The baker makes these three layers.

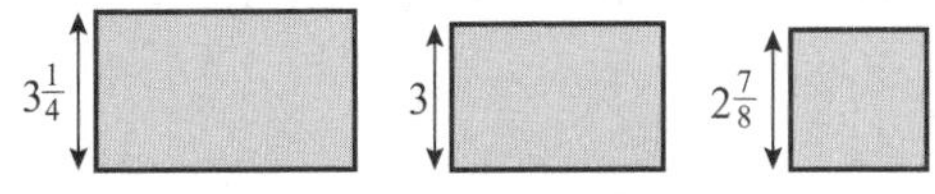

All measurements are in inches.

When the layers are placed on top of one another, will they make a cake that is between $8\frac{1}{2}$ and 9 inches high?
You must explain your answer. **(5 marks)**

Exam hint
Explain your answer by showing your calculations. Write a sentence, 'The cake will/will not be between $8\frac{1}{2}$ and 9 inches high because …'

11 Work out these subtractions.
a $5\frac{3}{4} - 2\frac{1}{6}$ b $6\frac{7}{8} - 1\frac{3}{5}$
c $2\frac{4}{5} - 3\frac{3}{10}$ d $2\frac{1}{6} - 5\frac{7}{12}$

Example

12 **P** Brenda earns a fixed amount each month.
She spends $\frac{3}{8}$ of this amount on her mortgage.
Out of the remaining amount, she saves $\frac{5}{12}$.
She spends the rest on supermarket shopping.
Brenda spends £350 per month on supermarket shopping.
How much does she spend on her mortgage each month?

13 **P** Tom sorted out bags of donated clothing at a homeless shelter.
Out of 6 bags, $2\frac{1}{4}$ were suitable for men, $1\frac{1}{3}$ were suitable for women and the rest were for children.
a How many bags of clothes were for children?
Tom spent $3\frac{1}{2}$ hours sorting out clothes and $2\frac{1}{3}$ hours working in the kitchen at the shelter.
He also took two 10-minute breaks and had $\frac{3}{4}$ hour for lunch.
b How many hours did he spend in total at the homeless shelter?

4.2 Ratios

1 Write each ratio in the form $1:n$.

a 5:30 b 32:16

c $\frac{1}{4}:3$ d $\frac{5}{8}:\frac{3}{5}$

2 Write each ratio in the form $n:1$.

a 16:4 b 15:60

c $4:\frac{1}{8}$ d $\frac{2}{5}:\frac{4}{9}$

3 Write these ratios in the form $1:n$.

a £2:80p

b 2 kg:400 g

c 3 hours:20 minutes

d 15 cm:6.6 m

4 **R** On Saturday at 12 pm, there were 45 staff members and 375 customers in a department store.

a Write the customer:staff ratio in the form $n:1$.

Another department store had 70 staff members and 637 customers at the same time.

b Which store had more customers per staff member?

5 **P** Cheryl and Jon both make hair dye by mixing hair colourant with peroxide.

Cheryl uses 750 ml of colourant and 450 ml of peroxide.

Jon uses 850 ml of colourant and 650 ml of peroxide.

Who has the greater concentration of peroxide in their hair dye?

6 **R** Jackie and Alan share the profits of their business in the ratio of the amounts they invested.

Jackie invested £170 and Alan invested £150.

Alan gets £18 000.

How much money does Jackie get?

7 To make concrete, Ali mixes 2 parts of cement with 5 parts of sand.

a Write down the ratio of sand to cement.

b To lay the foundation for an extension, Ali uses 60 kg of sand. How many kilograms of cement does she use?

c On another project, Ali used 75 kg of cement. How much sand did she use?

8 A scale model of a boat is 37 cm long. The real boat is 7.4 m long.

a Work out the scale of the model. Write it as the ratio of real length to model length.

The boat is 3.2 m wide in real life.

b How wide is the model?

9 The ratio of DVDs to CDs in Sam's collection is $1:\frac{5}{8}$.

Sam has 135 CDs.

How many DVDs does Sam have?

10 Jana splits £350 between her two nieces in the ratio of their ages.

Carlotta is 16 and Hannah is 12.

a What fraction does Carlotta get?

b What fraction does Hannah get?

c How much money does each niece get?

11 Share 342 sweets between Finn and Rosie in the ratio 5:4.

How many sweets does each child get?

12 Sahira needs to cut a length of wood into two pieces in the ratio 5:7.

The wood is 3600 mm long.

How long will each piece be?

13 Share each quantity in the given ratio.

a £451 in the ratio 4:5:2

b £52.60 in the ratio 4:1:3

c 930 mm in the ratio 2:3:5

d 885 kg in the ratio 8:2:2

14 **Exam-style question**

Jamal makes a batch of scones.

He needs to mix butter, sugar and flour in the ratio 3:2:15 by weight.

Jamal wants to make a scone mix weighing 500 g in total.

He has

85 g of butter

45 g of sugar

400 g of flour.

Does Jamal have enough butter, sugar and flour to make the scone mix? **(4 marks)**

15 Write each ratio as a whole number ratio in its simplest form.

a 40:22.5 b 83.25:11
c 14.4:73.5 d 60.2:2.56

> **Q15a hint** Multiply first by powers of 10 to make both sides of the ratios whole numbers, then simplify.
>
> ×10 (40 : 22.5) ×10
> 400 : 225
> ÷□ (16 : □) ÷□

16 **R** Jude wants to make lilac paint.
She is going to mix red paint, blue paint and white paint in the ratio 0.8:1.5:5.7.
Copy and complete the table to show how much of each colour Jude needs to make the paint quantities shown.

Size	Blue	Red	White
1 litre			
2.5 litres			

4.3 Ratio and proportion

1 The exchange rate between pounds and euros (€) is £1 = €1.28.
a Convert £500 to euros.
b Convert €576 to pounds.

2 **P** Cassie buys a smartphone in the UK for £449. She sees the same smartphone on sale in New York for US$499.
The exchange rate is £1 = US$1.53.
By how much is the smartphone cheaper in New York?

3 **R** Bron and Jill are both training to run a marathon. In one week Bron runs 83 miles and Jill runs 122 km. 5 miles = 8 km.
a Write the ratio of kilometres to miles in the form $1:n$.
b Work out who has run further in the week and by how much.

4 **R** Hal makes a fruit punch for a party by mixing orange juice with raspberry juice in the ratio 7:3. Lyn also makes a fruit punch by mixing raspberry juice and orange juice in the ratio 10:21.
Will the two fruit punches taste exactly the same? Explain your answer.

5 **P** Stuart is a plumber and is paid the same hourly rate for any work he does.
For one job he is paid £712.50 for 15 hours' work.
a What fraction of this amount is Stuart paid for 8 hours' work?
b Work out how much he is paid for 8 hours' work.
Karen also works as a plumber.
She gets £577.50 for doing 11 hours' work.
c Which plumber gets the better hourly rate?

6 A pattern is made from black squares, b, and white squares, w, in the ratio 2:5.
Copy and complete.
$w = b \times \square = \square b$
$b = w \times \square = \square w$

7 **R** Brian makes Mexican snacks by filling tortillas with a mixture of spiced meat and cheese. For every 4 tablespoons of meat, Brian uses 3 tablespoons of cheese.
a Write a formula for c, the number of tablespoons of grated cheese used with m tablespoons of spiced meat.
b Brian has 10 tablespoons of spiced meat. How many tablespoons of grated cheese does he need?
Brian wants to make the mixture twice as cheesy, so he doubles the number of tablespoons of cheese.
c Write a formula for the new recipe.

8 Are these pairs of quantities in direct proportion?
a 12 hotdogs cost £21.60, 15 hotdogs cost £27.00
b 5 apples cost £1.60, 9 apples cost £2.97
c Tom took 45 minutes to run 10 km, Ric took 1 hour 2 minutes to run 14 km

9 In a science experiment, the speed of a steel ball bearing is measured at different times after it is dropped.
The table shows the results.

Time, t (s)	1	2	3	4	5
Speed, s (m/s)	9.8	19.6	29.4	39.2	49

Are time and speed in direct proportion? Explain your answer.

10 The table gives readings A and B in a science experiment.

A	4	9	13
B	10	22.5	32.5

a Are A and B in direct proportion? Explain.

b Write a formula for B in terms of A.

c Write the ratio $A:B$ in its simplest form.

11 The values of X and Y are in direct proportion. Work out the missing values of p, q, r and s.

Value of X	p	20	45	r	108
Value of Y	10	q	24	54	s

12 The cost of wood is directly proportional to its length.
A 3000 mm piece of wood costs £58.50.
Work out the cost of 10 000 mm of this wood.

13 **P** The weight of a metal rod is directly proportional to its volume.
A metal rod weighs 350 g and has a volume of 15.71 cm^3.
Work out the volume of another rod made from the same metal that weighs 475 g.
Write your answer to 2 d.p.

14 **Exam-style question**

Bob is on holiday in Spain.
He pays €51.98 for 45 litres of petrol.
When Bob returns to the UK, he pays £44.03 for 37 litres of petrol.
The exchange rate is £1 = €1.28
In which country is it cheaper to buy petrol? **(4 marks)**

Q14 hint You only need to convert one of the prices into the other currency, not both, before you look at the amount of petrol.

4.4 Percentages

1 Tariq gets an electricity bill.
The cost of the electricity used before the VAT is added is £493.80.
VAT is charged at 5% on domestic fuel bills.
What is the cost of the electricity bill, including VAT?

2 **Exam-style question**

Monica booked a trip to Disneyland for herself and her family.
The total cost of the trip was £5800 *plus* VAT at 20%.
Monica paid £1300 of the total cost when she booked her holiday.
She paid the rest of the total cost in 6 equal monthly payments.
Work out the amount of each monthly payment. **(5 marks)**

3 A train ticket normally costs £182.
George gets a 30% discount.
How much does George pay for the train ticket?

4 **R** Dan has £2500 invested for 2 years.
The investment increases by 5% each year.
Work out the value of the investment after

a 1 year b 2 years.

5 a Work out the amount of simple interest earned in one year for each of these investments.

i £450 at 3% per year

ii £6000 at 7.5% per year

b Gary invests £23 750 for 3 years at 4.25% simple interest.
How much is his investment worth at the end of the 3 years?

6 **P** Income tax is paid on any money you earn over your personal tax allowance.
In 2014–15, the personal tax allowance was £10 000.
Above this amount, tax is paid at different rates, depending on how much you earn.
The table shows the rates for 2014–15.

Tax rate	Taxable income above your personal allowance
Basic rate 20%	£0 to £31 865
Higher rate 40%	£31 866 to £150 000

Work out the amount of income tax each of these people paid in the 2014–15 tax year.

a Harry earns £18 900 per annum.

b Sheila earns £27 950 p.a.

c Claire earns £57 560 p.a.

d Aylish earns £85 735 p.a.

Q6 hint Subtract the personal tax allowance before working out the tax owed.

7 Maddy saves £4500 in an account.
At the end of one year, the balance of her account is £4680.

a What is the actual increase?

b Work out the percentage increase.

8 In 2014, a local council recycled 14.28 million tonnes of refuse.
In 2004, they recycled 6.53 million tonnes of refuse.
What was the percentage increase over the decade?

9 Cerys bought an electric guitar for £380.
Two years later she sold it for £209.
What was her percentage loss?

10 Sean spent £91.35 buying ingredients to cook a charity dinner.
He sold tickets for the dinner for a total of £134.61.

a How much profit did Sean make?

b What was Sean's percentage profit?

11 **R** In 2009 Mike earned an annual salary of £11 400.
In 2015 Mike's salary had increased by 300%.
Mike says his salary in 2015 is 3 times the amount it was in 2009.
His boss says his salary in 2015 is 4 times the amount it was in 2009.
Who is correct?

12 Will pays £66.78 for a portable hard drive.
This price includes VAT at 20%.
What was the cost of the hard drive before VAT was added?

13 Jake invested an amount of money 2 years ago.
In the first year the amount increased by 5%.
In the second year, the amount decreased by 8%.
Copy and complete the calculation to work out the overall percentage decrease over these 2 years.

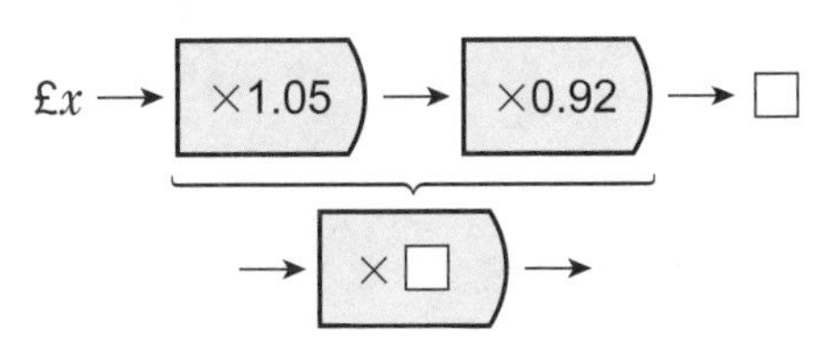

14 **P** In 2013 Sue bought a car. In 2014 the car depreciated by 15%. In 2015 it depreciated by 6%. The value of the car in 2015 was £7990.

a What was the total percentage decrease? Do not round your answer.

b Work out the value of the car in 2013.

15 **R** a Show that applying a 10% decrease followed by a 10% increase is the same as a 1% decrease overall.

b Will applying a decrease by a fixed amount and then an increase by the same fixed amount always result in a decrease overall?

4.5 Fractions, decimals and percentages

1 **R** Georgia bought 36 boxes of chocolates from the cash and carry.
Each box of chocolates cost her £1.50.
She sold $\frac{2}{3}$ of the boxes of chocolates for £7.50 each and another 7 boxes for £5 each.
She gave the rest away.

a How much profit did she make?

b Express this profit as a percentage of the total cost price.

2 **Exam-style question**

60 students took a maths test.
20% of the students scored 70 or more.
$\frac{3}{4}$ of the students scored between 55 and 69.
The rest scored less than 55.
Work out the number of students who scored less than 55. **(4 marks)**

3 **Exam-style question**

Tom works out he spends 0.75 hours out of every 3 hours he is awake playing computer games.
He also works out that, out of the time he plays computer games, he spends $\frac{3}{5}$ of the time playing a war game.
Tom is awake for $\frac{2}{3}$ of the day.
What percentage of each day does Tom spend playing the war game? **(3 marks)**

4 **Exam-style question**

Which is closer to 65%, $\frac{2}{3}$ or $\frac{9}{14}$?
You must show your working. **(3 marks)**

5 **P** Sally runs a hairdressing business.
The table shows the amount of profit she made in 2013 and 2014 for three different services she offers.

Service	2013	2014
Cut and blow dry	£4800	£6400
Colour and highlights	£7500	£5800
Wedding hair styles	£3200	£8500

Write three sentences comparing the profits in the two years.
Use fractions, decimals, percentages, ratio or proportion.

6 **P** Work out $\frac{2}{5}$ of 0.45 of 96% of £200.
Show all your working out.

7 **R** Work out
$9(\frac{1}{10} + \frac{1}{100} + \frac{1}{1000} + \ldots)$
as an integer to 1 s.f. (where … indicates that the sequence goes on forever).
Explain your answer.

8 **R** Two variables v and t are connected by the formula
$v = 10t$

a Are v and t in direct proportion? Explain.

b Write t as a fraction of v.

c Write the ratio $t:v$.

9 Write these recurring decimals as exact fractions.
Write each fraction in its simplest form.

Example

a $0.\dot{2}$ b $0.\dot{5}$

c $0.\dot{4}\dot{5}$ d 0.27272727…

e $0.\dot{8}5\dot{2}$ f $0.\dot{1}7\dot{1}$

Q9 hint Multiply by a power of ten.
If 1 decimal place recurs, multiply by 10.
If 2 decimal places recur, multiply by 100.
If 3 decimal places recur, multiply by 1000.

10 Which of these fractions are equivalent to recurring decimals?
Show your working out.

a $\frac{3}{20}$ b $\frac{3}{7}$ c $\frac{31}{40}$ d $\frac{4}{19}$

4 Problem-solving

Solve problems using these strategies where appropriate:

- **Use pictures or lists**
- **Use smaller numbers**
- **Use bar models.**

Example

1 **R** Malcolm finds the times (rounded to the nearest minute) that 10 people take to complete a half-marathon.
97, 108, 10, 77, 104, 107, 110, 106, 113, 89

a The outlier was recorded wrongly. Suggest what it might have been.

b Work out the range and the median time taking into account the corrected value.

2 The standard rate of value added tax (VAT) in the UK is 20%. VAT is reduced to 5% on some items, such as children's car seats.
If a child's car seat costs £48 before VAT is added, how much more would it cost if the standard VAT rate were charged instead of the reduced rate?

Q2 hint A bar model might help.

3 The official width-to-length ratio of the flag of Portugal is 2:3.
The official width-to-length ratio of the flag of Iceland is 18:25.
A gift shop sells official flags of both Portugal and Iceland.
Each flag has a width of 54 cm.

a Which flag is longer?

b How many cm longer?

4 **Exam-style question**

A company surveyed 380 people.
$\frac{3}{5}$ of the people surveyed were aged between 18 and 65 years.
10% of the people surveyed were aged under 18 years.
How many of the people surveyed were aged over 65 years? **(2 marks)**

5 Mirka is shopping online. She has £80. She is deciding whether to order an item from Australia or from Germany.

From Australia, the item she wants costs 115 Australian dollars (A$115) plus A$20 delivery. From Germany, the item she wants costs **€**80 plus **€**11 delivery.

The exchange rates are £1 = A$1.8 and £1 = **€**1.3.

a How much money will Mirka have left if she orders the item from Australia?

b How much money will Mirka have left if she orders the item from Germany?

6 One week Eoin paid £74.92 for his weekly groceries.

The next week he paid £61.32.

What is the percentage decrease in Eoin's grocery bill?

7 These rectangles have the same area.

Find x.

8 **R** Lionel draws three scatter graphs.

Graph A has a positive correlation, graph B has a negative correlation and graph C has no correlation.

One of Lionel's graphs shows the relationship between age (18–35 years) and income.

Which of the three graphs is this likely to be? Explain your choice.

9 **R** Shola and Paul are factorising $8pq^3 + 12p^2q$.

Shola says that the highest common factor of the terms is $4p$. Paul says it is $4q$.

a From just looking at the expression, is either of them correct? Explain your reason.

b What is the factorised answer?

10 Mike spends 450 hours working on a construction project. He spends some time on the construction site and some time in the office. When Mike calculates the fraction of the time he spent working on site as a decimal, he gets 0.666 666 666 …

a What fraction of the total time did Mike spend working on site?

b How many hours did Mike spend working on site?

5 ANGLES AND TRIGONOMETRY

5.1 Angle properties of triangles and quadrilaterals

1 ABC and CDE are straight lines.

AE is parallel to BD.

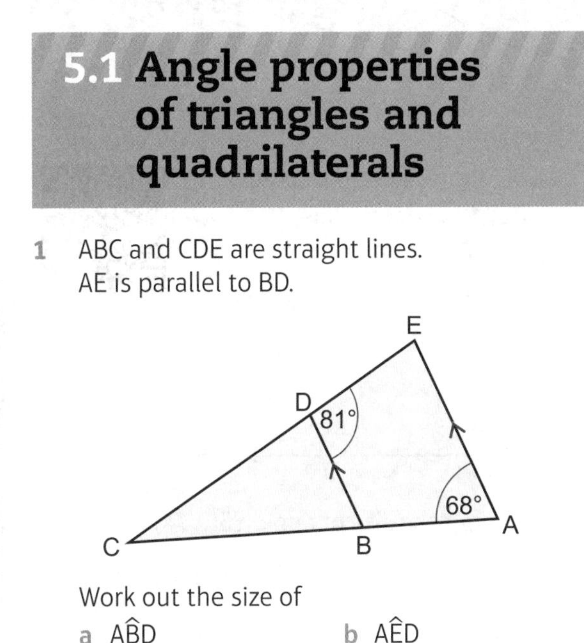

Work out the size of

a $A\hat{B}D$

b $A\hat{E}D$

c $B\hat{D}C$

d $A\hat{C}E$

2 **R** ABCD is a parallelogram.

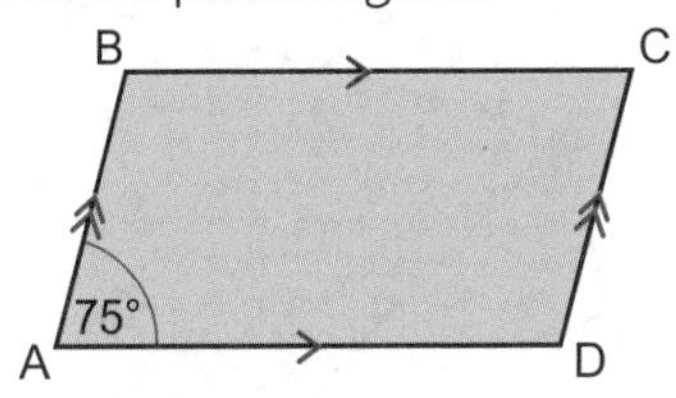

a Copy the parallelogram and extend each side as shown below.

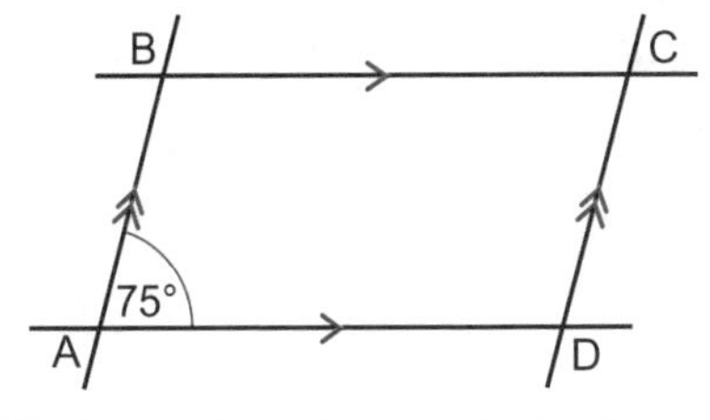

b Work out the other angles in the parallelogram.

c What do you notice about the opposite angles?

d Repeat with different parallelograms. Is your observation in part **c** still true?

3 Triangle BCF is shown. AD is parallel to EG.

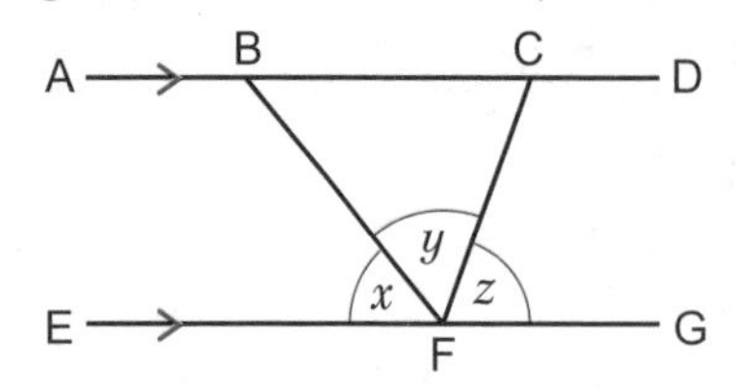

a What is the value of $x + y + z$?
Give a reason for your answer.

b Write the size of each of these angles in terms of x, y and z.
i angle FBC ii angle FCB
Give reasons for your answers.

c Use your answer to part **a** to derive the sum of the angles in a triangle.

4 **Exam-style question**

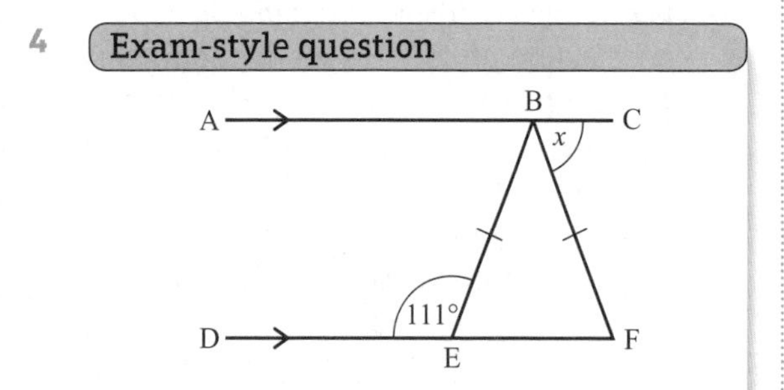

ABC and DEF are straight lines.
AC is parallel to DF. BE = BF.
Calculate the size of the angle marked x.
You must give reasons for your answer.

(4 marks)

5 **R** In this diagram a diagonal divides the quadrilateral into two triangles.

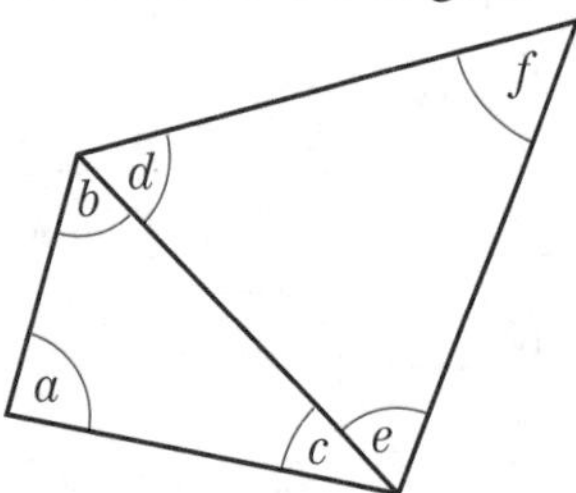

Use the diagram to prove that the angle sum of a quadrilateral is 360°.

6 Work out the size of each angle marked with a letter.

a

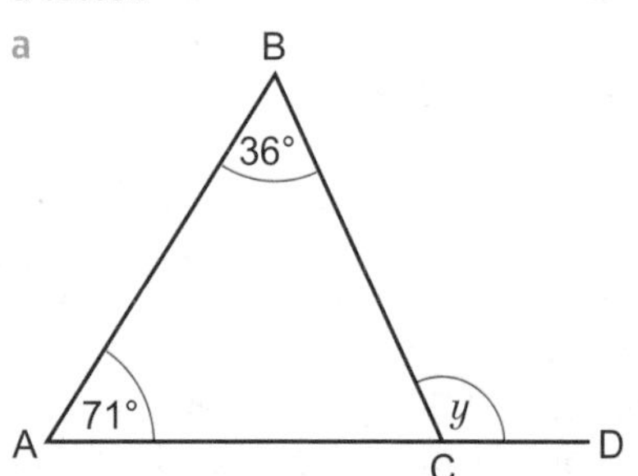

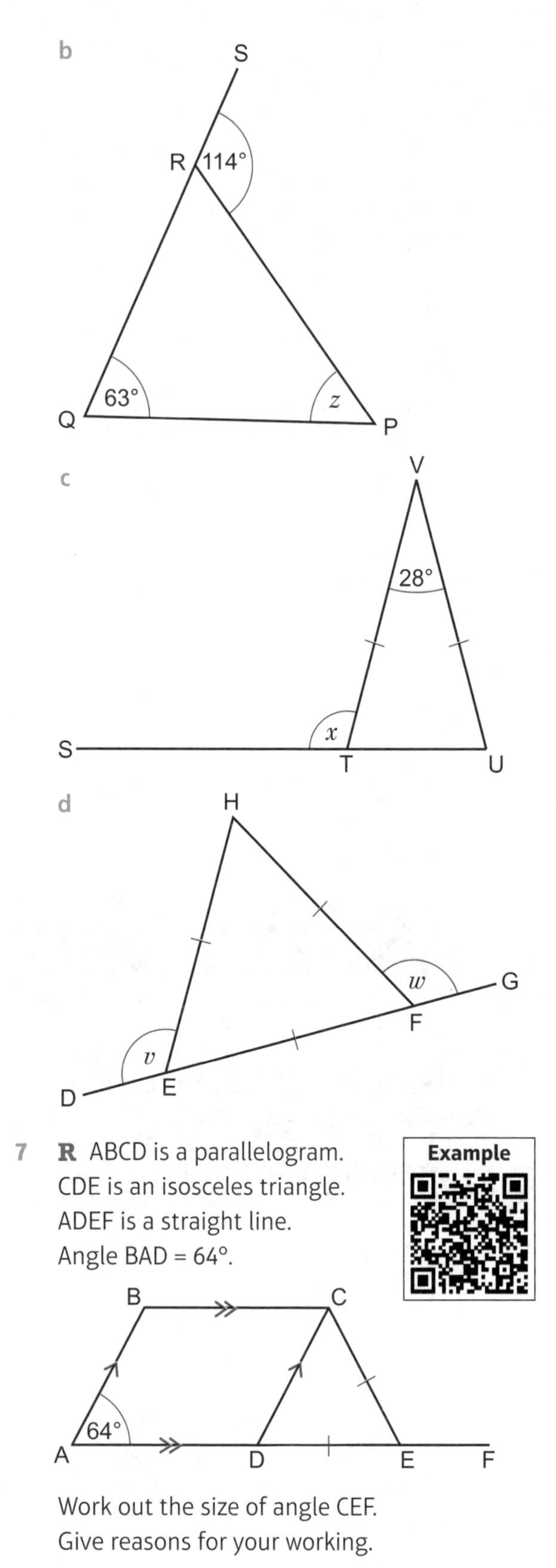

7 **R** ABCD is a parallelogram.
CDE is an isosceles triangle.
ADEF is a straight line.
Angle BAD = 64°.

Example

Work out the size of angle CEF.
Give reasons for your working.

8 **R** Work out the size of angle ABD.
Give reasons for your working.

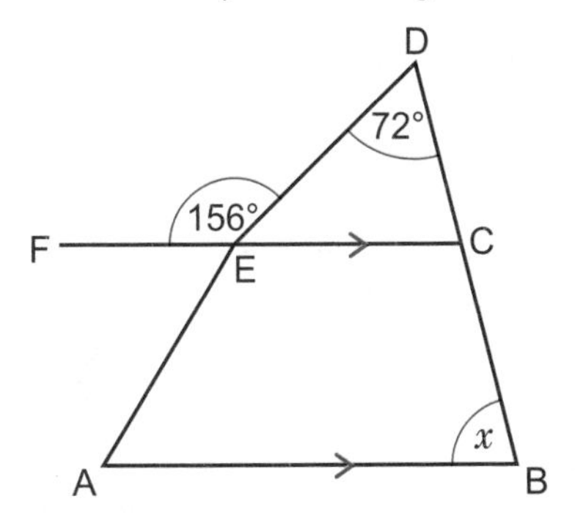

9 Work out the size of angle ADE.
Give reasons for your working.

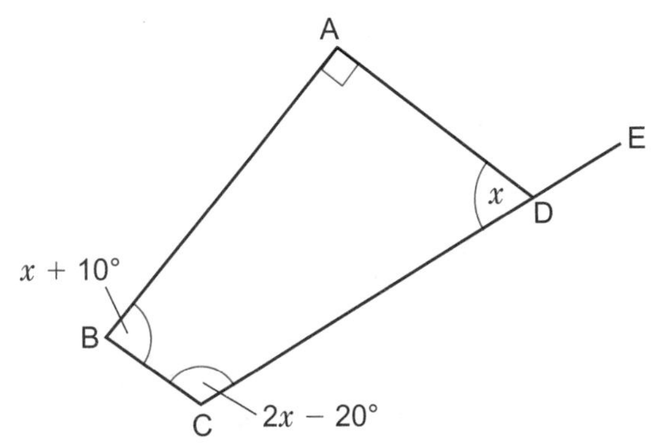

Q9 hint Use the fact that the angles in a quadrilateral add up to 360° to write an equation.

5.2 Interior angles of a polygon

1 Work out the sum of the interior angles of a pentagon.

Example

2 **R** Copy and complete the table.

Polygon	Number of sides (n)	Number of triangles formed	Sum of interior angles
Triangle	3	1	180°
Quadrilateral	4		
Pentagon	5		
Hexagon	6	4	720°
Heptagon	7		

3 A regular polygon has 16 sides.
a Work out the sum of the interior angles of the polygon.
b Work out the size of each interior angle.

Q3a hint Substitute into $(n - 2) \times 180°$.

4 **R** Work out the size of each interior angle of
a a regular hexagon
b a regular nonagon
c a regular decagon
d a regular polygon with 18 sides.

5 **R** Work out the size of each unknown interior angle.

a
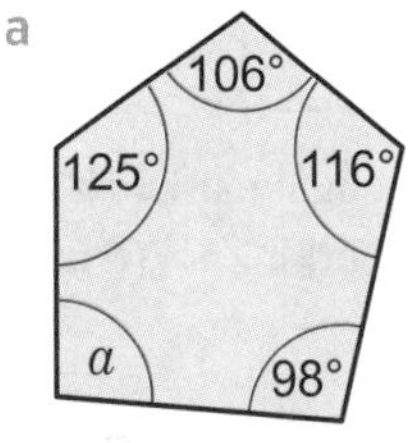

b
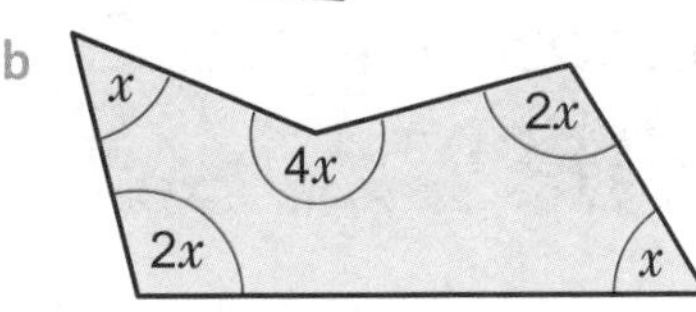

6 The sum of the interior angles of a polygon is 1980°.
How many sides does the polygon have?

Example

7 **P** A regular octagon is divided into 8 isosceles triangles.

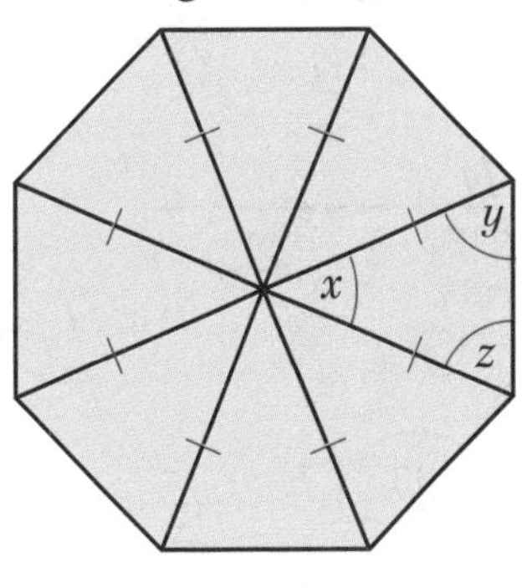

Work out the size of
a angle x b angle y c angle z.

8 **R** **Q7** shows an octagon made from isosceles triangles.
What polygon can you make from equilateral triangles?

9 **Exam-style question**

The diagram shows a regular pentagon and a square.

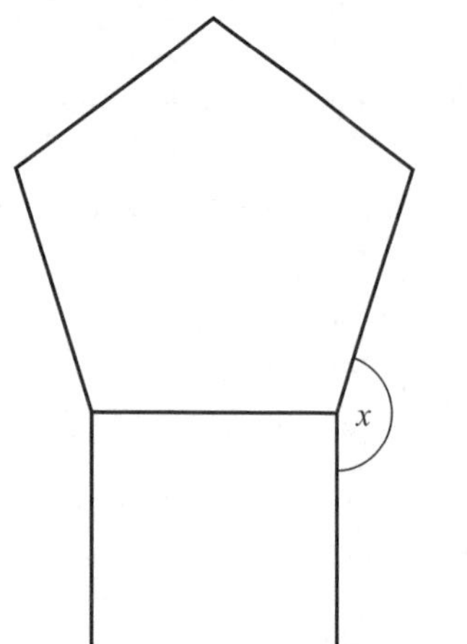

Calculate the size of the angle marked x.
You must show all your working. **(4 marks)**

5.3 Exterior angles of a polygon

1 **R** A pentagon and a hexagon are shown.

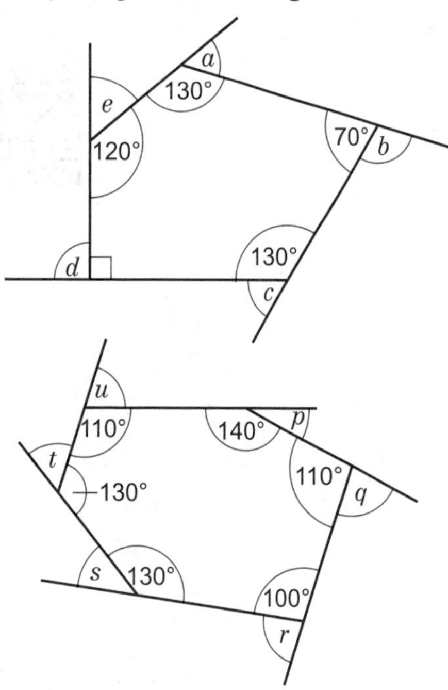

a Work out the sizes of the angles marked with letters.

b Work out the sum of the exterior angle for each polygon.

c What do you notice about the sum of the exterior angles?

2 Work out the size of an exterior angle of a regular octagon.

3 Work out the sizes of the angles marked with letters.

a

b

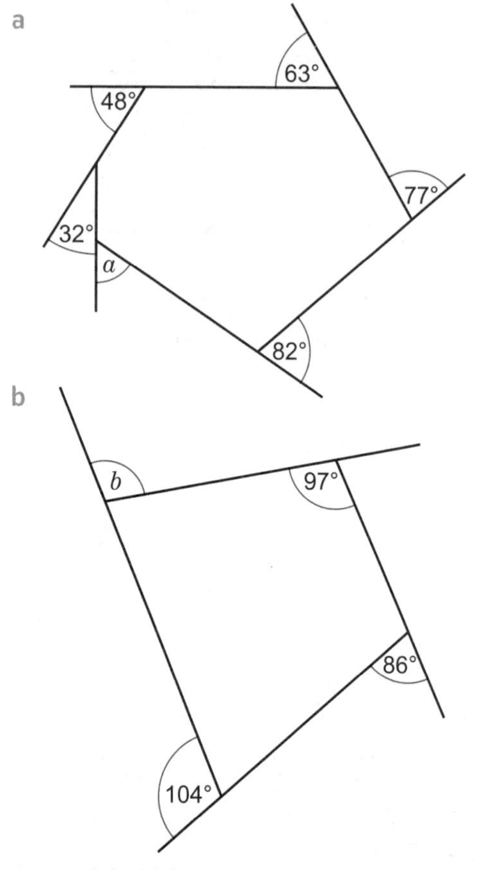

Q3a hint $a + 32° + 48° + 63° + 77° + 82° = 360°$
Q3b hint To find angle b, first work out the exterior angle not marked with a letter.

4 **R** The sizes of five of the exterior angles of a hexagon are 35°, 79°, 21°, 95° and 54°.
Work out the size of each interior angle.

5 **R** Work out the size of each unknown exterior angle in this polygon.

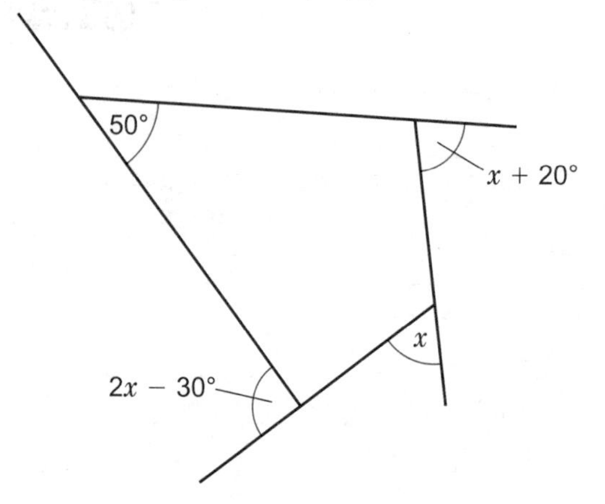

6 How many sides does a regular polygon have if its exterior angle is
a 36° b 15° c 9°?

7 How many sides does a regular polygon have if its interior angle is
a 60° b 156° c 170°?

8 Can the exterior angle of a regular polygon be 50°? Explain.

9 **P** One side of a regular pentagon ABCDE forms the side of a regular polygon with n sides.

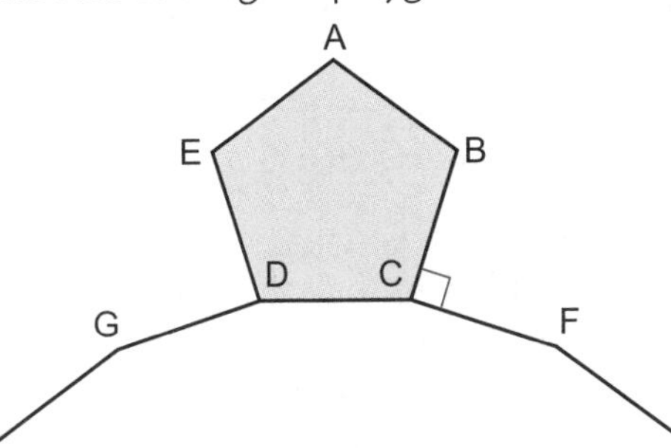

Angle BCF = 90°. Work out the value of n.

10 **P** The interior angle of a regular polygon is 3 times the size of its exterior angle.
How many sides does the polygon have?

5.4 Pythagoras' theorem 1

1 Calculate the length of the hypotenuse in each triangle.
Give your answers correct to 1 d.p.

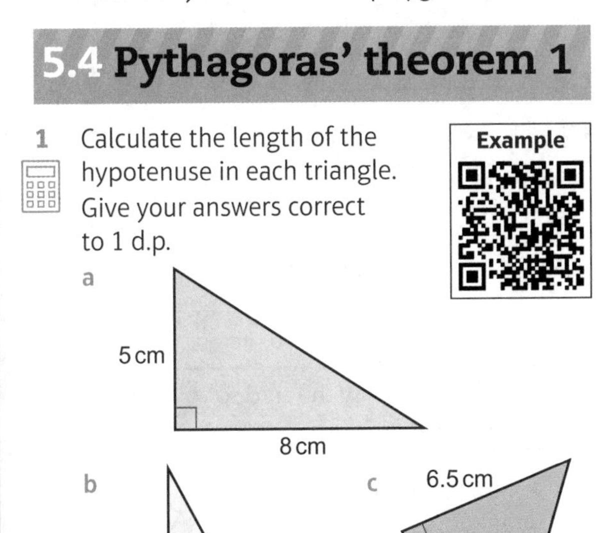

Q1 hint Do not round *before* taking the square root. Use all the figures on your calculator display.

2 ABC is a right-angled triangle.

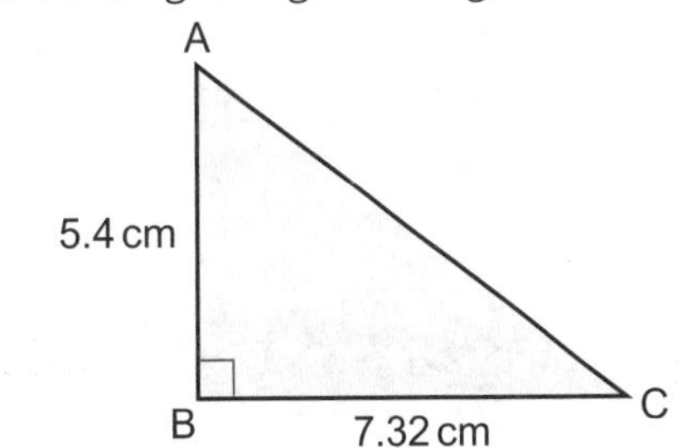

Calculate the length of AC.
Give your answer correct to 3 significant figures.

3 Calculate the length of the diagonal of a square with side length 6.3 cm.
Give your answer correct to 3 significant figures.

4 A guy rope is attached to the top of a 10 m flagpole. It is anchored to the ground 4 m from the base of the flagpole.
What is the length of the guy rope?

5 **P** Ashley walks 1.8 miles east and then 2.4 miles north. He then walks straight back to his starting point.
How far does Ashley walk in total?

6 **P** A garden is in the shape of an isosceles triangle.
Calculate the perimeter of the garden.

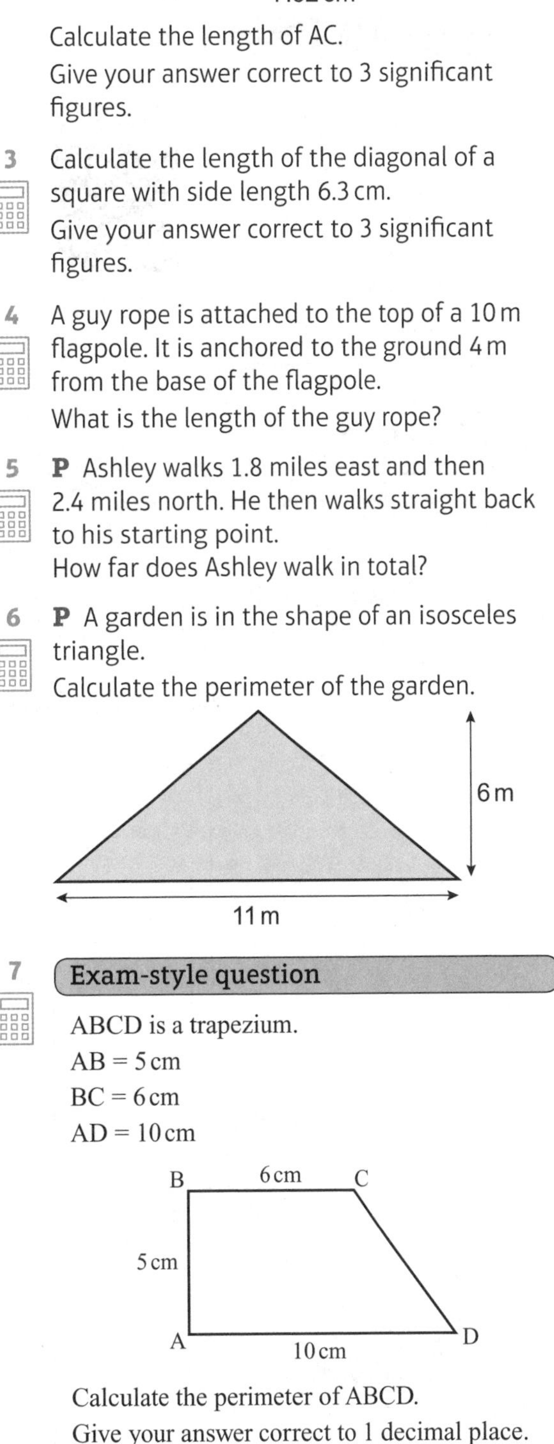

7 **Exam-style question**

ABCD is a trapezium.
AB = 5 cm
BC = 6 cm
AD = 10 cm

Calculate the perimeter of ABCD.
Give your answer correct to 1 decimal place.
(3 marks)

8 **R** Can a right-angled triangle have sides of length

a 5 cm, 7 cm, 9 cm b 6 cm, 8 cm, 10 cm

c 7 cm, 8 cm, 11 cm?

Explain your answers.

5.5 Pythagoras' theorem 2

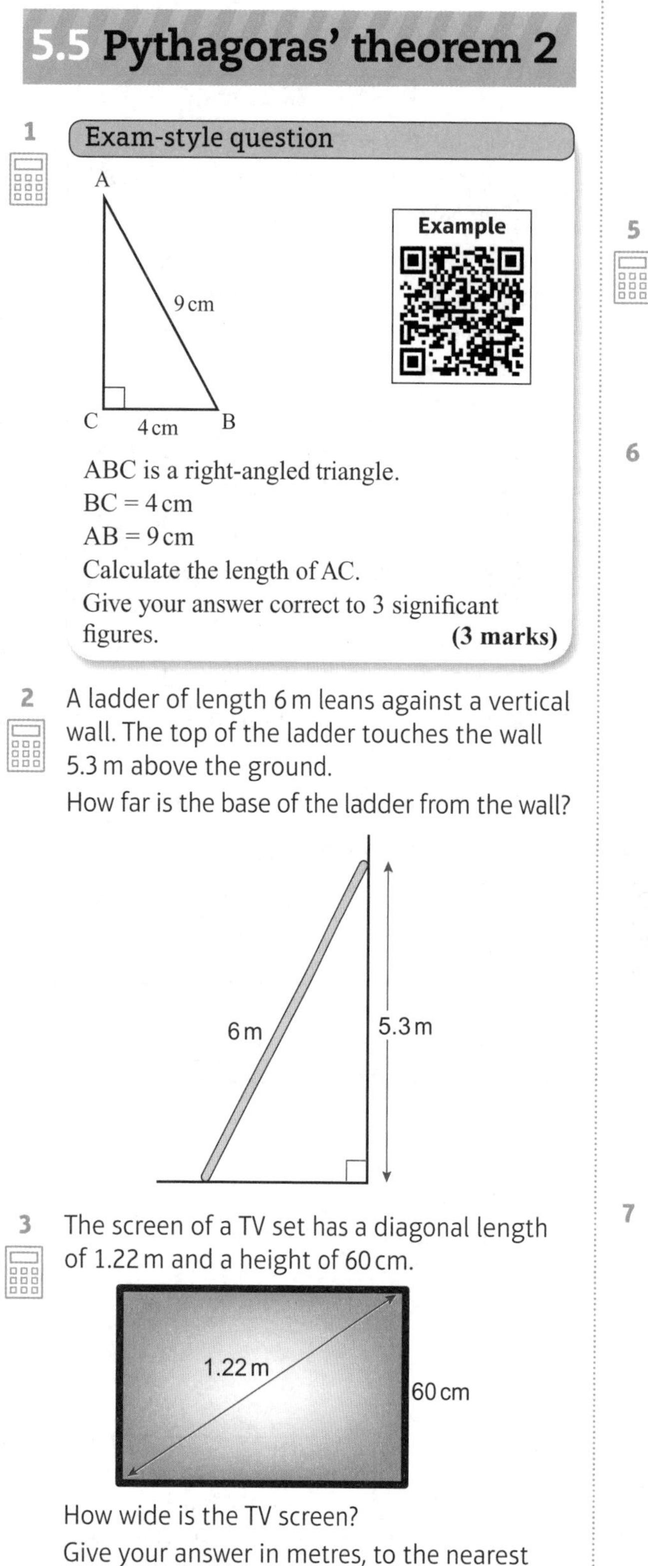

1 **Exam-style question**

Example

ABC is a right-angled triangle.
BC = 4 cm
AB = 9 cm
Calculate the length of AC.
Give your answer correct to 3 significant figures. **(3 marks)**

2 A ladder of length 6 m leans against a vertical wall. The top of the ladder touches the wall 5.3 m above the ground.

How far is the base of the ladder from the wall?

3 The screen of a TV set has a diagonal length of 1.22 m and a height of 60 cm.

How wide is the TV screen?

Give your answer in metres, to the nearest centimetre.

4 Calculate the vertical height of trapezium ABCD.

Give your answer in centimetres, to the nearest millimetre.

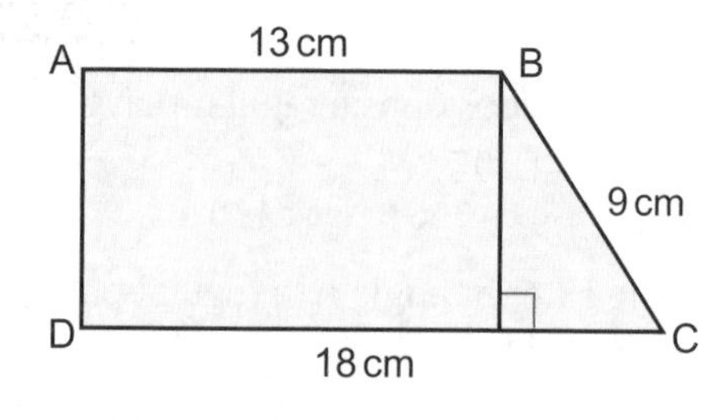

5 **P** a Calculate the length of the side of the largest square that *fits inside* an 8 cm diameter circle.

b Work out the length of the side of the smallest square that *surrounds* an 8 cm diameter circle.

6 Work out the length of the unknown side in each right-angled triangle.

Give your answers in surd form.

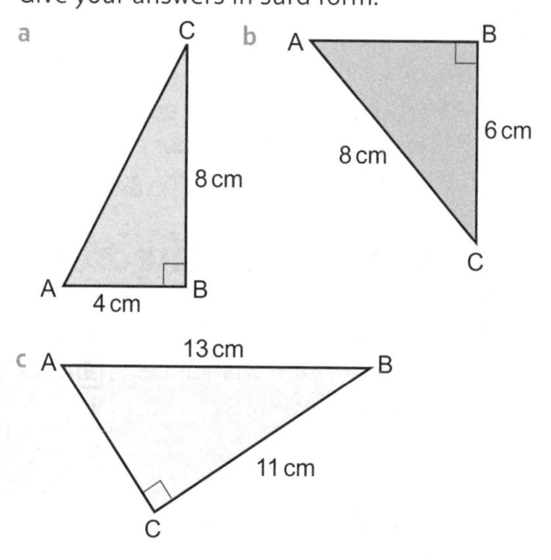

Q6a hint Simplify the surd so your answer looks like this: AC = $\square\sqrt{\square}$ cm

7 **P** ABC is an isosceles triangle.

D is the midpoint of BC.

Work out the height of the triangle.

Give your answer in surd form.

A
8 cm
8 cm
h
B
D
C
6 cm

8 **P** Work out

a the length of BD b the length of CD

c the perimeter of the quadrilateral.

Give all your answers to 3 s.f.

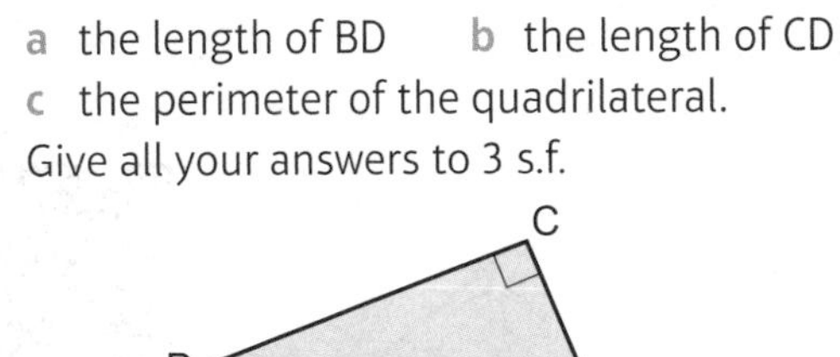

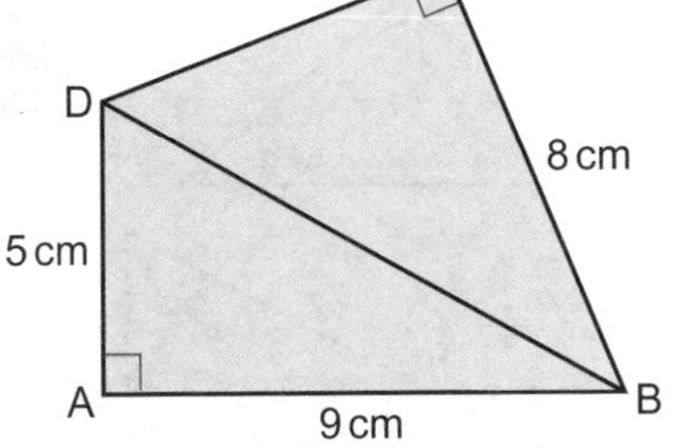

5.6 Trigonometry 1

1 **R** Draw triangle ABC accurately using a ruler and protractor.

Angle A = 90°, angle B = 40° and AB = 4 cm.

a Label the **hyp**otenuse (**hyp**), **opp**osite side (**opp**) and **adj**acent side (**adj**).

b Measure each unknown side to the nearest millimetre.

c Write the fraction

i $\frac{\text{opposite}}{\text{hypotenuse}}$ ii $\frac{\text{adjacent}}{\text{hypotenuse}}$

iii $\frac{\text{opposite}}{\text{adjacent}}$

Convert each fraction to a decimal.

Give your answers correct to 1 decimal place.

d Repeat parts **a** to **c** for triangle ABC with

i angle A = 90°, angle B = 40° and AB = 5 cm

ii angle A = 90°, angle B = 40° and AB = 7 cm.

2 Use your calculator to find, correct to 2 d.p. where necessary

a sin 25° b cos 15° c tan 38°

d cos 76° e tan 83° f sin 63°

3 Calculate the length of the side marked x in each triangle.

Give your answers correct to 3 significant figures.

Example

a

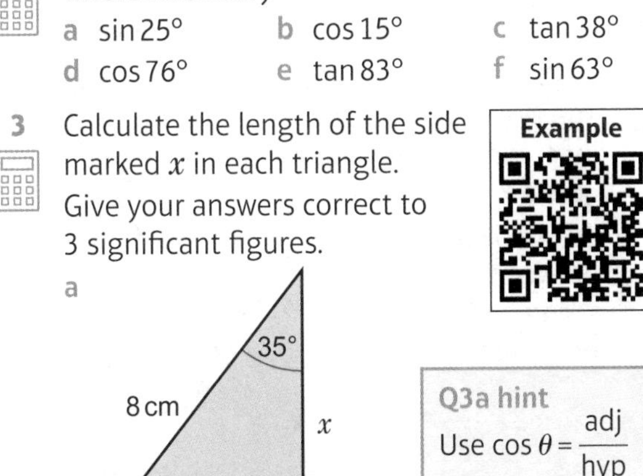

Q3a hint

Use $\cos\theta = \frac{\text{adj}}{\text{hyp}}$

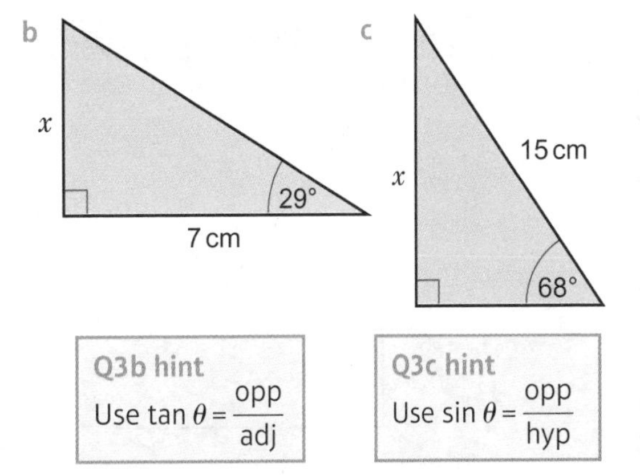

Q3b hint

Use $\tan\theta = \frac{\text{opp}}{\text{adj}}$

Q3c hint

Use $\sin\theta = \frac{\text{opp}}{\text{hyp}}$

4 Calculate the length of the side marked x.

Give your answer correct to 1 decimal place.

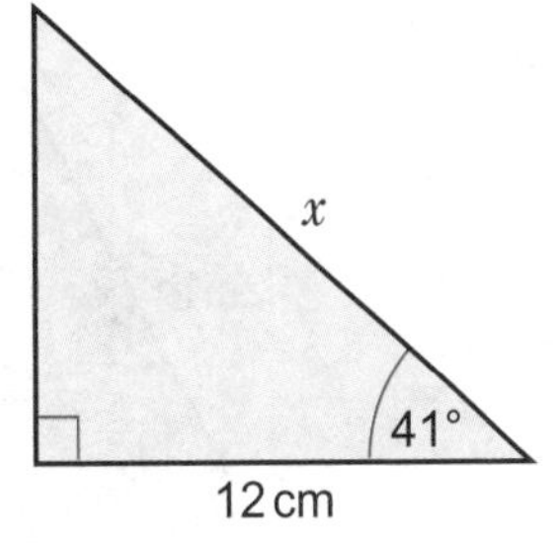

5 Calculate the length of the side marked x in each triangle.

Give your answers correct to 1 d.p.

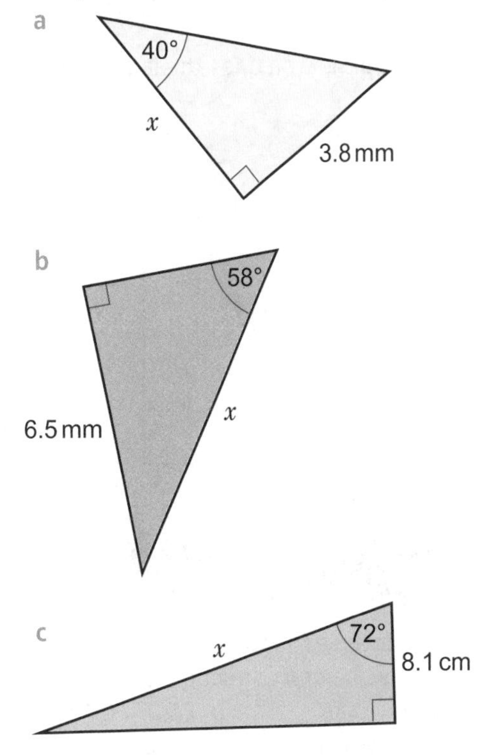

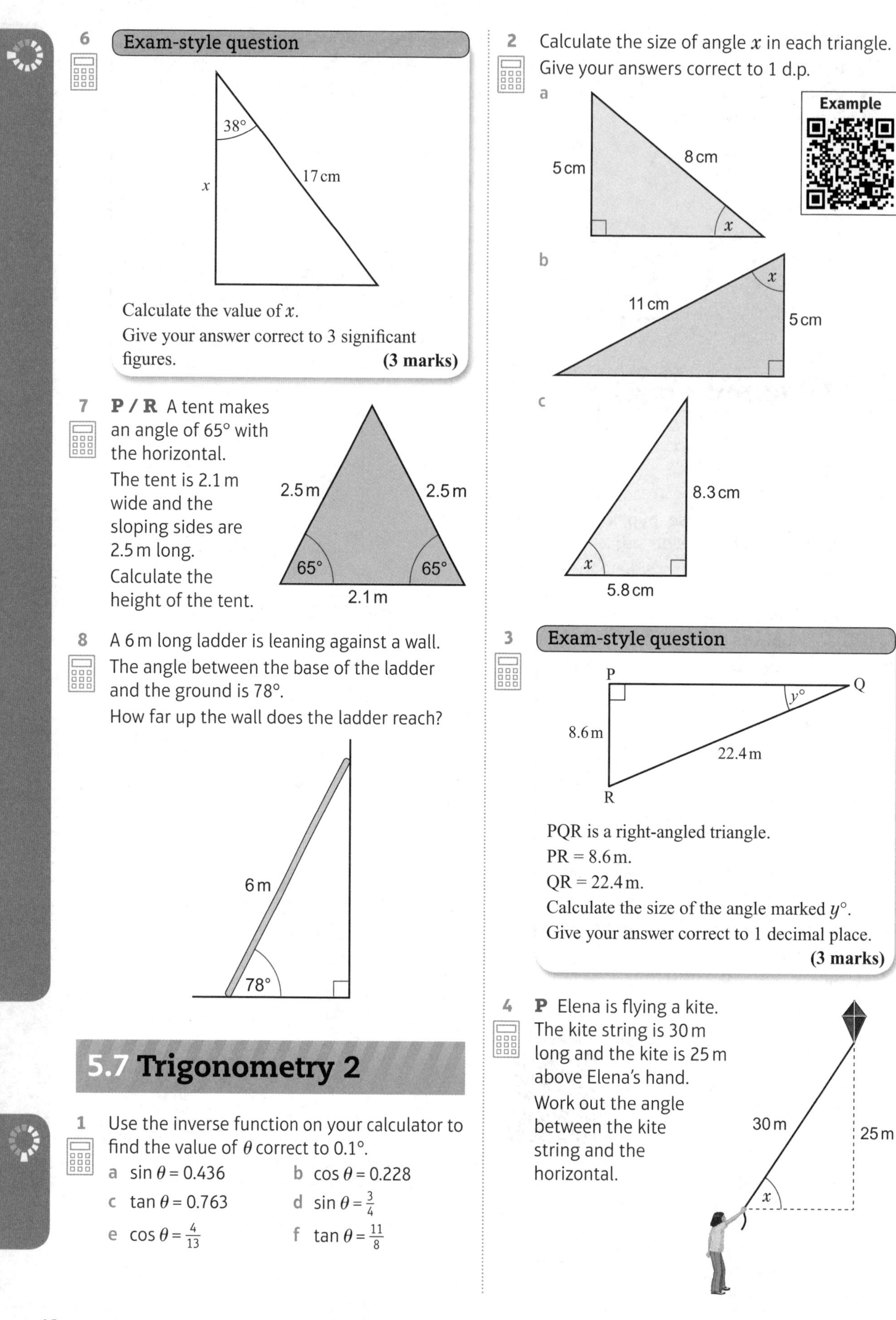

6 **Exam-style question**

Calculate the value of x.

Give your answer correct to 3 significant figures. **(3 marks)**

7 **P / R** A tent makes an angle of 65° with the horizontal.

The tent is 2.1 m wide and the sloping sides are 2.5 m long.

Calculate the height of the tent.

8 A 6 m long ladder is leaning against a wall.

The angle between the base of the ladder and the ground is 78°.

How far up the wall does the ladder reach?

5.7 Trigonometry 2

1 Use the inverse function on your calculator to find the value of θ correct to 0.1°.

a $\sin\theta = 0.436$ b $\cos\theta = 0.228$

c $\tan\theta = 0.763$ d $\sin\theta = \frac{3}{4}$

e $\cos\theta = \frac{4}{13}$ f $\tan\theta = \frac{11}{8}$

2 Calculate the size of angle x in each triangle.

Give your answers correct to 1 d.p.

Example

a

b

c

3 **Exam-style question**

PQR is a right-angled triangle.

PR = 8.6 m.

QR = 22.4 m.

Calculate the size of the angle marked y°.

Give your answer correct to 1 decimal place. **(3 marks)**

4 **P** Elena is flying a kite.

The kite string is 30 m long and the kite is 25 m above Elena's hand.

Work out the angle between the kite string and the horizontal.

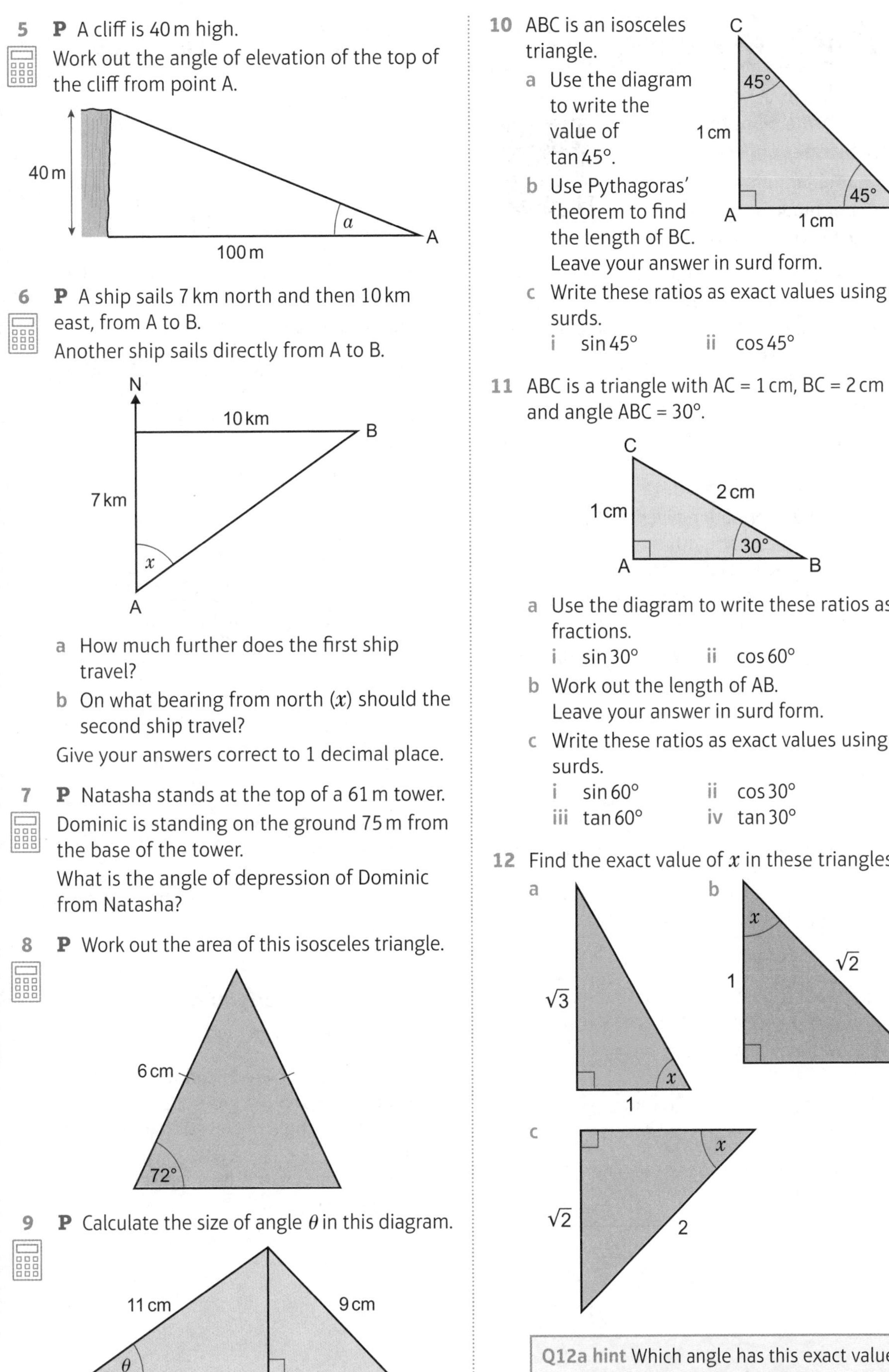

5 **P** A cliff is 40 m high.
Work out the angle of elevation of the top of the cliff from point A.

40 m, 100 m, a, A

6 **P** A ship sails 7 km north and then 10 km east, from A to B.
Another ship sails directly from A to B.

N, 10 km, B, 7 km, x, A

a How much further does the first ship travel?
b On what bearing from north (x) should the second ship travel?

Give your answers correct to 1 decimal place.

7 **P** Natasha stands at the top of a 61 m tower.
Dominic is standing on the ground 75 m from the base of the tower.
What is the angle of depression of Dominic from Natasha?

8 **P** Work out the area of this isosceles triangle.

6 cm, 72°

9 **P** Calculate the size of angle θ in this diagram.

11 cm, 9 cm, θ, 6 cm

10 ABC is an isosceles triangle.

a Use the diagram to write the value of tan 45°.
b Use Pythagoras' theorem to find the length of BC.
Leave your answer in surd form.
c Write these ratios as exact values using surds.
i sin 45° ii cos 45°

C, 45°, 1 cm, 45°, A, 1 cm, B

11 ABC is a triangle with AC = 1 cm, BC = 2 cm and angle ABC = 30°.

C, 2 cm, 1 cm, 30°, A, B

a Use the diagram to write these ratios as fractions.
i sin 30° ii cos 60°
b Work out the length of AB.
Leave your answer in surd form.
c Write these ratios as exact values using surds.
i sin 60° ii cos 30°
iii tan 60° iv tan 30°

12 Find the exact value of x in these triangles.

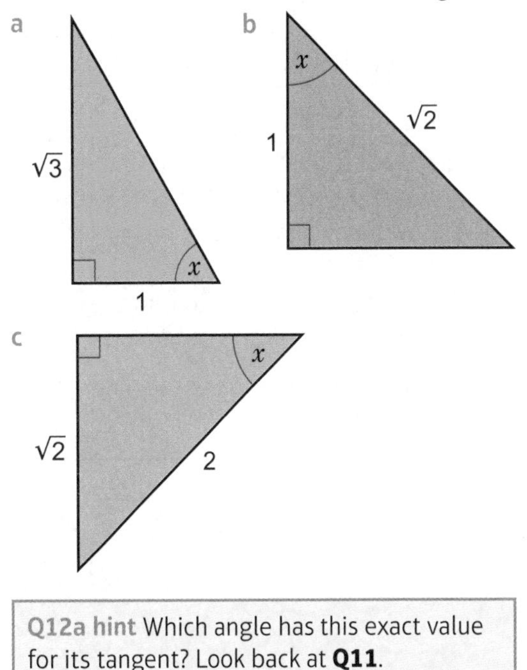

Q12a hint Which angle has this exact value for its tangent? Look back at **Q11**.

5 Problem-solving

Solve problems using these strategies where appropriate:

- **Use pictures or lists**
- **Use smaller numbers**
- **Use bar models**
- **Use x for the unknown.**

Example

1 **R** Neil is building a rabbit hutch in the shape of a triangular prism. The end view of the hutch makes an isosceles triangle with a base of 80 cm. The angle of the roof of the hutch is 76°. Part of the side panel folds down to be flat on the ground.

Neil wants to know the exterior angle that is made when the side panel is open, to make sure he has the correct hinge.

a Use the information given to draw an end view of the hutch with the door down.
Use a scale of 1 cm : 10 cm.
Label all known interior angles of the triangle.
Use the diagram below as a starting point.

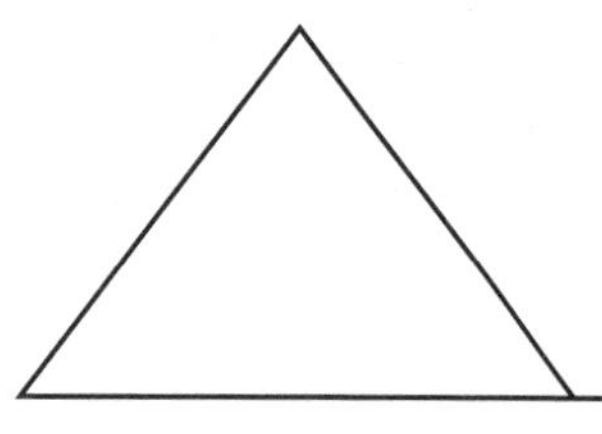

b Find the angle the side panel makes when it is fully open.

Q1 hint Use x for the exterior angle.

2 **R** Pia is part of a 'calling tree'. The first person calls three people, then each person in this second round calls three people and so on.

a How many people have been called after the fifth round?

b Write the expression to find the number of people called in the nth round.

3 **R** Use the information in the diagram to find angles a, b and c.

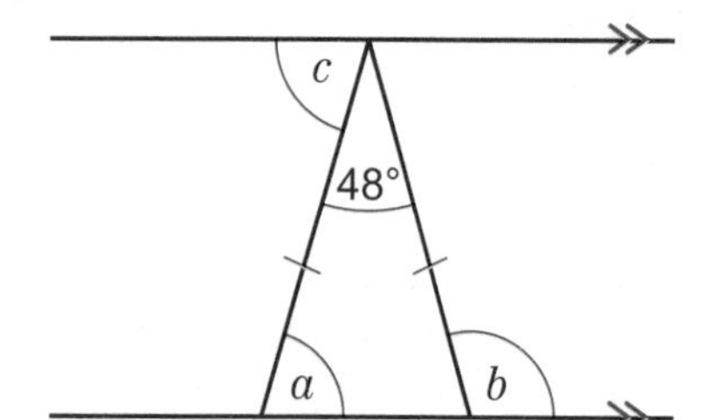

4 **R** The local election results were displayed in the newspaper using this bar chart.

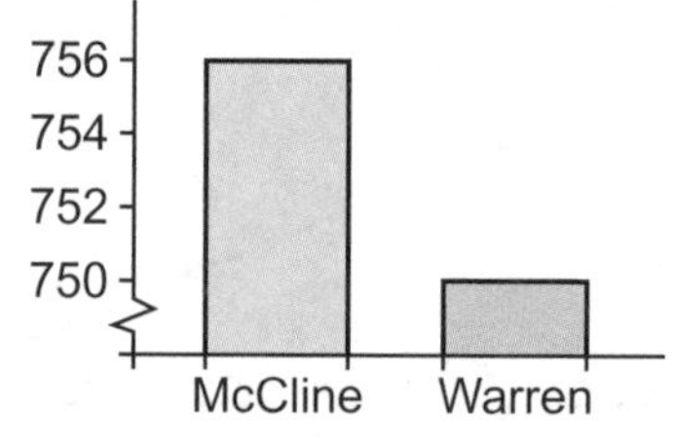

The headline read, 'McCline storms ahead in local election.'
How is this bar chart misleading?

5 **R** Geometric tiling is used all over the world to create patterns. A geometric pattern has been started below with a regular hexagon and a square.

a Find angles a and b.

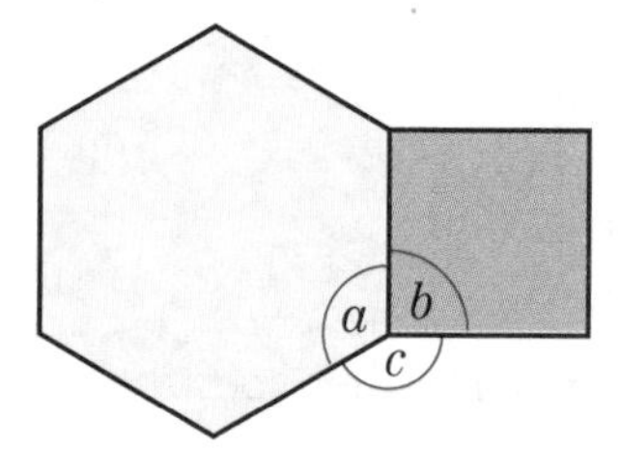

b Use angles a and b to find angle c.

c Use angle c to find which regular polygon will fit to join the two shapes.

6 Ben is putting up his tent.
The front centre pole is 1.2 m high.
Ben attaches a 2 m guy line to the top of the pole and pegs it out.
How far from the bottom of the pole does Ben put the peg for the guy line?

7 Juan pours $6\frac{1}{6}$ litres of water into some jugs.
Each jug can hold $1\frac{1}{4}$ litres.
How many jugs can Juan fill?

8 The Elizabeth Tower in London is 96.3 m tall.
From a point on the ground 25 m away, what is the angle of elevation to the top of the tower?
Give your answer to the nearest whole number.

9 Sketch a right-angled triangle.
Label the triangle ABC with the hypotenuse AC.
AB = 4 cm
BC = 6.5 cm
a Find the length of the hypotenuse to 1 d.p.
b Find the angles of the triangle to 1 d.p.

Q9 hint We know one of the angles is 90° as it is a right-angled triangle. Use SOHCAHTOA to find one of the missing angles and then subtract from 180°.

10 **Exam-style question**

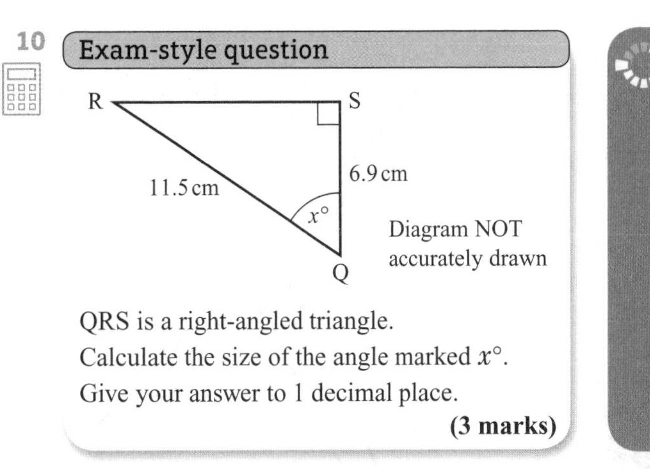

QRS is a right-angled triangle.
Calculate the size of the angle marked x°.
Give your answer to 1 decimal place.
(3 marks)

6 GRAPHS

6.1 Linear graphs

1 On squared paper, draw a line with gradient
a 2
b $\frac{1}{3}$
c −1

2 Copy and complete this table for the graphs on the grid.

Equation of line	Gradient	y-intercept
$y = 3x - 2$		
$y = 3x$		
$y = 3x + 1$		

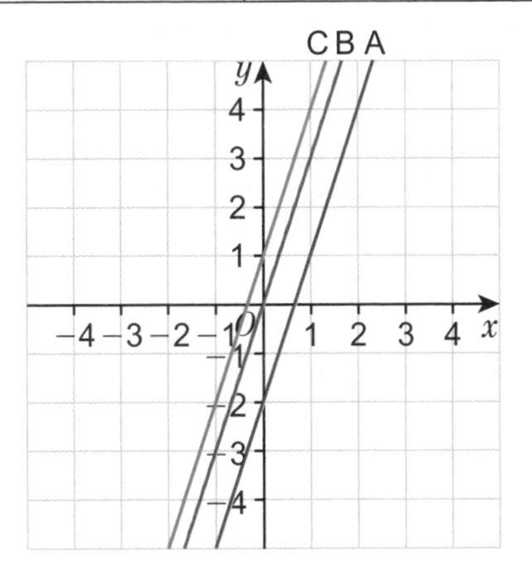

3 a Match each line to an equation.

Example

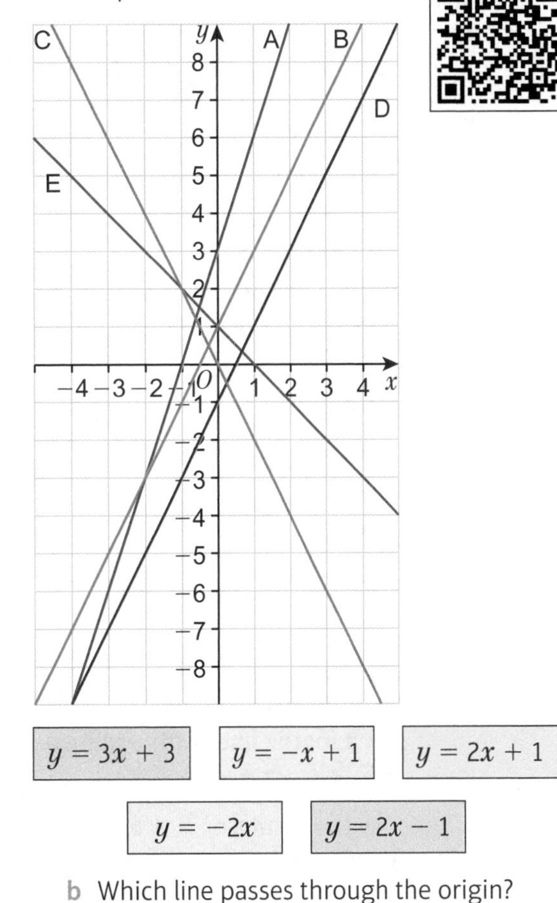

$y = 3x + 3$ $y = -x + 1$ $y = 2x + 1$ $y = -2x$ $y = 2x - 1$

b Which line passes through the origin?
c Which line is the steepest?
d Which lines have the same y-intercept?
e Which lines are parallel?

4 Write the equations of these lines.

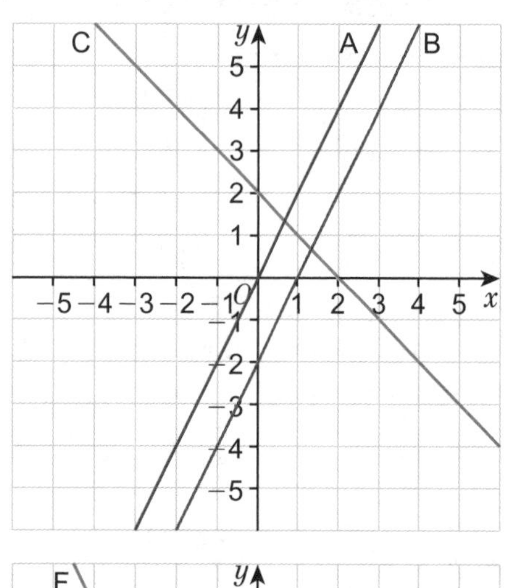

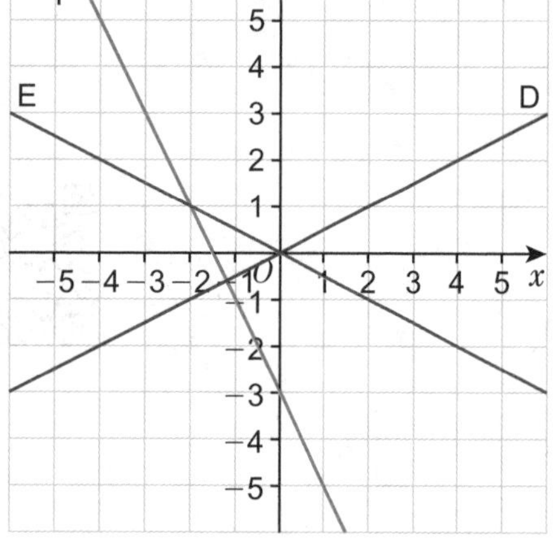

5 Here are the equations of some linear graphs.

A $y = -x + 2$ **B** $y = -x$ **C** $y = 4x + 2$

D $y = 4x - 2$ **E** $y = -4x - 2$

Which of these graphs

a cross the y-axis at the same point

b are parallel?

6 a For the equation $3y - 6x = 12$

i copy and complete the table of values

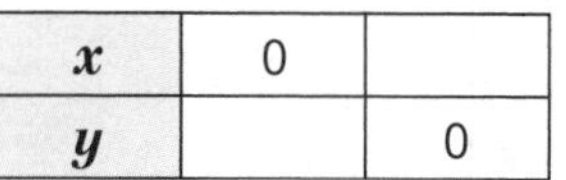

x	0	
y		0

ii plot the graph on suitable axes.

b Repeat part **a** for the lines with equation

i $x - y = 5$ ii $2x + y = 6$

7 In **Q6** you drew the graphs of $3y - 6x = 12$, $x - y = 5$ and $2x + y = 6$

a Rearrange each equation to make y the subject.

b Read the gradient and y-intercept from each.

c Look back at your graphs in **Q6** to check the gradients and y-intercepts are correct.

8 **R** Which is the steepest line?

A $y = 10 - 2x$ **B** $2 + 3x = 3y$

C $y = \frac{1}{2}x$ **D** $6x - y = 12$

E $x + y = 15$

Q8 hint Rearrange to $y = mx + c$ if necessary.

9 **P** Which of these lines pass through (0, 5)? Show how you worked it out.

A $y = 2x - 5$ **B** $2y - 3x = 10$

C $3y = 15 + 6x$ **D** $x + y = 5$

E $2x - 10 = y$

6.2 More linear graphs

1 Draw these graphs from their equations.

Use a coordinate grid from −10 to +10 on both axes.

Example

a $y = 2x + 1$ b $y = -x + 2$

c $y = 5$ d $y = 2 + x$

e $y = 3 - x$ f $y = -3x + 1$

2 **R** Match each equation to one of the sketch graphs below.

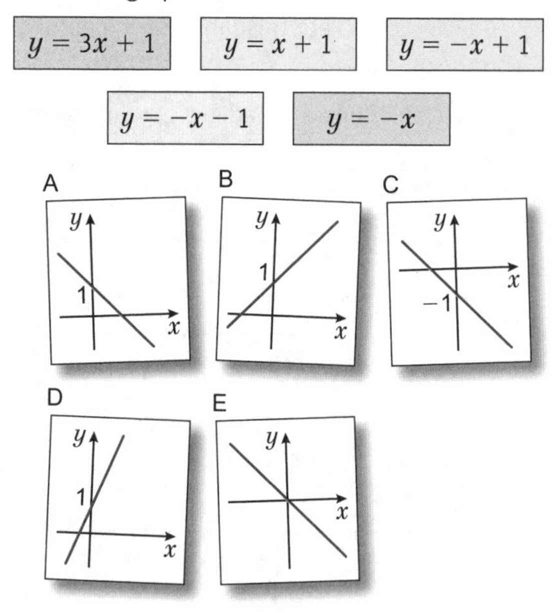

3 Sketch the graphs of

a $y = 3x$ b $y = 2x - 4$ c $x + y = 6$

4 a Find the x-intercept and y-intercept of the graph with equation

i $x + y = 7$ ii $x - y = 4$

iii $2x - y = 6$ iv $y - x = 3$

b Sketch the graphs.

5 **R** Which of these are linear functions?

A $y = x^2$ B $x = y + 4$ C $3x = 2y - 6$

D $y = \frac{10}{x}$ E $xy = 5$ F $2y + 7x + 10 = 0$

6 Copy and complete the definition.

A linear function has a graph that is a ……………… line.

7 **R** a Does the point (1, 2) lie on the line $y = 2x$?

b Does the point (1, 4) lie on the line $y = 2x - 3$?

c Does the point (−1, 5) lie on the line $y = -3x + 2$?

8 **P** A straight line has gradient −2.

The point (3, 7) lies on the line.

Find the equation of the line.

9 **P** Work out the equations of these straight-line graphs.

a The line with gradient 2 that passes through the point (0, −3)

b The line with gradient −1 that passes through the point (−4, 0)

c The line with gradient $-\frac{1}{2}$ that passes through the point (2, 5)

d The line with gradient 4 that passes through the point (−1, −1)

10 Find the gradient of the line joining points A (−3, −1) and B (−6, 5)

a by drawing the graph and using the formula

$$\text{gradient} = \frac{\text{difference in } y\text{-coordinates}}{\text{difference in } x\text{-coordinates}}$$

b using the formula $m = \frac{y_2 - y_1}{x_2 - x_1}$ where

A = (x_1, y_1) and B = (x_2, y_2)

(−3, −1) (−6, 5)

11 **R** P is the point (2, −4). Q is the point (4, 0).

a Find the gradient of line PQ.

b Write $y = mx + c$ using your gradient from part **a**. Substitute the coordinates of Q into this equation. Solve to find c.

c Write the equation of the line PQ.

12 To find the coordinates of the point where these graphs intersect

$y = 2x - 1$ and $y = -x - 4$

a write the two equations equal to each other

b solve to find x

c substitute x into one of the first equations to find y

d write the coordinates (x, y).

13 Find the coordinates of the point where these graphs intersect:

$y = -2x + 12$ and $2y - x = 14$

Q13 hint Write both as $y = mx + c$

6.3 Graphing rates of change

1 Aahil cycles from his house to his friend's house.

The distance–time graph shows his journey.

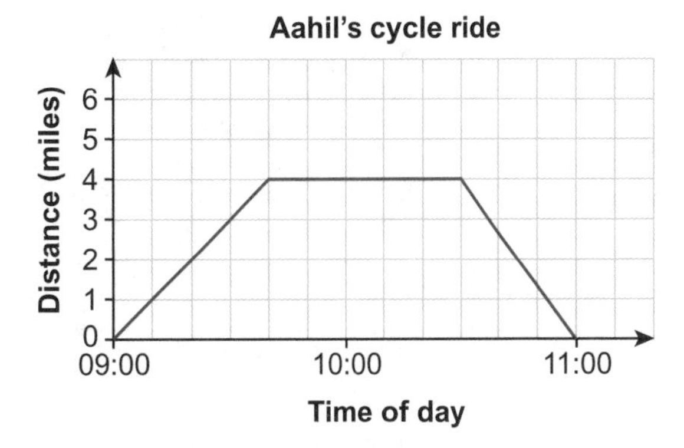

a How far is Aahil's house from his friend's house?

b What time does he arrive at his friend's house?

c How long does he take to cycle there?

d How long does he stay at his friend's house?

e What was his speed on the way to his friend's house?

f Work out the gradient for his cycle ride to his friend's house.

What do you notice?

Q1f hint Find the time as a fraction of an hour when finding the gradient.

2 Fadeelah goes for a run.

She runs for 1 hour and travels 5 miles.

She then stops for a half-hour break.

Then she runs a further 3 miles in 40 minutes.

a On graph paper draw a horizontal axis from 0 to 3 hours and a vertical axis from 0 to 8 miles.

Draw a distance–time graph to show Fadeelah's run.

b Work out her speed for the last part of her run.

3 Dillon is practising for a swimming race. He swims 1.5 km along the coast in 30 minutes, rests for 5 minutes, then returns at a speed of 2 km per hour.

a Draw a distance–time graph to show Dillon's swim.

b Work out the fastest speed he swam.

4 **Exam-style question**

James drove from Petersfield to London.

He left home at 11:00.

The graph represents part of James's journey.

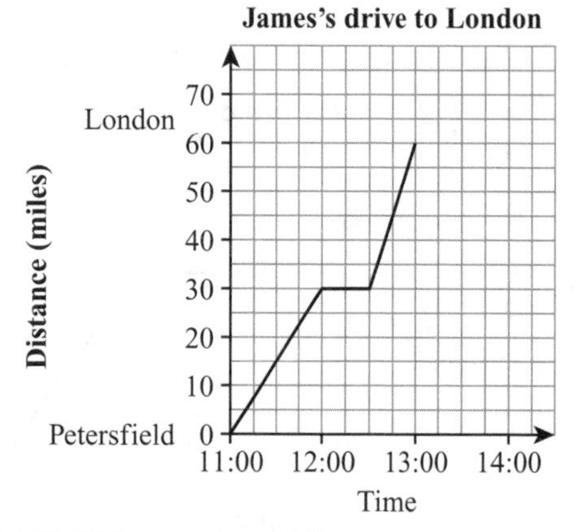

At 12:00 James stopped for a rest.

a How many minutes did he rest? **(1 mark)**

b How far was James from home at 13:00? **(1 mark)**

The last 5 miles took James 30 minutes.

c Copy and complete the graph. **(1 mark)**

5 The table shows a bus journey from Bournemouth to Salisbury. The bus stops at Ringwood and Fordingbridge on the way.

Bus stop	Time
Bournemouth (departing)	0705
Ringwood (arriving)	0740
Fordingbridge (arriving)	0755
Salisbury (arriving)	0825

When the bus stops it waits for 5 minutes at the stop before leaving.

The distances between stops are

- Bournemouth to Ringwood – 12 miles
- Ringwood to Fordingbridge – 7 miles
- Fordingbridge to Salisbury – 11 miles.

a Assuming a steady speed between stops, draw a distance–time graph for the journey.

b Work out the speed of the bus between Fordingbridge and Salisbury.

c What was the average speed for the whole journey?

6 Look at the graph you drew for **Q3**. What was Dillon's average speed for the whole swim?

7 Ferry A travels from Ryde to Portsmouth. Ferry B travels from Portsmouth to Ryde.

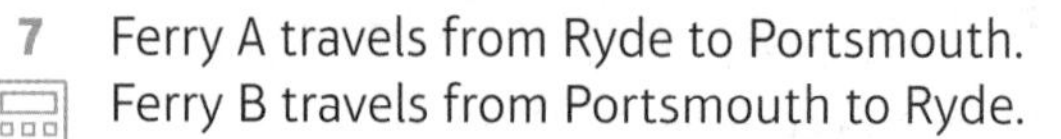

Ferry journeys to and from Portsmouth

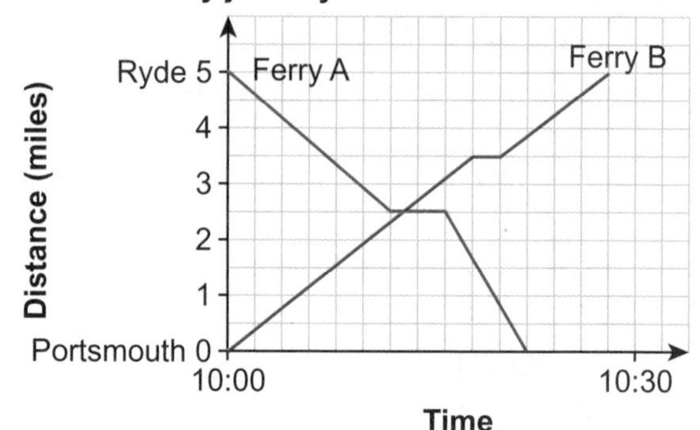

a Use the graph to estimate how far the ferries are from Portsmouth when they pass each other.

b Work out the speed for each part of the journey for Ferry A.

c When was Ferry B travelling fastest? How can you tell this from the graph?

d Which ferry travelled faster on average?

8 **R** These three vases are filled with water at a constant rate.

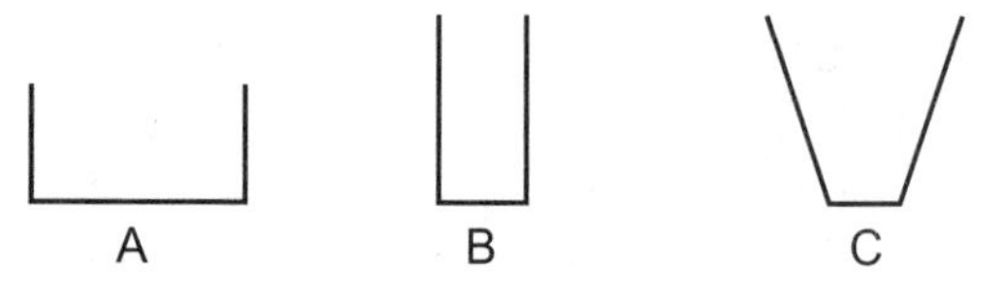

a In which container does the depth of water not increase at a constant rate?

b Which graph shows this?

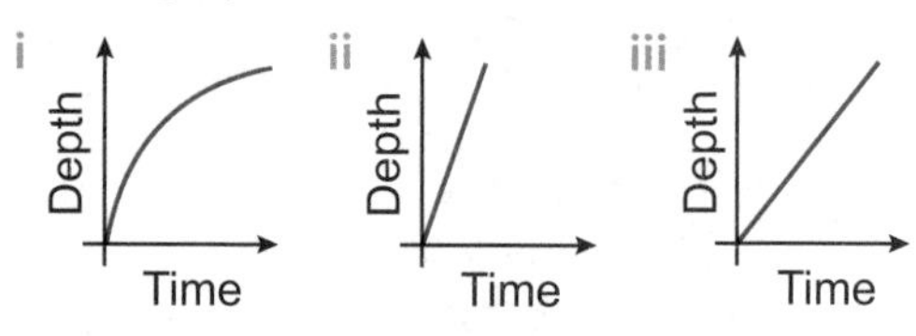

c Match each graph to a different container.

9 **R** Here are three baths.

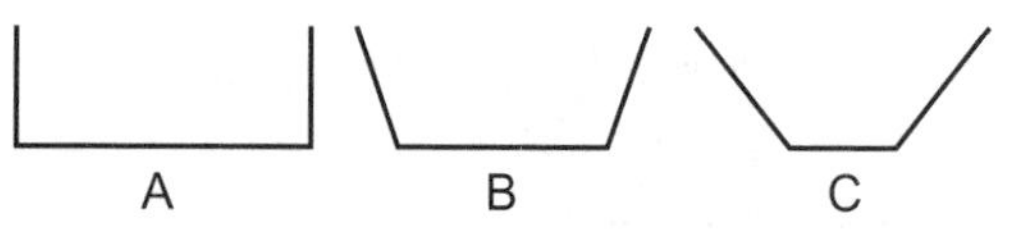

They are filled with water at a constant rate.

The baths are all the same depth.

On the same axes, sketch and label three graphs showing the rate at which water fills the baths.

10 Donald takes his boat out on a lake.
The graph shows his journey.

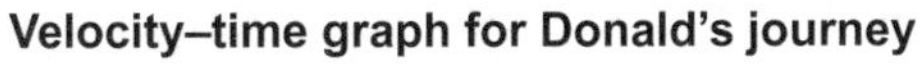

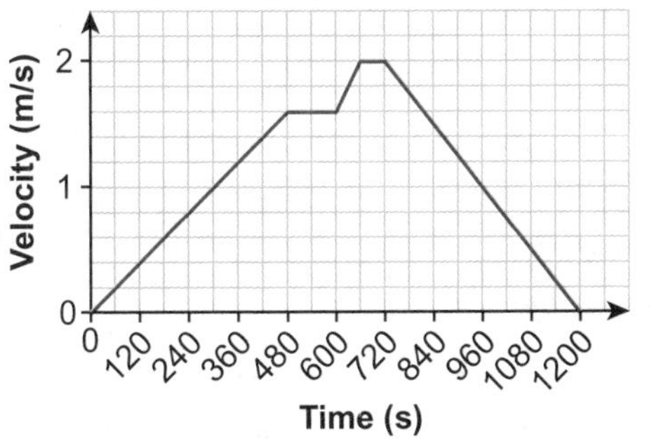

Work out

a Donald's maximum velocity

b how many minutes his velocity was greater than 1 m/s

c his acceleration for the first part of the journey

d the distance Donald travelled during the last 8 minutes of the journey.

e Copy and complete this description of Donald's journey.
He accelerated at □ m/s^2 for the first □ minutes, then travelled at a constant velocity of □ m/s for □ minutes. Next, …

> **Q10c and e hint** Acceleration is given by the gradient: $\frac{\text{change in velocity (m/s)}}{\text{time (s)}}$

6.4 Real-life graphs

1 **P** A company is buying branded memory sticks to give to its customers.
The graph shows the price per stick depending on how many are ordered.

a How much would a single memory stick cost?

b The company orders 150 memory sticks. How much does this cost altogether?

c Next year the company has a budget of £500.
How many memory sticks can they order?

2 **R** The graph shows the conversion from miles (m) to kilometres (km).

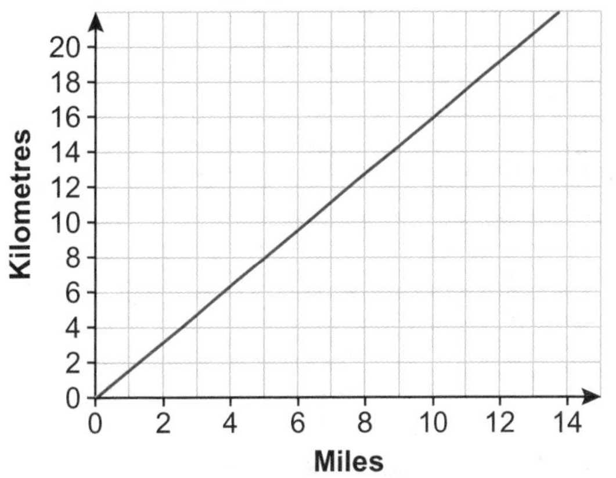

a How many km are equivalent to 10 miles?

b How many miles are equivalent to 14 km?

c Work out the gradient of the graph.

3 **R** The graph shows the cost of renting a car in Spain for a number of days.

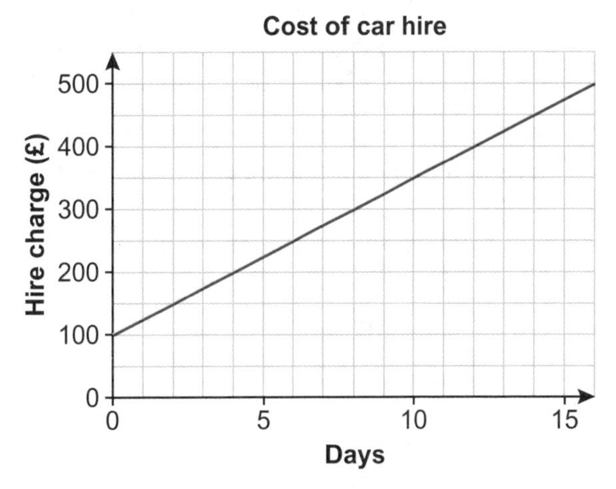

a Calculate the gradient of the line.

b What is the initial charge before you add on the daily hire charge?

c Write down the equation of the line.

d Amy is going on a two-week holiday. How much will her car hire cost?

4 Which of these graphs show one variable in direct proportion to another?

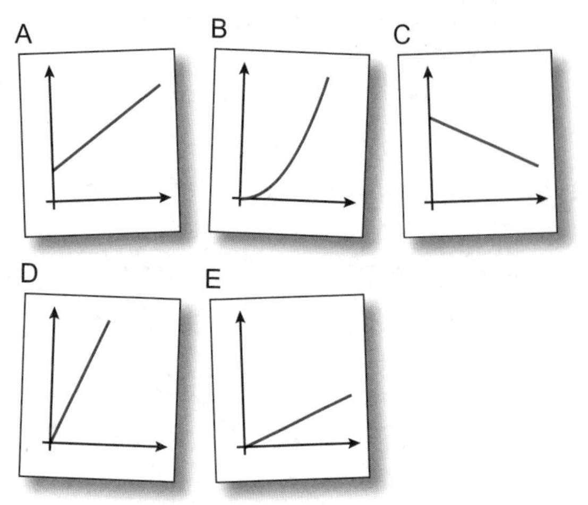

5 Look at the graphs in **Q1**, **Q2** and **Q3**. Which shows direct proportion?

6 **R** A recipe for meringues uses eggs and sugar in the ratio 1:60.

a Copy and complete this table.

Number of eggs	1	5	10
Sugar (grams)			

b Draw a graph showing grams of sugar (y) against number of eggs (x).

c Write the equation linking x and y.

d Each egg used makes 6 meringues. What is the maximum number of meringues you can make using 400 g of sugar?

7 **R** Angelica carries out an experiment recording the temperature of a cup of tea as it cools.

The graph gives information about the temperature, T°C, of the tea over time.

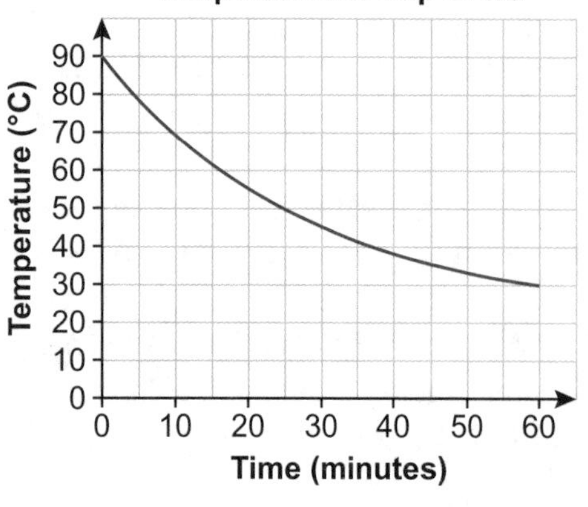

a What does the y-intercept tell you?

b Use the graph to estimate the tea's temperature half an hour after it is made.

c Will the graph intercept the x-axis? Explain your answer.

d Is the rate of decrease of temperature constant? How can you tell from the graph?

8 **R** The table shows the sales of DVDs for different prices.

Price (£)	5	7	9	11
Sales (000s)	20	15	10	5

a On a suitable grid, plot the points and draw a graph to illustrate this information.

b Use your graph to find the price at which sales will be zero.

The equation of the graph is in the form $y = mx + c$

c Use your graph to find the values of m and c.

d Explain whether your answer to part **b** is sensible.

9 **R** The graph shows lease packages for two different cars, Car A and Car B.

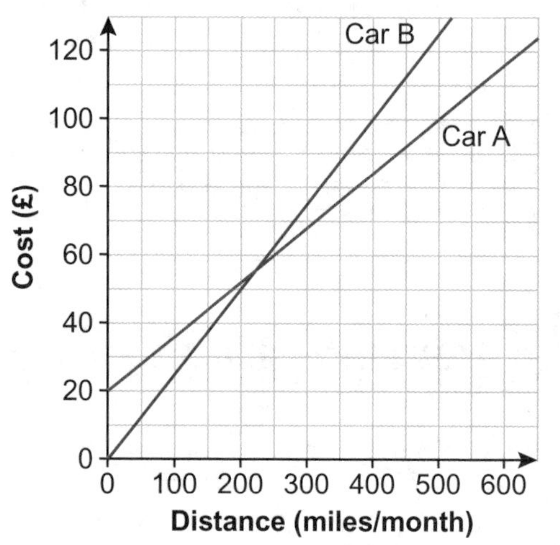

a How much does 400 miles per month cost with

i Car A ii Car B?

What is the practical meaning of

b the y-intercept value for Car A

c the point where the two graphs intersect?

d Another lease package, for Car C, is introduced. With Car C you will pay £90 per month for unlimited miles.

Which package should each person choose?

Danny: Average 250 miles per month.

Sally: Average 500 miles per month.

10 **R** The heights and weights of six students are recorded.

Height (m)	1.2	1.3	1.5	1.7	1.8	1.3
Weight (kg)	32	38	50	68	70	42

a Plot a scatter graph of the data.

b What type of correlation does this graph show?

c Draw in a line of best fit.

d Write the equation of your line of best fit.

e Rebecca is 1.35 m tall. What would you expect her to weigh?

11 **Exam-style question**

The tables show the age and value of 12 cars of the same make and type.

Age (years)	1	1	1.4	1.6	1.8	2.0
Value (£)	8200	9000	8000	7200	7400	6400

Age (years)	2.4	3.0	3.0	3.5	3.5	4.0
Value (£)	6600	6000	5200	5000	4400	3800

a From this data, work out the value of

i a 5-year old car

ii a 7-year old car. **(4 marks)**

b Which value is more reliable? Why? **(2 marks)**

Q11 hint You could draw a graph to show this information and extend it.

6.5 Line segments

1 Work out the midpoint of a line segment AB, where

a A is (6, 0) and B is (10, 4)

b A is (1, 5) and B is (2, 8)

c A is (7, 2) and B is (4, −2)

d A is (0, 0) and B is (−7, −5).

2 Work out the midpoint of a line segment PQ, where

a P is (2, 0) and Q is (6, 4)

b P is (5, 7) and Q is (−1, 9)

c P is (−4, 8) and Q is (2, −1)

d P is (5, 3) and Q is (−2, −6).

3 Work out the gradient of each line segment in **Q2**.

4 What is the length of the line segment with end points

a E (3, −2) and F (−1, −6)

b G (3, 5) and H (−2, 0)

c J (6, −3) and K (−2, 4)?

5 **R** A line is parallel to the line $y = 2x + 1$ and passes through the point (3, −2).

a Substitute the value of m for this line into $y = mx + c$

b Substitute the coordinates of the known point to work out the equation of the line.

6 **R** The graph shows the prices of two different mobile phone packages. Write the equation for the price with Plan B.

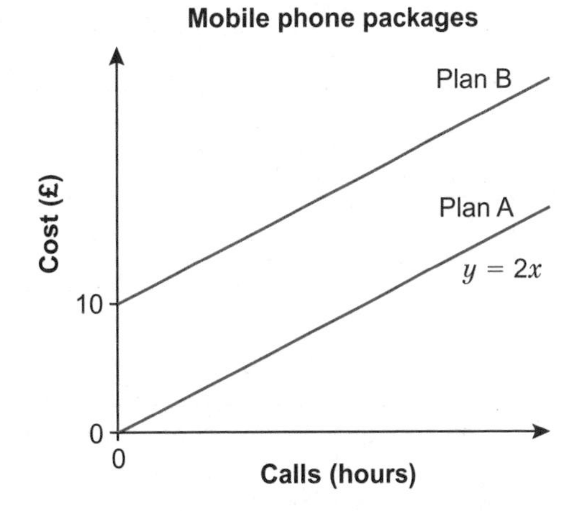

7 **P** Write the equation of a line parallel to $y = \frac{1}{2}x - 3$, which passes through the point (−1, 0).

8 **P** Find the equation of a line that passes through the point (−3, 3) and is parallel to the line with equation $y - x = 4$

9 Here are three pairs of perpendicular lines.

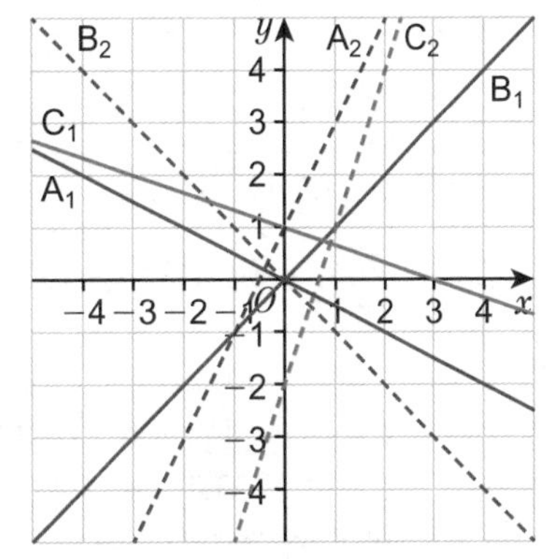

a Write down the gradient of each line.

b Multiply the gradients in each pair together. What do you notice?

10 Write down the gradient of a line perpendicular to

a $y = 2x + 5$ b $y = -x + 1$ c $y = \frac{1}{3}x - 4$

11 **Exam-style question**

Find the equation of a line

a parallel to the line $y = 2x + 1$ that passes through (0, 5) **(2 marks)**

b perpendicular to $y = 2x + 1$ that passes through (−2, 7). **(2 marks)**

6.6 Quadratic graphs

1 Which points lie on the graph of $y = x^2$?

A (0, 0)   **B** (1, 2)   **C** (2, 4)   **D** (3, 6)
E (4, 16)   **F** (−1, 1)   **G** (−2, −4)   **H** (−3, 9)

2 a Copy and complete this table of values for $y = x^2 + 2$

x	−3	−2	−1	0	1	2	3
x^2							
+ 2	+ 2	+ 2	+ 2	+ 2	+ 2	+ 2	+ 2
y							

b Plot the graph of $y = x^2 + 2$

Q2a hint For quadratic functions with more than one step, you can include a row for each step in the table.

3 **Exam-style question**

Draw the graph of $y = (x + 1)^2$ for $-5 \leqslant x \leqslant 3$ **(4 marks)**

4 a Copy and complete this table of values for $y = -2x^2$

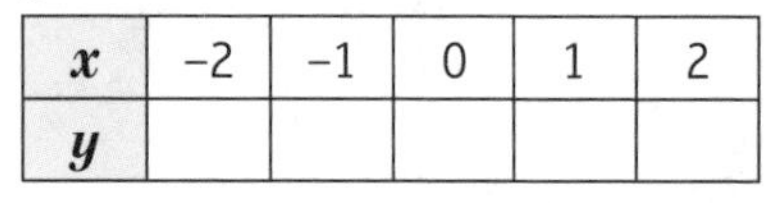

x	−2	−1	0	1	2
y					

b Plot the graph of $y = -2x^2$

5 **R** Compare your graphs from **Q2**, **Q3** and **Q4**.

a What is the same about these graphs?

b Which ones have a minimum point? Which ones have a maximum point?

c Find the coordinates of the minimum/ maximum point for each graph.

d Describe the symmetry of each graph by giving the equation of its mirror line.

6 A maths student uses a graph to model the trajectory of a ball.
The graph shows the height, h metres, of the ball at a time t seconds after it is thrown straight up.

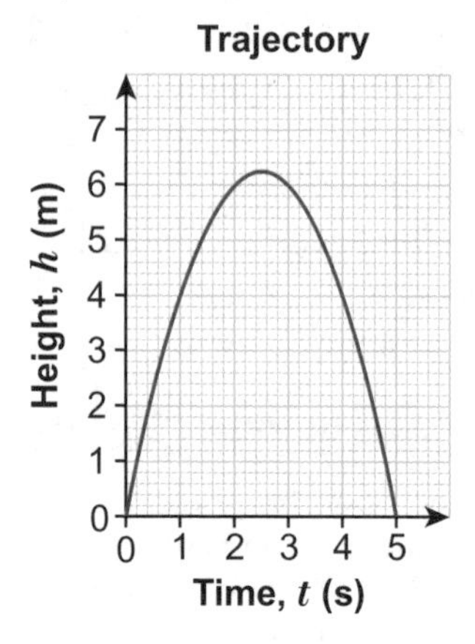

a What type of graph is this?

b When is the ball travelling fastest?

c When is the ball's speed zero?

d What is the maximum height that the ball reaches?

e How long is the ball in the air?

f The ball was caught when it was 1.5 m above the ground. Use the graph to estimate the times this could have occurred.

7 Here is the graph of $y = x^2 - 4$
Use the graph to solve the equation $x^2 - 4 =$ 0

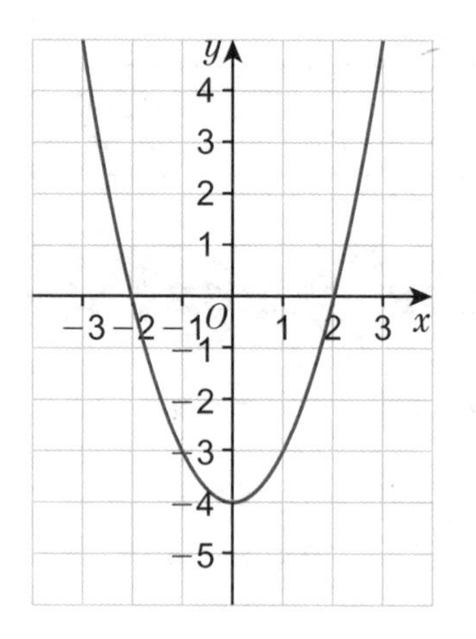

8 Here are four graphs. Use these graphs to solve the equations

a $x^2 - 3x = 0$ b $x^2 + 3x = 0$

c $-x^2 + 5x - 4 = 0$ d $(x - 2)^2 = 0$

a $y = x^2 - 3x$

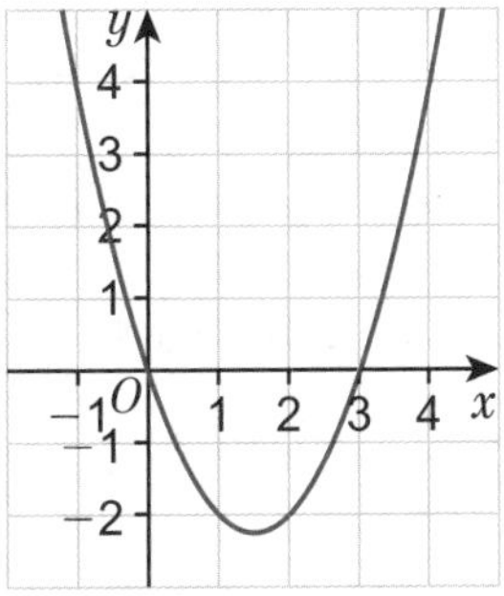

b $y = x^2 + 3x$

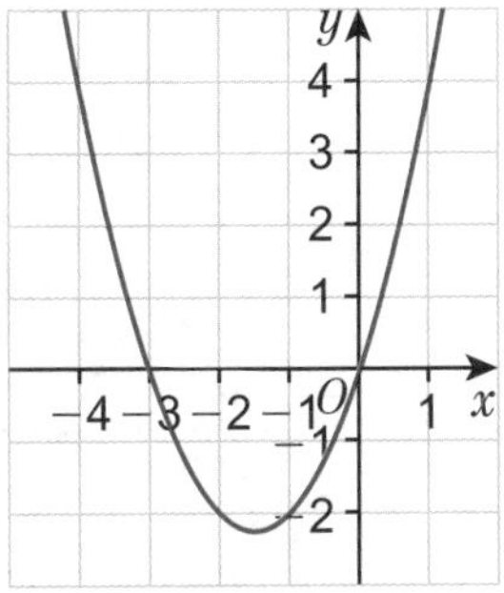

c $y = -x^2 + 5x - 4$

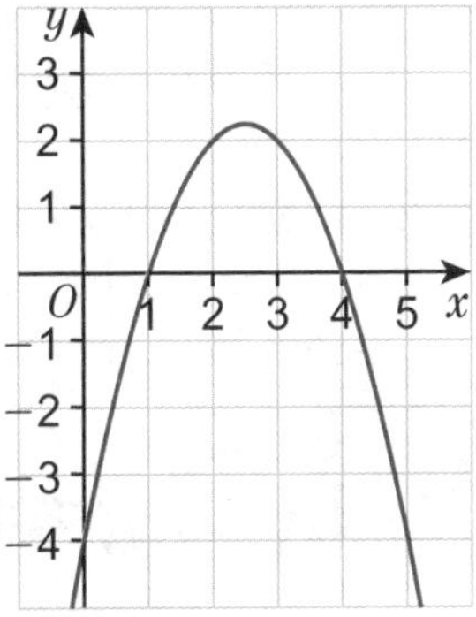

d $y = (x - 2)^2$

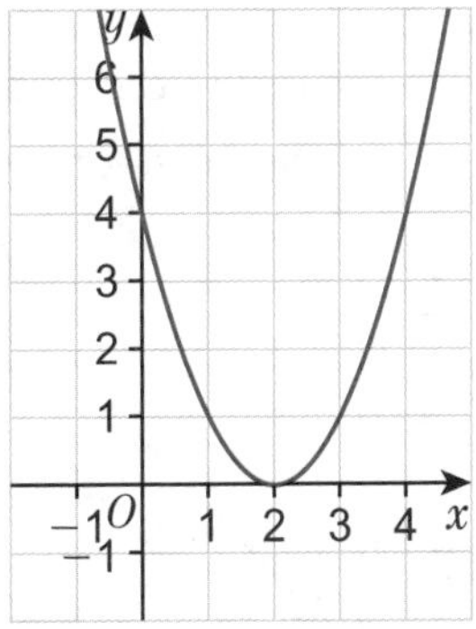

9 **R** Use the graphs in **Q8** to solve the equations

a $x^2 - 3x - 2 = 0$

b $x^2 + 3x - 4 = 0$

c $-x^2 + 4x - 4 = 0$

d $(x - 2)^2 - \frac{1}{2}x = 0$

e Explain why $x^2 - 3x = -3$ has no solutions.

Example

10 Exam-style question

a Copy and complete the table for $y = x^2 + 2x + 4$

x	−4	−3	−2	−1	0	1
y		7				

(2 marks)

b Draw the graph of $y = x^2 + 2x + 4$ **(2 marks)**

c By drawing a suitable line on your graph, solve the equation $x^2 + x - 3 = 0$ **(2 marks)**

11 An ornithologist records the height, h metres, of a bird during flight at time, t seconds, after it is released.
The table gives the data he records.

Time, t (seconds)	0	1	2	3	4
Height, h (metres)	0	3.4	6.4	9	11.2

a Use this data to draw a graph showing the trajectory of the bird.

b Continue the graph to predict the height of the bird after 8 seconds.

6.7 Cubic and reciprocal graphs

1 Which of the points are on the graph of $y = x^3$?

A (0, 0) **B** (1, 1) **C** (−1, −1) **D** (2, 6)

E (−2, −8) **F** (1, 3) **G** (−1, 1) **H** (2, −8)

2 **R** Here is the graph of $y = x^3$

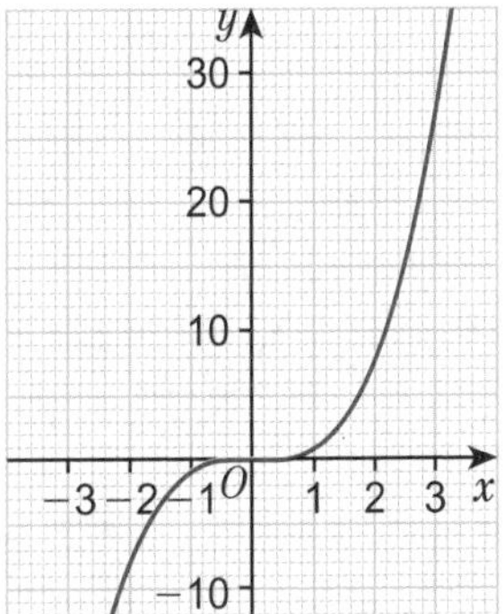

a Use the graph to estimate

i 2.5^3 ii $\sqrt[3]{-2}$

b Use a calculator to work out

i 2.5^3 ii $\sqrt[3]{-2}$

3 a Copy and complete the table of values for $y = x^3 + 1$

x	-3	-2	-1	0	1	2	3
y	-26						

b Draw the graph of $y = x^3 + 1$ for $-3 \leqslant x \leqslant 3$.

c What is the same and what is different about this graph and the one in **Q2**?

4 a Plot graphs of $y = x^3 - 1$ and $y = x^3 + 2$ for $-3 \leqslant x \leqslant 3$.

b Compare these two graphs and the graphs from **Q2** and **Q3**.
What similarities can you see?
What are the differences?

5 **R** a Draw a table of values for $y = \frac{2}{x}$, where $x \neq 0$, for $-3 \leqslant x \leqslant 3$.

b Draw the graph of $y = \frac{2}{x}$

c Draw a table of values for $y = -\frac{2}{x}$, where $x \neq 0$, for $-3 \leqslant x \leqslant 3$.

d Draw the graph of $y = -\frac{2}{x}$

e What is the same and what is different about $y = \frac{2}{x}$ and $y = -\frac{2}{x}$?

6 a Draw the graph of $y = \frac{1}{x}$, where $x \neq 0$, for $-4 \leqslant x \leqslant 4$.

b Use your graph to estimate the value of y when

i $x = 2.5$ ii $x = -3.5$ iii $x = -1.2$

7 **R** Match each equation to a graph.

a $y = x^2$ b $y = -\frac{1}{x}$ c $y = x^3$

d $y = -2x$ e $y = -x$ f $y = -\frac{1}{2}x$

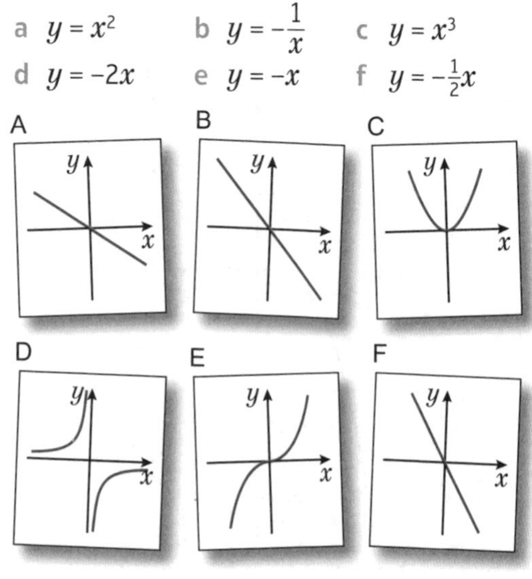

8 Use the graphs you drew in **Q4** to solve the equations

a $x^3 - 1 = 0$ b $x^3 + 2 = 0$ c $x^3 - 1 = -5$

9 **Exam-style question**

a Complete the table of values for $y = \frac{6}{x}$

x	0.5	1	2	3	4	5	6
y		6	3		1.5		1

(2 marks)

b Draw the graph of $y = \frac{6}{x}$ for $0.5 \leqslant x \leqslant 6$ **(2 marks)**

c Hence or otherwise solve $\frac{6}{x} = 4$ **(2 marks)**

10 This is the graph of $y = 2x^3 + 7x^2 - 4$

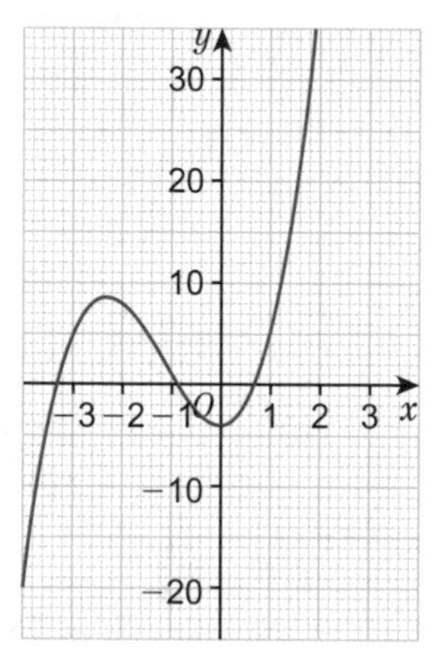

By drawing suitable lines on the graph,

a solve the equation $2x^3 + 7x^2 - 4 = 0$

b solve the equation $2x^3 + 7x^2 - 4 = 10$

6.8 More graphs

1 **R** The distance–time graphs represent Mr Murphy's cycle rides on two different days.

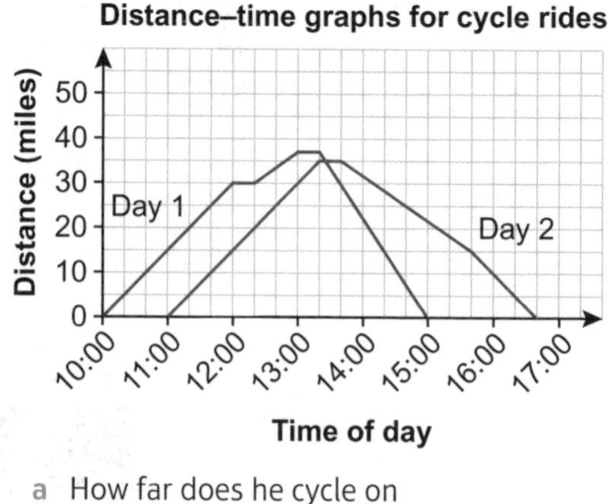

a How far does he cycle on

i Day 1 ii Day 2?

b Which day does his cycle ride take longer, and by how many minutes is it longer?

c Work out his greatest speed on Day 2.

d How long did he stop for in total on Day 1?

e Work out his average speed during the whole journey on Day 1.

f What does the change in gradient between 15:00 and 16:00 on Day 2 show?

2 The graph shows the height of an object dropped from the top of a 100 m tower.
When does the object reach terminal velocity?

Height of falling object

Q2 hint Any object when dropped will reach terminal velocity. This is the point where the velocity will not increase any more and is constant.

3 **R** Here are two sets of data.

Data set A

x	3	3	4	6	6	8	9	9
y	7	6	10	13	14	16	19	20

Data set B

x	3	3.1	3.3	3.4	3.4	3.7	3.8	3.8
y	10.5	10.7	12.0	12.6	12.5	14.9	15.5	15.4

a Plot each set of data on a scatter graph.

b Describe the correlation for each set.

c Draw a line of best fit for the graph for data set A. What does this show?

d Draw the graph of $y = x^2 + 1$ on the same grid as the graph for data set B.
Copy and complete this table of values to help you.

x	3	3.1	3.2	3.3	3.4	3.5
y						

x	3.6	3.7	3.8	3.9	4
y					

e What do you think the relationship is between x and y in data set B?

4 **R** The table gives some information about the value of a car over 6 years from new.

Age (years)	0	1	2	3	4	5	6
Value (£000s)	22	18.5	15.5	13	11	10	9.5

a Draw axes on graph paper using 1 cm to represent 1 year on the horizontal axis and 1 cm to represent £2000 on the vertical axis. Plot the values from the table and join them with a smooth curve.

From the graph, estimate

b the value of the car after 3.5 years

c the age at which the car is worth 75% of its original price.

5 The graph shows the number of bacteria in a Petri dish.

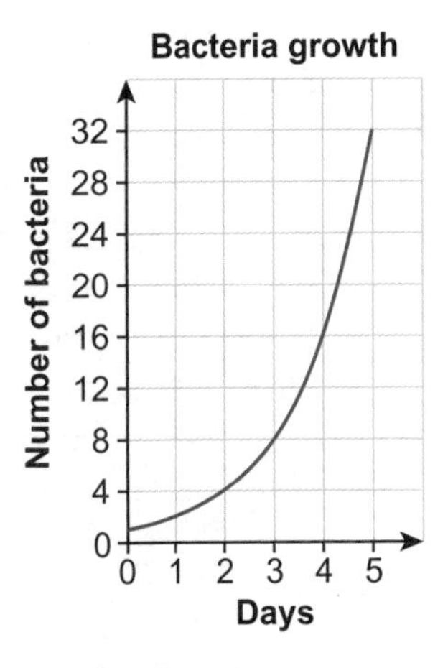

a How many bacteria were there at the start of the experiment?

b Explain how you found your answer to part **a**.

c Estimate the number of bacteria after
i 2 days ii 5 days.

d Describe the change in the number of bacteria from
i day 1 to day 2
ii day 2 to day 3
iii day 3 to day 4.

6 **R** The graph shows the temperature of a liquid during an experiment.

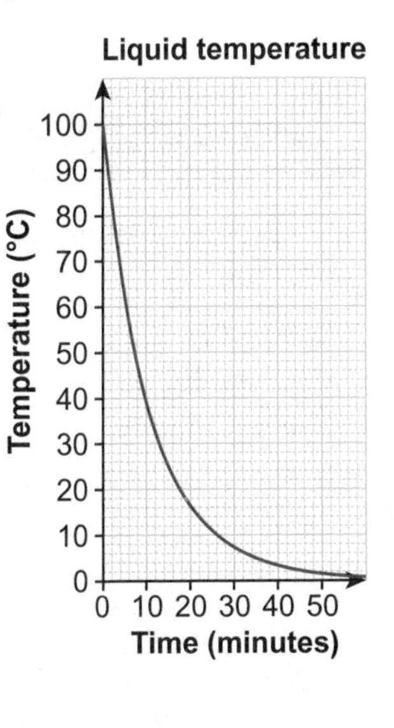

a Estimate the temperature after 10 minutes.

b After how many minutes will the temperature have halved?

c Does the temperature ever reach zero?

7 **P** The graph shows the value of a child's investment.

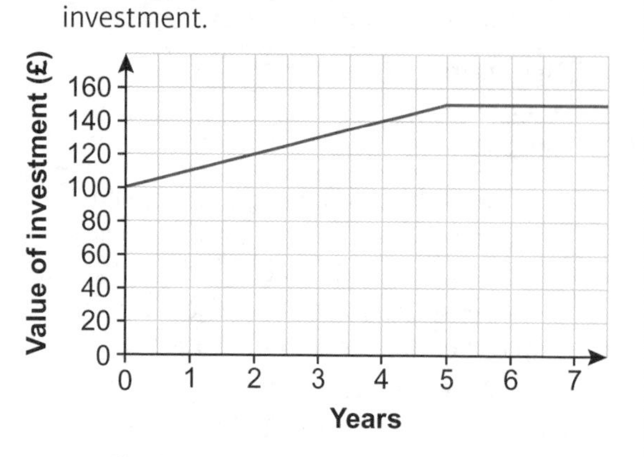

a What was the initial value of the investment?

b What is the value of the investment after 5 years?

c How much did the value increase each year?

d The rate of interest remained the same for the first five years.
Work out the percentage interest rate.

e What happened after 5 years?

> **Q7d hint**
> Percentage interest rate
> $= \frac{\text{actual change}}{\text{original amount}} \times 100$

8 On graph paper, draw the graphs of

a $x^2 + y^2 = 4$

b $x^2 + y^2 = 9$

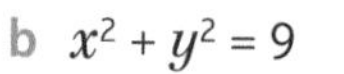

c $x^2 + y^2 = 100$

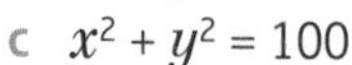

d $x^2 + y^2 = 25$

6 Problem-solving

Solve problems using these strategies where appropriate:

- **Use pictures**
- **Use smaller numbers**
- **Use bar models**
- **Use x for the unknown.**

1 This distance–time graph shows part of Jason's journey.
Lee made the same journey with an average speed that was 2 mph faster than Jason's.
Yasmin made the same journey at an average speed of 27 mph.

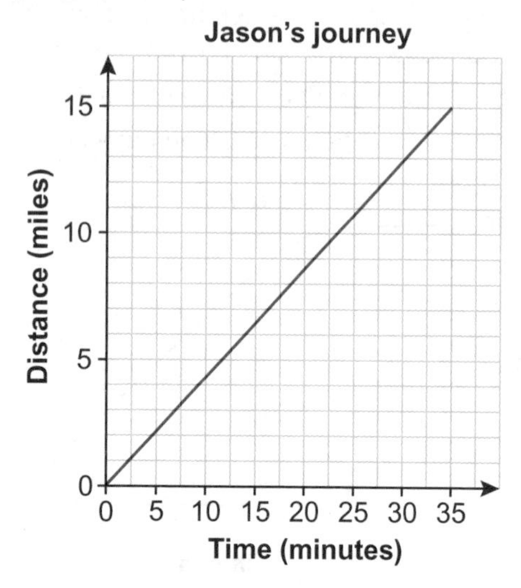

a Which of Lee or Yasmin was travelling at a faster average speed?

b How much faster?

2 **R** Look at this scatter graph.
It shows the attendance at an outdoor pool against the maximum daily temperature over two weeks in the summer.

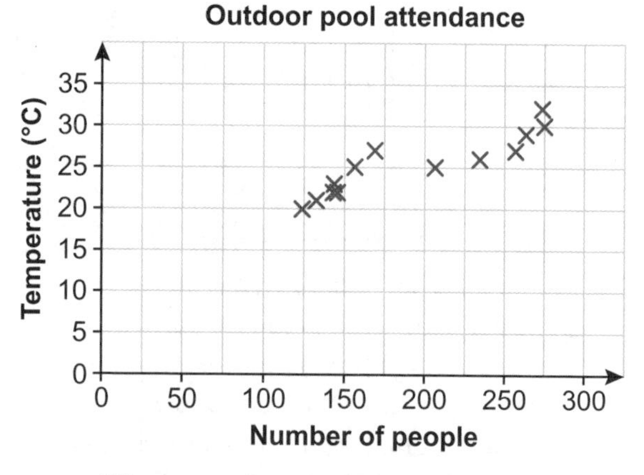

a What were the approximate highest and lowest maximum daily temperatures during the two weeks?

b What were the approximate highest and lowest numbers of people at the pool over the two weeks?

c Explain how you know if there is a correlation or not.

d Make a statement summarising the results shown on this graph.

3 **R** A vase holds 800 cm³ of water.
The warmth of the room makes 2.5 cm³ of water evaporate each day.

a Find an expression, in terms of n, for the amount of water in the vase at the end of the nth day.

b How many days before the vase is completely empty?
Explain your answer.

4 **Exam-style question**

The scatter graph shows information about the height and arm length of each of 8 students in Year 11.

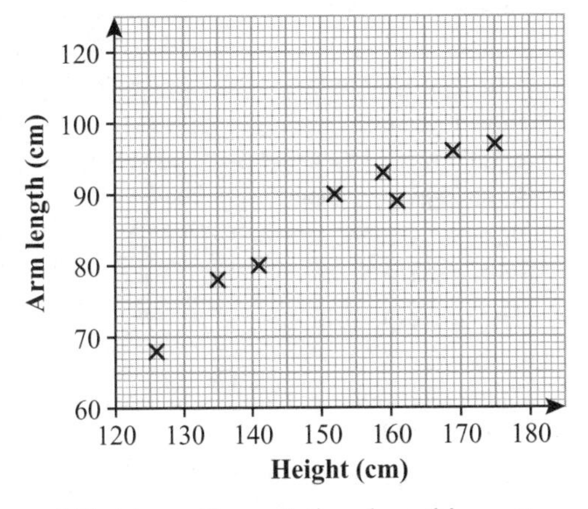

a What type of correlation does this scatter graph show? **(1 mark)**

Another student in Year 11 has a height of 148 cm.

b Estimate the arm length of this student. **(2 marks)**

November 2012, Q2, 1MA0/1H

5 Jan, Rowena and Atiq share £420 in the ratio 4 : 3 : 5.

a How much does each person get?

Rowena shares her amount equally with Flo.

b Write the new ratio Jan : Rowena : Flo : Atiq using integers.

6 **R** Daniel and Maria are using these linear equations to draw graphs.

$y = x - 2$ $\quad$ $y = 8x$ $\quad$ $y = 2x + 1$

$y = 3x - 3$ $\quad$ $y = 4x$

The first line Daniel draws is parallel to $y = 2x - 6$. The first line Maria draws passes through the origin.

a Which equation did Daniel use?

b Can you tell which equation Maria used? Explain.

7 **R** Ryan sketches a graph to show that 1 British pound = 2.06 Singapore dollars.
Which of these graphs could be Ryan's sketch?
Explain how you know.

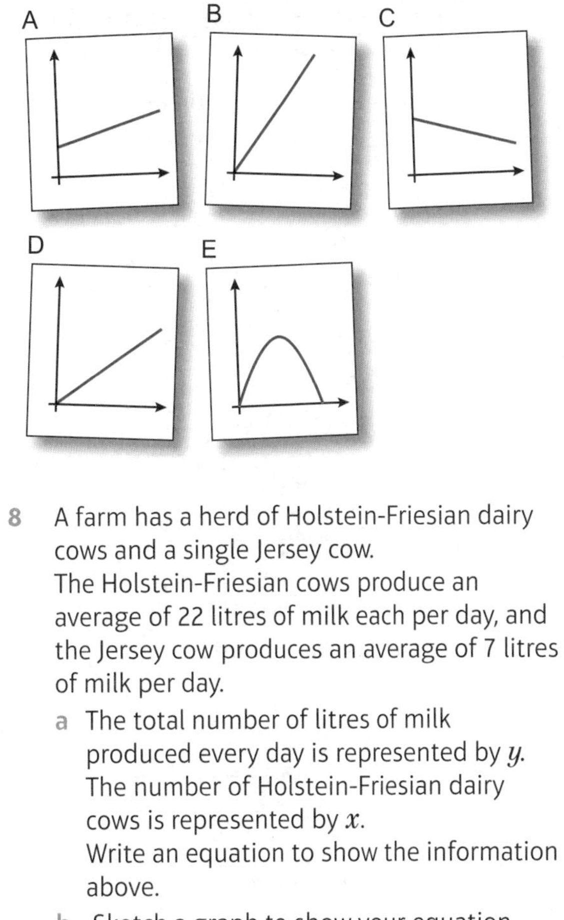

8 A farm has a herd of Holstein-Friesian dairy cows and a single Jersey cow.
The Holstein-Friesian cows produce an average of 22 litres of milk each per day, and the Jersey cow produces an average of 7 litres of milk per day.

a The total number of litres of milk produced every day is represented by y.
The number of Holstein-Friesian dairy cows is represented by x.
Write an equation to show the information above.

b Sketch a graph to show your equation.

9 A road ascends from sea level at an angle of 36° for 2.2 miles.
How much higher is the top of the road than the bottom?
Give your answer to 1 d.p.

10 Work out the equation of a line perpendicular to $y = 4x - 5$, and passing through the point (8, 0).

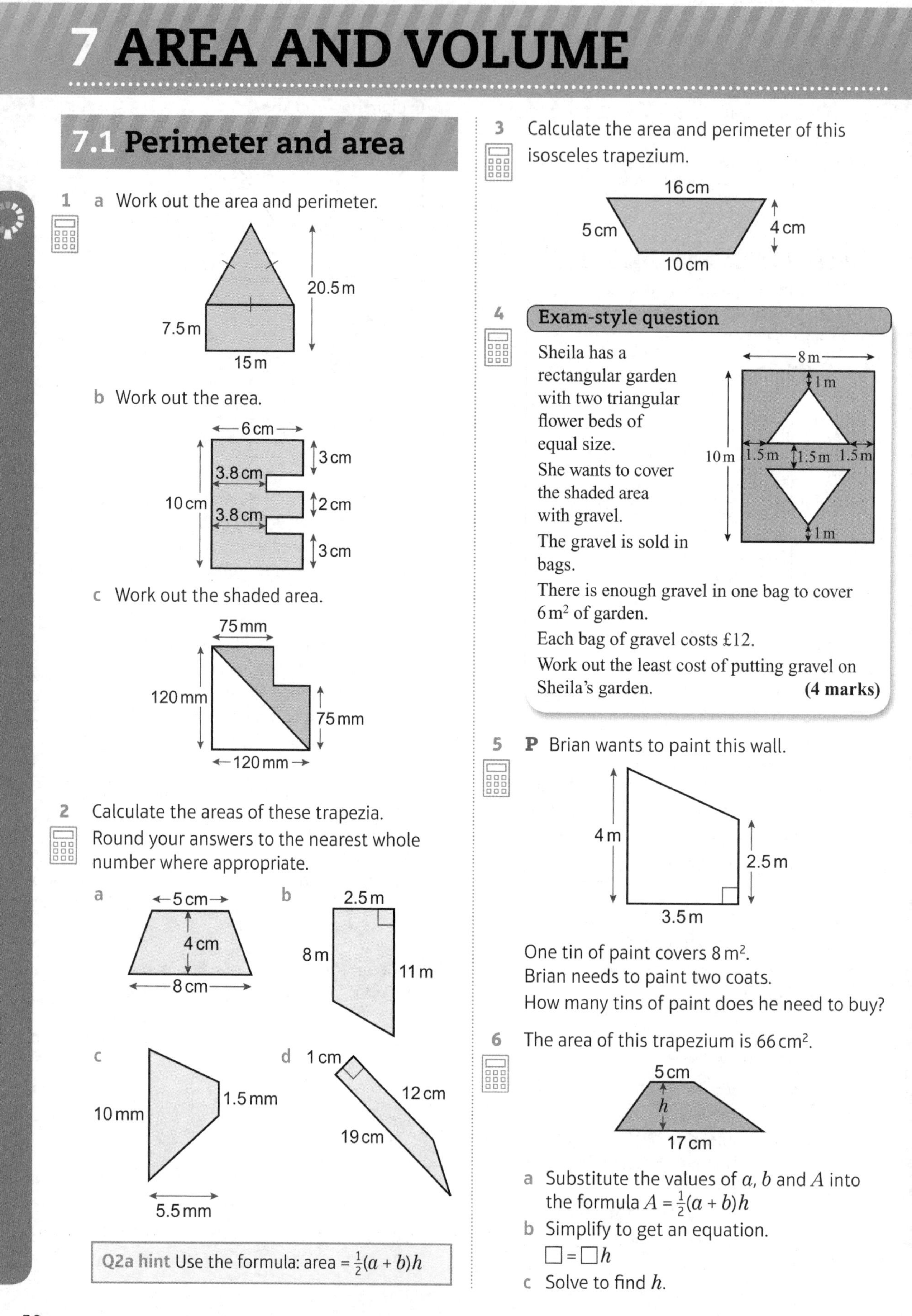

7 AREA AND VOLUME

7.1 Perimeter and area

1 **a** Work out the area and perimeter.

b Work out the area.

c Work out the shaded area.

2 Calculate the areas of these trapezia.
Round your answers to the nearest whole number where appropriate.

a

b

c

d

Q2a hint Use the formula: area = $\frac{1}{2}(a + b)h$

3 Calculate the area and perimeter of this isosceles trapezium.

4 **Exam-style question**

Sheila has a rectangular garden with two triangular flower beds of equal size.

She wants to cover the shaded area with gravel.

The gravel is sold in bags.

There is enough gravel in one bag to cover $6\,m^2$ of garden.

Each bag of gravel costs £12.

Work out the least cost of putting gravel on Sheila's garden. **(4 marks)**

5 **P** Brian wants to paint this wall.

One tin of paint covers $8\,m^2$.
Brian needs to paint two coats.
How many tins of paint does he need to buy?

6 The area of this trapezium is $66\,cm^2$.

a Substitute the values of a, b and A into the formula $A = \frac{1}{2}(a + b)h$

b Simplify to get an equation.
$\square = \square h$

c Solve to find h.

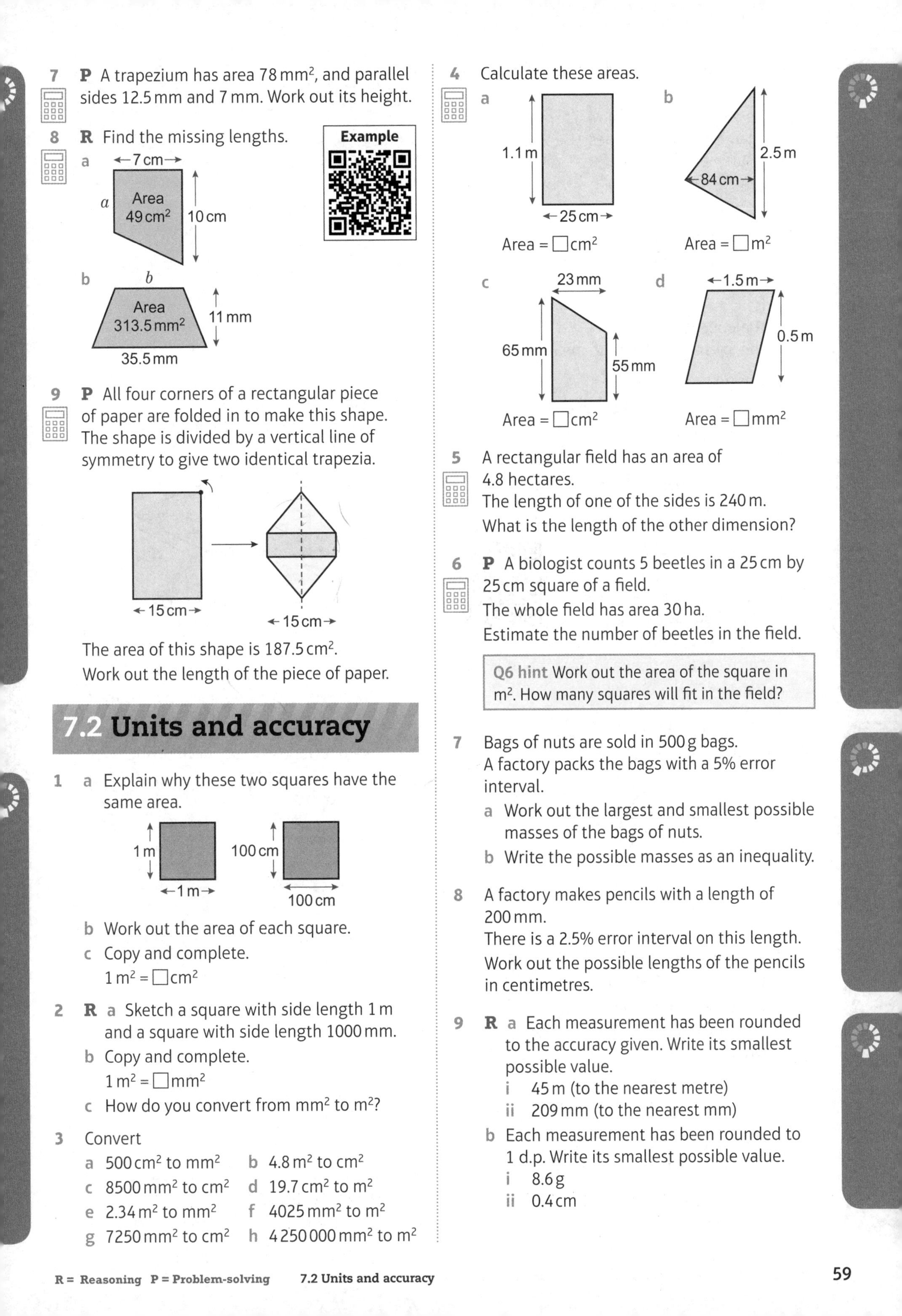

7 **P** A trapezium has area 78 mm², and parallel sides 12.5 mm and 7 mm. Work out its height.

8 **R** Find the missing lengths.

Example

a

b

9 **P** All four corners of a rectangular piece of paper are folded in to make this shape. The shape is divided by a vertical line of symmetry to give two identical trapezia.

The area of this shape is 187.5 cm².
Work out the length of the piece of paper.

7.2 Units and accuracy

1 a Explain why these two squares have the same area.

b Work out the area of each square.

c Copy and complete.
1 m² = □ cm²

2 **R** a Sketch a square with side length 1 m and a square with side length 1000 mm.

b Copy and complete.
1 m² = □ mm²

c How do you convert from mm² to m²?

3 Convert

a 500 cm² to mm²
b 4.8 m² to cm²
c 8500 mm² to cm²
d 19.7 cm² to m²
e 2.34 m² to mm²
f 4025 mm² to m²
g 7250 mm² to cm²
h 4 250 000 mm² to m²

4 Calculate these areas.

a

Area = □ cm²

b

Area = □ m²

c

Area = □ cm²

d

Area = □ mm²

5 A rectangular field has an area of 4.8 hectares.
The length of one of the sides is 240 m.
What is the length of the other dimension?

6 **P** A biologist counts 5 beetles in a 25 cm by 25 cm square of a field.
The whole field has area 30 ha.
Estimate the number of beetles in the field.

Q6 hint Work out the area of the square in m². How many squares will fit in the field?

7 Bags of nuts are sold in 500 g bags.
A factory packs the bags with a 5% error interval.

a Work out the largest and smallest possible masses of the bags of nuts.

b Write the possible masses as an inequality.

8 A factory makes pencils with a length of 200 mm.
There is a 2.5% error interval on this length.
Work out the possible lengths of the pencils in centimetres.

9 **R** a Each measurement has been rounded to the accuracy given. Write its smallest possible value.

i 45 m (to the nearest metre)
ii 209 mm (to the nearest mm)

b Each measurement has been rounded to 1 d.p. Write its smallest possible value.

i 8.6 g
ii 0.4 cm

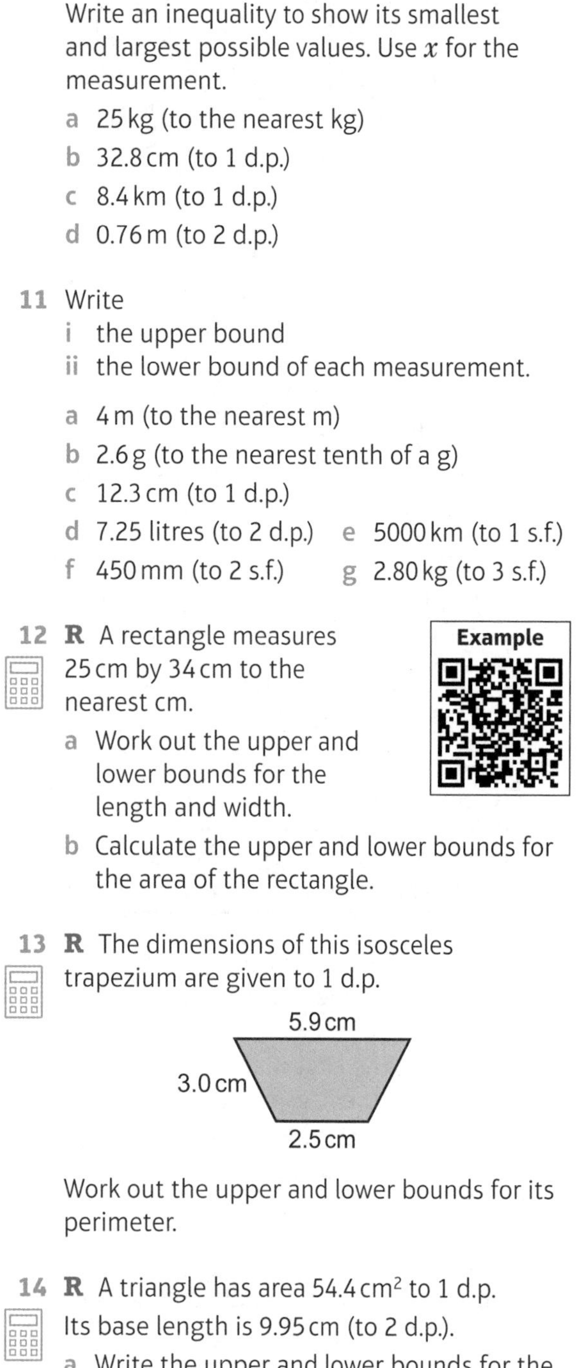

10 Each measurement has been rounded to the accuracy given.

Write an inequality to show its smallest and largest possible values. Use x for the measurement.

a 25 kg (to the nearest kg)
b 32.8 cm (to 1 d.p.)
c 8.4 km (to 1 d.p.)
d 0.76 m (to 2 d.p.)

11 Write
i the upper bound
ii the lower bound of each measurement.

a 4 m (to the nearest m)
b 2.6 g (to the nearest tenth of a g)
c 12.3 cm (to 1 d.p.)
d 7.25 litres (to 2 d.p.) e 5000 km (to 1 s.f.)
f 450 mm (to 2 s.f.) g 2.80 kg (to 3 s.f.)

12 R A rectangle measures 25 cm by 34 cm to the nearest cm.

Example

a Work out the upper and lower bounds for the length and width.
b Calculate the upper and lower bounds for the area of the rectangle.

13 R The dimensions of this isosceles trapezium are given to 1 d.p.

Work out the upper and lower bounds for its perimeter.

14 R A triangle has area 54.4 cm² to 1 d.p.
Its base length is 9.95 cm (to 2 d.p.).

a Write the upper and lower bounds for the area and the base length.
b Work out

i $\dfrac{\text{lower bound for area} \times 2}{\text{upper bound for base length}}$

ii $\dfrac{\text{upper bound for area} \times 2}{\text{lower bound for base length}}$

c What is the lower bound for the height of the triangle to 2 d.p.?

7.3 Prisms

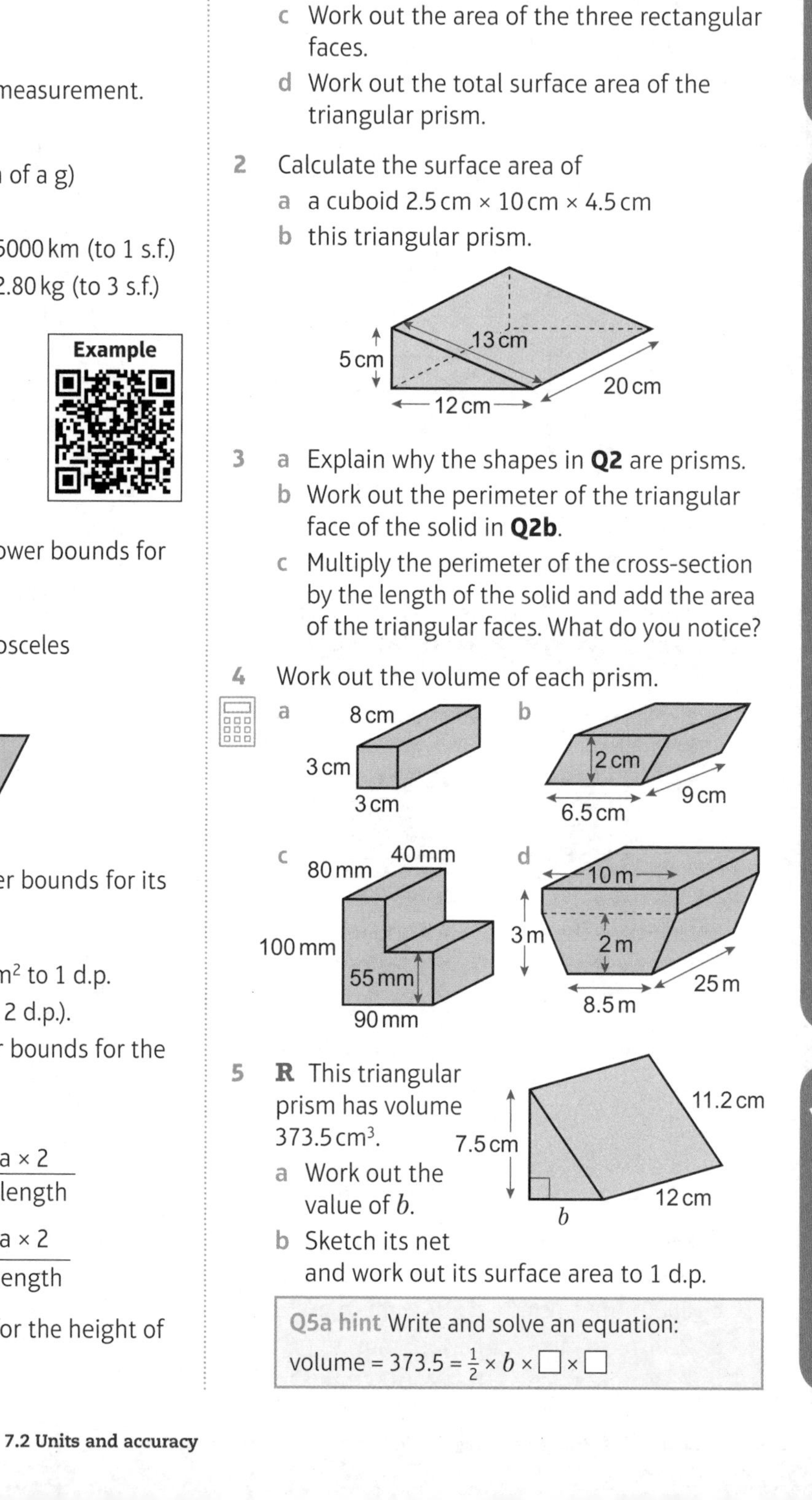

1 R a Make a sketch of this triangular prism.

b Work out the area of the triangular face. How many faces of the triangular prism also have this area?
c Work out the area of the three rectangular faces.
d Work out the total surface area of the triangular prism.

2 Calculate the surface area of
a a cuboid 2.5 cm × 10 cm × 4.5 cm
b this triangular prism.

3 a Explain why the shapes in **Q2** are prisms.
b Work out the perimeter of the triangular face of the solid in **Q2b**.
c Multiply the perimeter of the cross-section by the length of the solid and add the area of the triangular faces. What do you notice?

4 Work out the volume of each prism.

5 R This triangular prism has volume 373.5 cm³.

a Work out the value of b.
b Sketch its net and work out its surface area to 1 d.p.

Q5a hint Write and solve an equation:
volume = 373.5 = $\frac{1}{2} \times b \times \square \times \square$

6 Exam-style question

Maggie has a petrol tank.

The tank is in the shape of a cuboid.

The depth of the petrol in the tank is 12 cm.

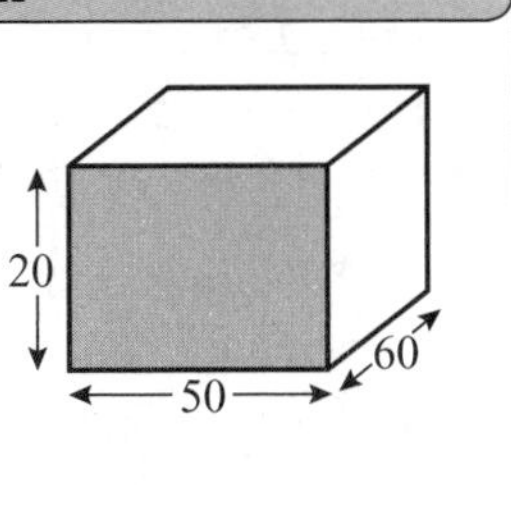

The tank is turned on its end, so it is standing on the shaded face.

Work out the depth, in cm, of the petrol in the tank now. **(3 marks)**

7 **R** a Sketch a cube with side length 1 m and a cube with side length 100 cm.

b Copy and complete. $1\text{ m}^3 = \square\text{ cm}^3$

c How do you convert from cm^3 to m^3?

8 **R** a Work out the volumes of a cube with side length 1 m and a cube with side length 1000 mm.

b How do you convert from m^3 to mm^3?

9 Convert

a 8.2 m^3 into cm^3

b 4800 mm^3 into cm^3

c $1\,650\,000\text{ cm}^3$ into m^3

d 12.55 cm^3 into mm^3

e $87\,500\text{ cm}^3$ into litres

f 6.05 litres into cm^3

g 0.25 m^3 into litres

h 98 cm^3 into ml

10 **P** Sam wants to paint the walls and ceiling of his loft conversion. He will paint two coats on all the walls and the ceiling. What is the total area Sam will paint to the nearest m^2?

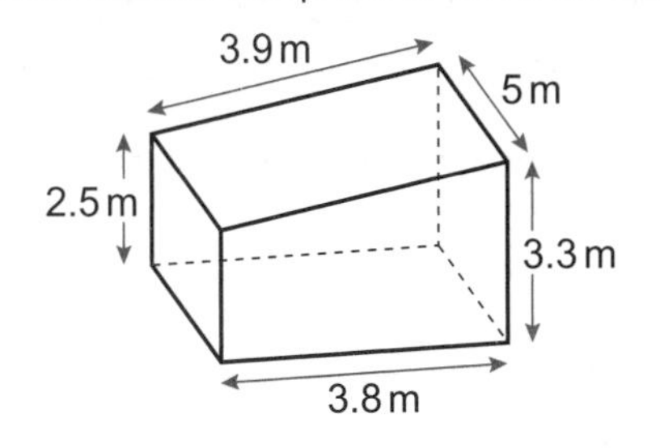

11 **P / R** A cuboid has 8 edges of the same length and 4 shorter edges which are 3.5 cm each.

Its volume is 143.5 cm^3.

What is the length of the longer edge (to 1 d.p.)?

12 Mike owns a factory which makes fence posts.

He needs to treat the fence posts with wood preservative.

He will treat all the surface area of each post.

Each post is modelled as a cuboid 2 m in length with base dimensions 40 mm by 40 mm.

a Work out the surface area of each post.

b Mike will treat 500 posts. He needs 30 ml of preservative for every square metre of wood.

Estimate the amount of preservative he needs to the nearest litre.

13 Show that the surface area of this prism is $12x(7x - 1)$.

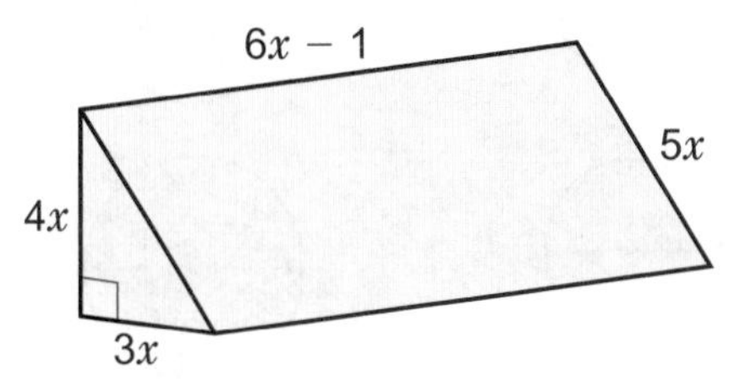

14 **R** The dimensions of this triangular prism are measured to 1 d.p.

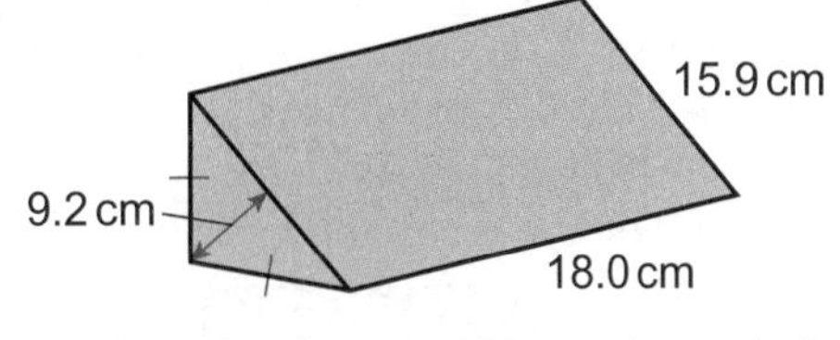

Calculate the upper and lower bounds for the volume of the triangular prism.

7.4 Circles

1 **R** The table gives the diameter and circumference of some circles.

Diameter	Circumference
5 cm	15.7 cm
10 m	31.4 m
42 mm	131.9 mm

a Work out the ratio $\frac{\text{circumference}}{\text{diameter}}$ for each one. What do you notice?

b The ratio $\frac{\text{circumference}}{\text{diameter}}$ of a circle is represented by the Greek letter π (pi).

Find the π key on your calculator.

Write the value of π to 8 decimal places (8 d.p.).

2 Work out the circumference of each circle. Give your answers to 1 d.p. and the units of measurement.

a 8 cm b 5.2 m

c 47 mm d 8.6 cm

3 Find the area of each circle.

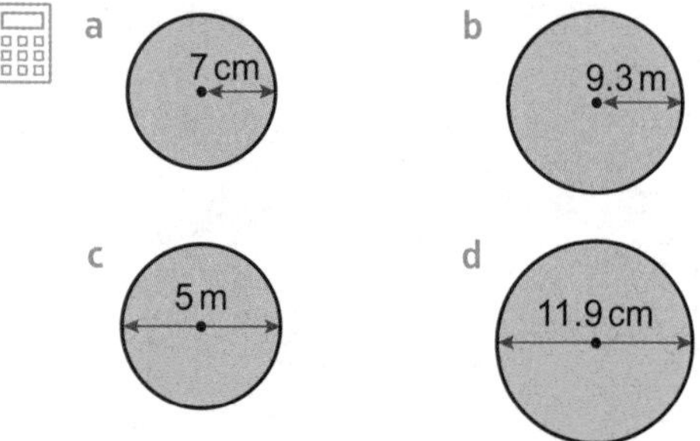

4

Exam-style question

Tariq designs a pattern for a park square in the city centre.

He draws this design to be made from dark grey and light grey gravel.

Each bag of gravel covers an area of 8 m².

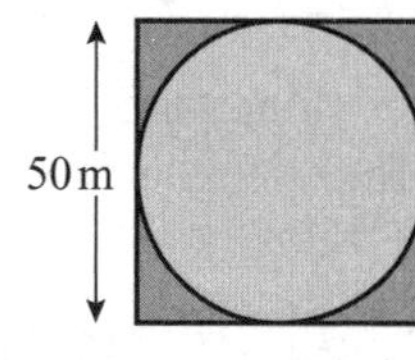

How many bags of dark grey and light grey gravel are needed to complete the design?

You must show your working. **(5 marks)**

5 Georgie owns a factory which produces aluminium bars for climbing gear.

The diagram shows the cross-section of one of the bars.

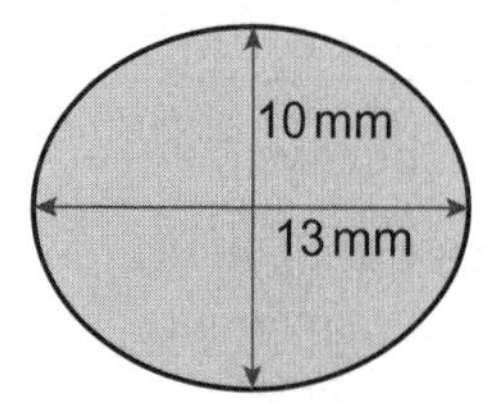

a Work out the mean distance across the bar.

b By modelling the cross-section as a circle, calculate an estimate for its area to the nearest square millimetre.

6 P The diameter of a wheelchair wheel measures 63 cm.

How many complete revolutions does the wheel make in an 800 m wheelchair race?

7 R The areas and circumferences of these circles are given in terms of π.

Match each circle to its area and circumference.

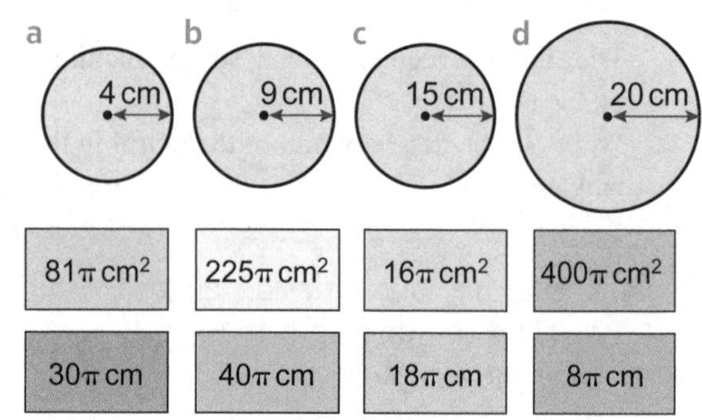

8 R a Work out the area and circumference of a circle with radius 7 cm

i in terms of π ii to 2 s.f.

b Which values for the area and circumference are the most accurate?

9 The circumference of a circle is 215 cm.

a Substitute the value for C into the formula $C = \pi d$

b Solve the equation to find the diameter to 1 d.p.

10 Find the radius of a circle with circumference 20 m. Give your answer to 1 d.p.

11 a Find the radius of a circle with area 484 cm². Give your answer to the nearest cm.

b Find the diameter of a circle with area 952 cm². Give your answer to the nearest mm.

Example

12 a Copy and complete to make r the subject of the formula for the area of a circle.

$$A = \pi r^2$$

$$\frac{A}{\square} = r^2$$

$$\sqrt{\frac{A}{\square}} = \square$$

b Use your formula for r from part **a** to work out the radii of these circles to 2 d.p.

i circle 1 with area 120 cm²

ii circle 2 with area 750 mm²

iii circle 3 with area 6.7 m²

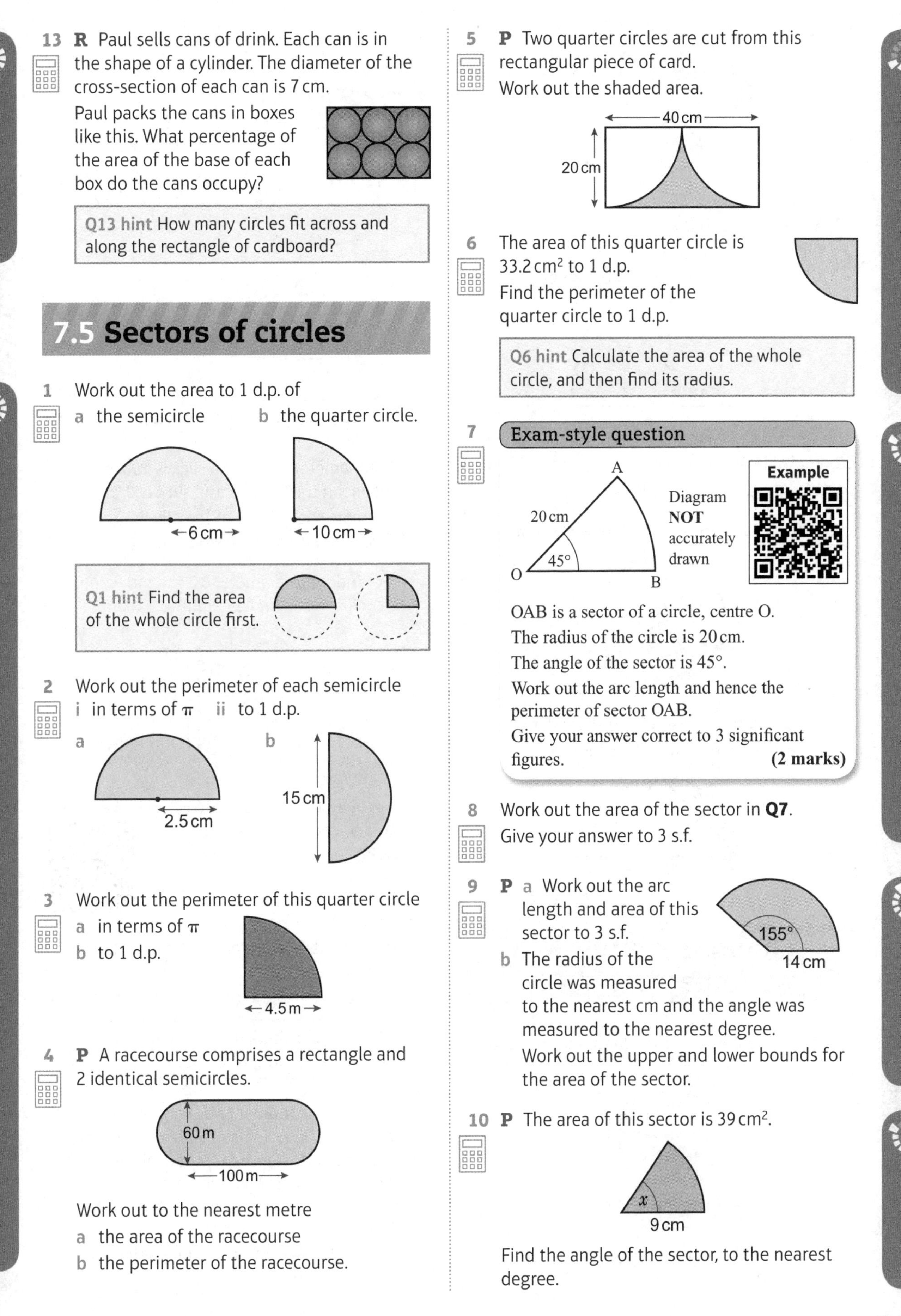

13 **R** Paul sells cans of drink. Each can is in the shape of a cylinder. The diameter of the cross-section of each can is 7 cm.

Paul packs the cans in boxes like this. What percentage of the area of the base of each box do the cans occupy?

Q13 hint How many circles fit across and along the rectangle of cardboard?

7.5 Sectors of circles

1 Work out the area to 1 d.p. of

a the semicircle b the quarter circle.

Q1 hint Find the area of the whole circle first.

2 Work out the perimeter of each semicircle

i in terms of π ii to 1 d.p.

3 Work out the perimeter of this quarter circle

a in terms of π

b to 1 d.p.

4 **P** A racecourse comprises a rectangle and 2 identical semicircles.

Work out to the nearest metre

a the area of the racecourse

b the perimeter of the racecourse.

5 **P** Two quarter circles are cut from this rectangular piece of card.

Work out the shaded area.

6 The area of this quarter circle is 33.2 cm^2 to 1 d.p.

Find the perimeter of the quarter circle to 1 d.p.

Q6 hint Calculate the area of the whole circle, and then find its radius.

7 **Exam-style question**

OAB is a sector of a circle, centre O.

The radius of the circle is 20 cm.

The angle of the sector is 45°.

Work out the arc length and hence the perimeter of sector OAB.

Give your answer correct to 3 significant figures. **(2 marks)**

8 Work out the area of the sector in **Q7**.

Give your answer to 3 s.f.

9 **P** a Work out the arc length and area of this sector to 3 s.f.

b The radius of the circle was measured to the nearest cm and the angle was measured to the nearest degree.

Work out the upper and lower bounds for the area of the sector.

10 **P** The area of this sector is 39 cm^2.

Find the angle of the sector, to the nearest degree.

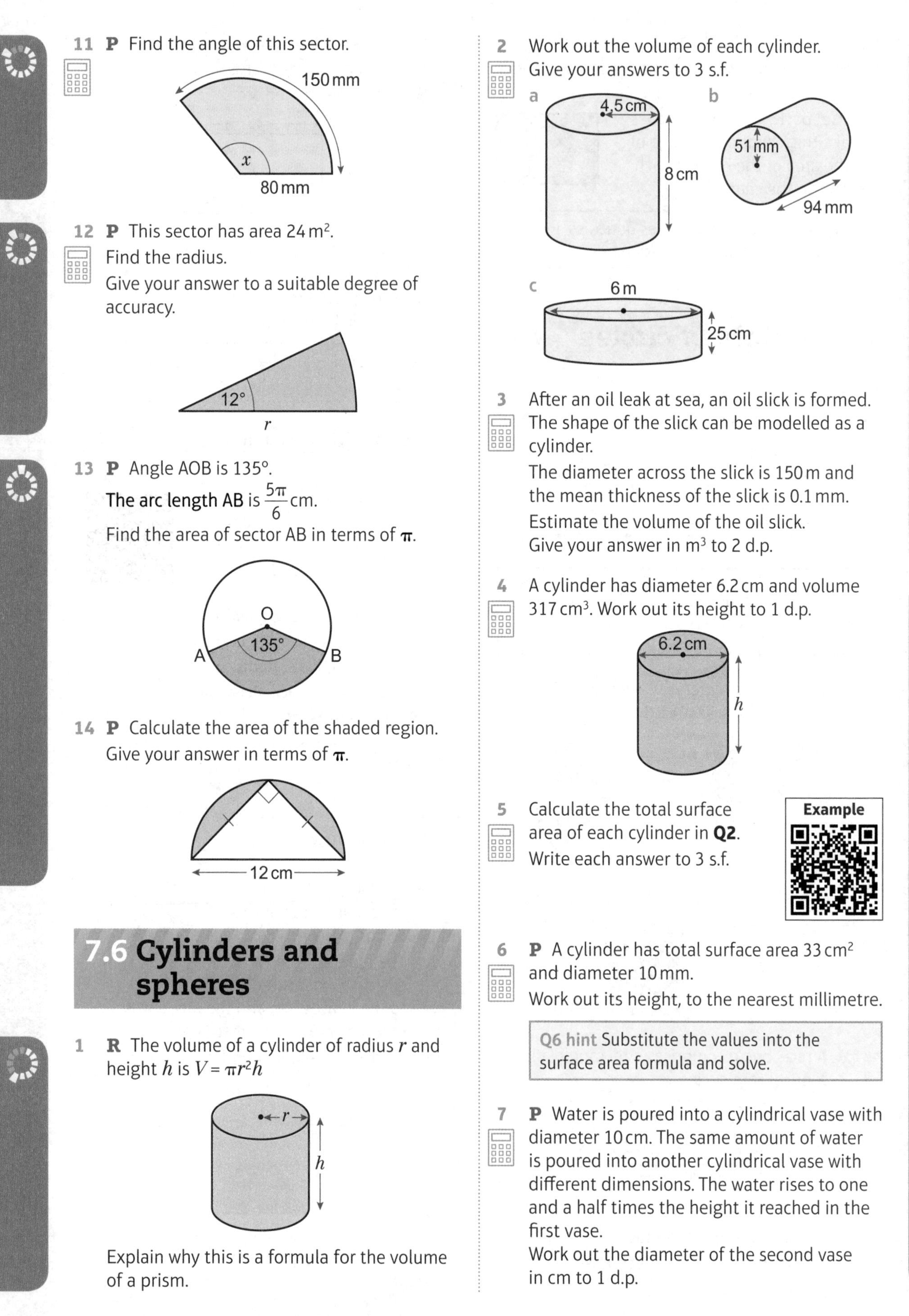

11 **P** Find the angle of this sector.

12 **P** This sector has area 24 m^2.

Find the radius.

Give your answer to a suitable degree of accuracy.

13 **P** Angle AOB is 135°.

The arc length AB is $\frac{5\pi}{6}$ cm.

Find the area of sector AB in terms of π.

14 **P** Calculate the area of the shaded region.

Give your answer in terms of π.

7.6 Cylinders and spheres

1 **R** The volume of a cylinder of radius r and height h is $V = \pi r^2 h$

Explain why this is a formula for the volume of a prism.

2 Work out the volume of each cylinder.

Give your answers to 3 s.f.

3 After an oil leak at sea, an oil slick is formed. The shape of the slick can be modelled as a cylinder.

The diameter across the slick is 150 m and the mean thickness of the slick is 0.1 mm.

Estimate the volume of the oil slick.

Give your answer in m^3 to 2 d.p.

4 A cylinder has diameter 6.2 cm and volume 317 cm^3. Work out its height to 1 d.p.

5 Calculate the total surface area of each cylinder in **Q2**.

Write each answer to 3 s.f.

Example

6 **P** A cylinder has total surface area 33 cm^2 and diameter 10 mm.

Work out its height, to the nearest millimetre.

> **Q6 hint** Substitute the values into the surface area formula and solve.

7 **P** Water is poured into a cylindrical vase with diameter 10 cm. The same amount of water is poured into another cylindrical vase with different dimensions. The water rises to one and a half times the height it reached in the first vase.

Work out the diameter of the second vase in cm to 1 d.p.

8 Calculate the surface area and volume of each sphere. Give your answers in terms of π.

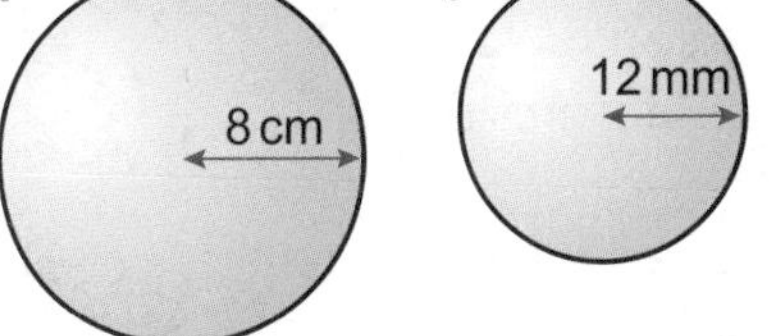

9 A factory makes hollow metal hemispheres as garden ornaments. The factory produces the hemispheres with external diameter 20 cm and a thickness of 3 mm.
Work out the volume of metal used for each hemisphere, to the nearest cubic millimetre.

10 **Exam-style question**

A manufacturing company makes plastic containers for children's toys.
Each container comprises a cylinder with identical hemispheres at each end.
The length of the cylinder is 2.5 cm.
The diameter of the cylinder is 1.75 cm.

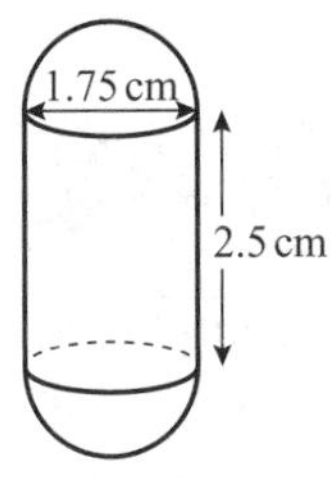

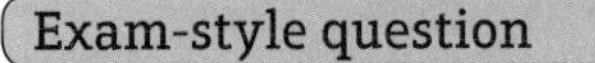

a Calculate the total volume of the container. Give your answer correct to 3 significant figures. **(3 marks)**

b Calculate the surface area of the container. Give your answer correct to 3 significant figures. **(3 marks)**

11 **P** A cuboid-shaped block of metal has dimensions 20 cm by 10 cm by 4 cm.
The block is melted down and made into spherical ball bearings, each with a radius of 5 mm.
How many of these ball bearings can be made from the block?

12 **P** What is the radius of a sphere with volume 6000 cm^3?

13 **P** A cylinder has a volume of 450 cm^3 measured to the nearest 10 cm^3, and a radius of 5 cm measured to the nearest cm.
What are the upper and lower bounds of its height?

7.7 Pyramids and cones

1 Here is a square-based pyramid.

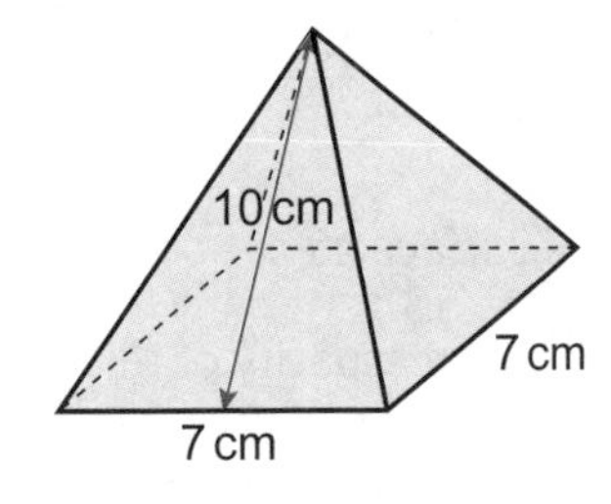

a Work out the area of each triangular face.

b Calculate the total surface area of the pyramid.

2 This pyramid has a square base of side 5 cm, and vertical height 9 cm.
Calculate its volume.

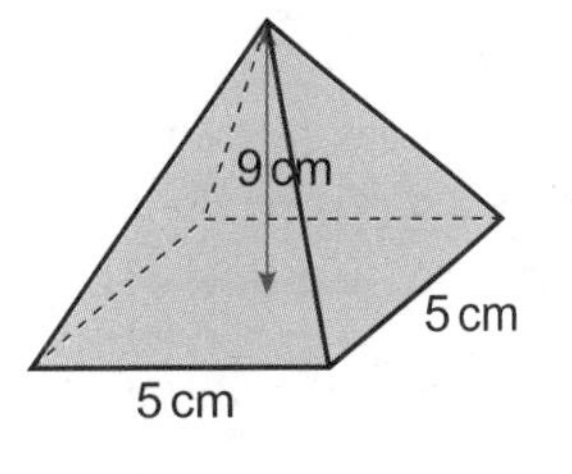

3 **Exam-style question**

This solid is made from a triangular-based pyramid and a triangular prism.
The triangular prism has a right-angled triangular cross-section and a length of 10 m.
The height and base of the triangle are 12 m and 5 m respectively.

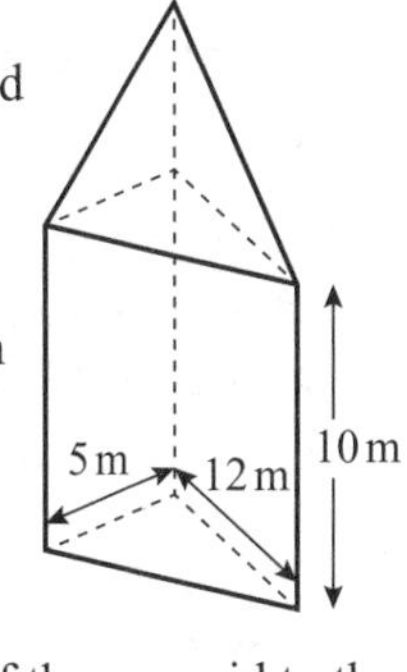

The ratio of the volume of the pyramid to the volume of the prism is 1 : 4
Find

a the volume of the prism

b the height of the pyramid. **(6 marks)**

4 A cone has base radius 9 cm and height 12 cm.
Calculate its volume

a in terms of π

b to 3 s.f.

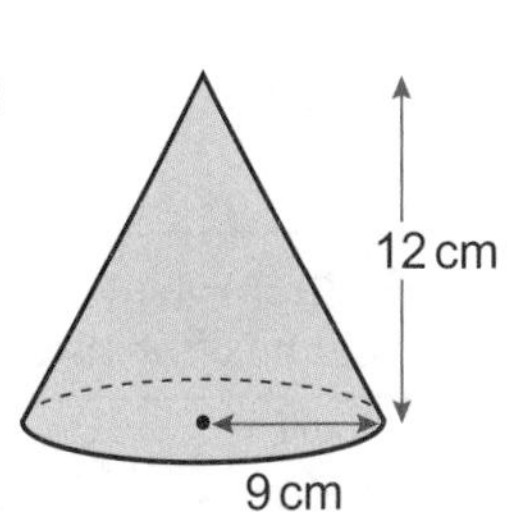

5 A cone has base radius 70 mm and slant height 250 mm.

Calculate, in terms of π

a the area of its base

b its curved surface area

c its total surface area.

6 **P** Work out the capacity of this disposable cup in ml.

Give your answer to 3 s.f.

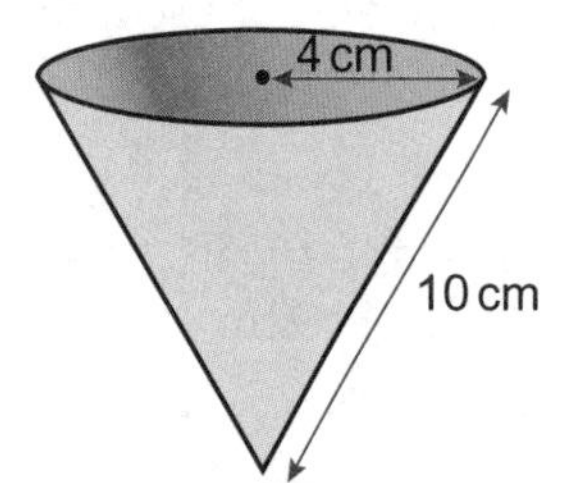

7 **P** Work out the total surface area of a cone with radius 3.6 cm and volume 129 cm^3.

Give your answer to 3 s.f.

8 **P** A solid cone is made from 144 000 mm^3 of plastic.

The diameter of the cone is 50 mm.

a What is the height of the cone?

b What is the surface area of the cone?

Give your answers to 3 s.f.

9 **P** A cone and a sphere have identical total surface areas. The height of the cone is 24 cm and the slant height is 26 cm.

Work out the radius of the sphere to the nearest millimetre.

10 The top half of this cone is cut off to leave a 3D solid called a frustum.

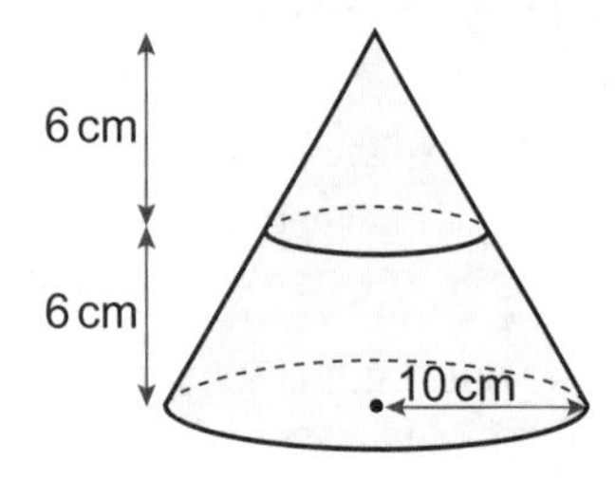

Work out the volume of the frustum, in terms of π.

> **Q10 hint** Volume of frustum
> = volume of whole cone – volume of top cone

11 This 3D solid is made from a cylinder and a cone.

Write an expression in terms of x and π for

a the surface area of the cylinder

b the curved surface area of the cone

c the total surface area of the solid.

12 Calculate the volume and surface area of this 3D solid. Give your answers to 3 s.f.

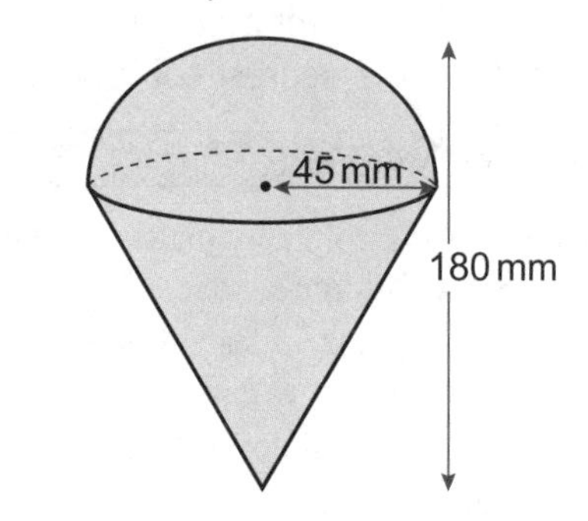

7 Problem-solving

Solve problems using these strategies where appropriate:

Example

- **Use pictures or lists**
- **Use smaller numbers**
- **Use bar models**
- **Use x for the unknown**
- **Use a flow diagram.**

1 **R** Gillian sells items at a craft fair.

The prices of the items sold at her last craft fair were

£3.50, £12.50, £25, £8, £5, £4.50, £5, £16, £5, £2.50, £5, £12.50, £2.50, £5, £3.50

a Find the mean, median, mode and range of the prices of the items sold.

b Why might Gillian want to know the mode?

c What might Gillian say about the range?

2 The first floor of an office occupies 1200 m^2.

$\frac{3}{8}$ of the space is taken up by work stations and $\frac{1}{3}$ of the space is taken up by the canteen.

The rest is taken up by conference rooms.

What is the floor area of the conference rooms?

3 **R** A local school has been given an area of a memorial park to create a mosaic.
The shape and measurements of the area are shown below.

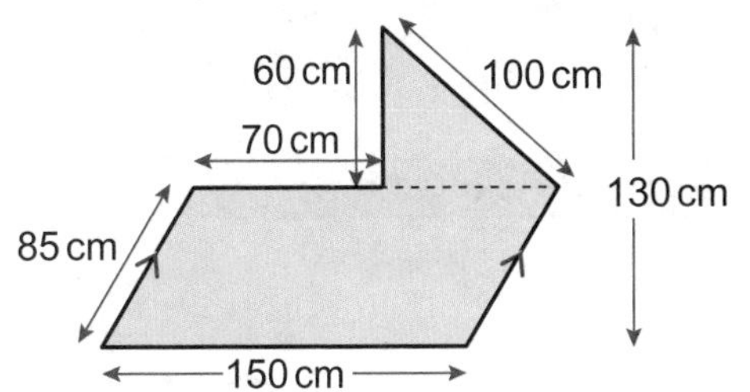

a What is the perimeter of the shape for the mosaic?

b What is the area of the mosaic?

c How did you work out the area of the mosaic?

4 **R** A farmer rents two rectangular fields.
One field measures 350 m × 620 m and the other measures 430 m × 860 m.
How many hectares of field does the farmer rent?

Q4 hint 1 ha = 100 m × 100 m

5 A company makes measuring wheels where the circumference of the wheel is used to measure distance.
What would the radius of the wheel need to be in cm for each turn to measure

a 1 metre

b 1.5 metres

c 5 metres?

Give your answers to 1 d.p.

6 **R** The sum of the interior angles of a polygon is 1080°.
What is the name of the polygon?
Explain how you know.

7 **R** A class have been asked to design a container to hold exactly 8 litres of liquid.
Knowing that 1 litre = 1000 cm^3, what possible dimensions could the container have if it was

a a cube

b a cuboid

c a triangular prism?

8 **Exam-style question**

The cabins on a ship need repainting.
The walls overlooking the sea will be painted blue. There are enough cans of blue paint to cover 50 m^2.

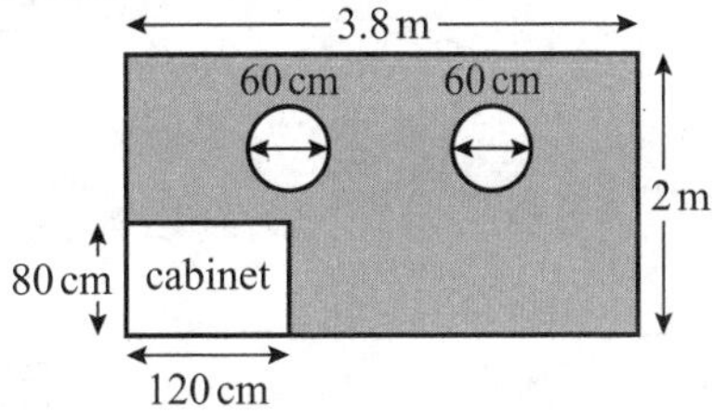

3.8 m
60 cm 60 cm
2 m
80 cm cabinet
120 cm

The shaded area is the part of the wall to be painted. If all the cabins are the same, how many complete cabin walls can be painted with the paint they already have?
Show all your working.
Round all answers to 2 d.p. **(5 marks)**

9 An open-air stage is being built in a park.
A track needs to be placed along the arc of the sector for the scenery.

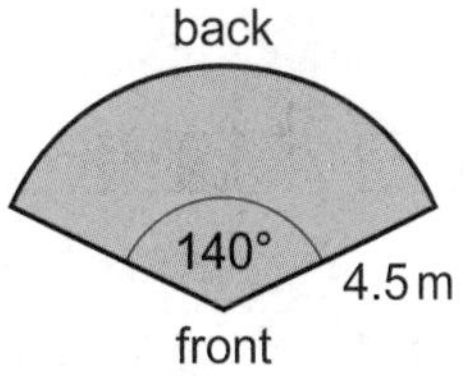

back
140°
4.5 m
front

a What length does the track need to be?
Give your answer to 1 d.p.

b What is the area of the stage?
Give your answer to 2 d.p.

10 **R** Philip and Craig are building model castles for a project.
Philip uses cylinders for his towers and needs to make cones for the turrets.
Craig uses prisms with square-faced ends and wants to make square-based pyramids for the turrets.

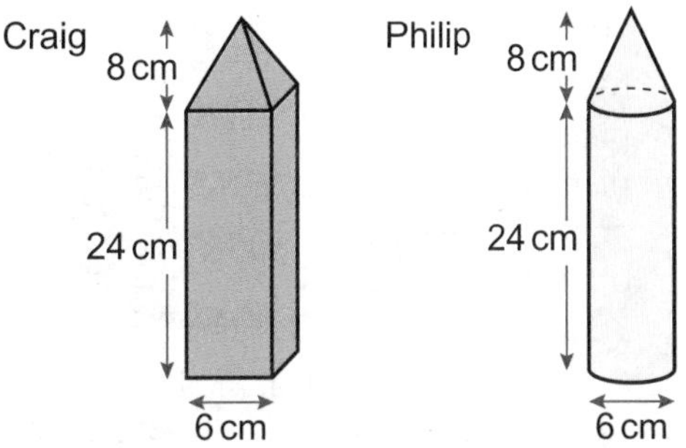

Craig 8 cm 24 cm 6 cm
Philip 8 cm 24 cm 6 cm

If both types of tower are the same height and width/diameter, which tower has the larger surface area? Show your working.

8 TRANSFORMATIONS AND CONSTRUCTIONS

8.1 3D solids

1 On squared paper, draw and label the plan, front elevation and side elevation of these solids.

Example

a

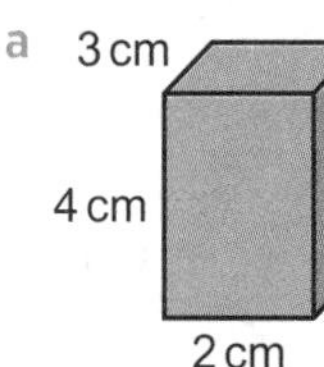

b

c

2 **R** Sketch the solids represented by these plans and elevations.

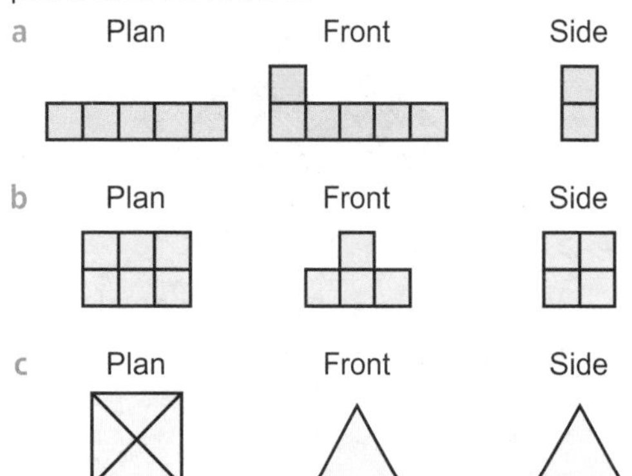

3 **P** Here is the side elevation of a 3D solid. Sketch three possible 3D solids it could belong to.

4 **P / R** Here is a cube.

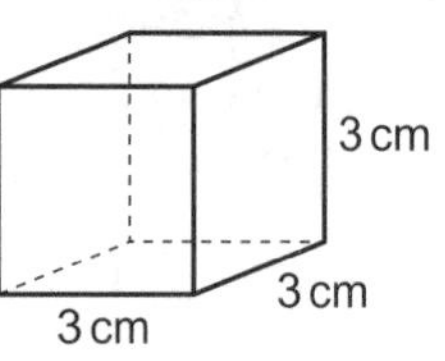

a Calculate the surface area of the cube.

b The cube is cut in half along the red plane. Sketch the plan, front elevation and side elevation of each of the new 3D solids.

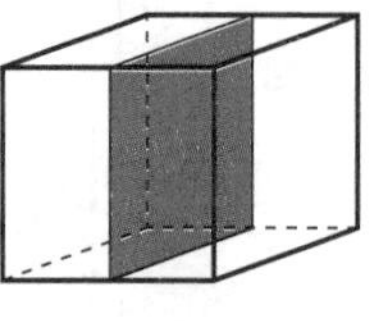

c Calculate the surface area of each of the new solids.

d Repeat parts **b** and **c** for the cube cut along this red plane.

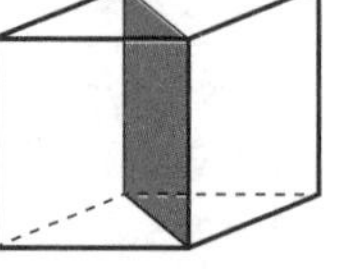

5 Exam-style question

Here is a solid prism.

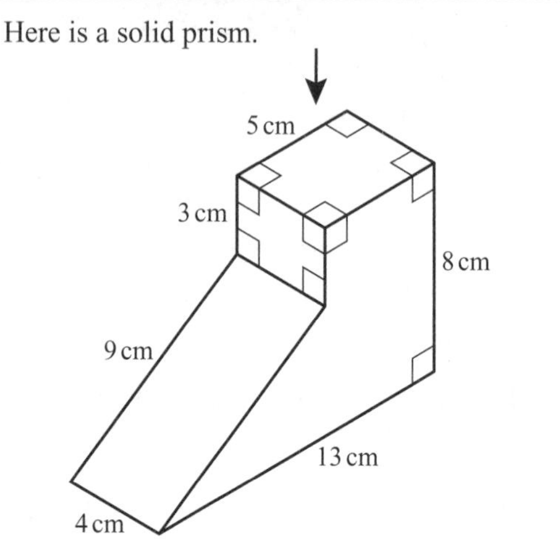

On a centimetre-square grid, draw an accurate plan view of the solid prism from the direction of the arrow. **(2 marks)**

8.2 Reflection and rotation

1 Draw a coordinate grid from −8 to +8 on both axes.

a Draw a rectangle Q with coordinates A(3, 2), B(1, 2), C(1, 5) and D(3, 5).

b Reflect rectangle Q in the y-axis. Label the image R.

c Reflect rectangle R in $y = -1$. Label the image S.

d Reflect rectangle S in the y-axis. Label the image T.

e Describe the single reflection that maps rectangle T onto rectangle Q.

Q1 hint Use tracing paper to help.

2 **R** Draw a coordinate grid from −6 to +6 on both axes.

a Draw triangle A with coordinates (0, 2), (2, 2) and (2, 4).

b Reflect triangle A in the line $y = x$.

c Reflect triangle A in the line $x = -2$.

3 Describe the reflection that maps

a P onto Q b P onto R
c P onto S d P onto T.

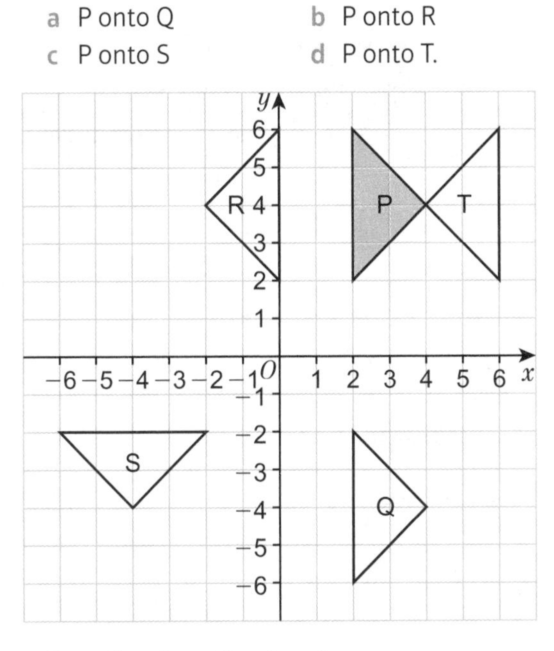

4 Describe the reflection that maps

a A to B b A to C.

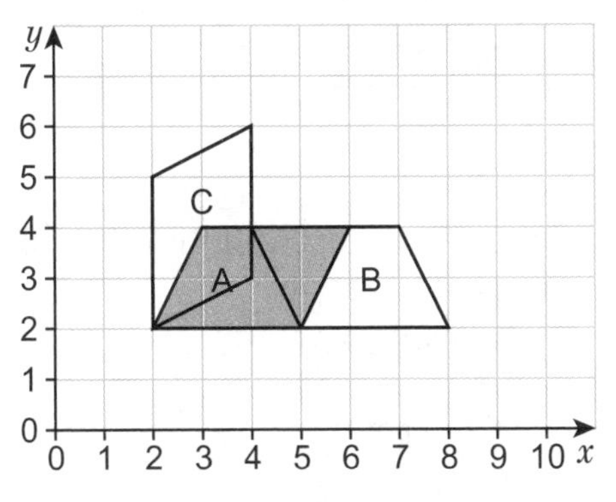

5 Draw a coordinate grid from –5 to +5 on both axes.

a Draw shape A with coordinates (–2, 1), (–4, 1) and (–3, 3).
b Reflect shape A in the line $y = x$. Label the image B.
c Reflect shape A in the line $y = -x$. Label the image C.
d Reflect shape C in the x-axis. Label the image D.
e Describe the reflection that maps shape D to shape B.

6 **R** Draw a coordinate grid from –8 to +5 on both axes.

a Draw shape A with coordinates (2, 1), (4, 1), (4, 4) and (2, 4).
b Rotate triangle A
 i 90° anticlockwise about (2, 1)
 ii 180° about (0, 0)
 iii 90° clockwise about (3, 1)
 iv 180° about (–1, 4)
 v 90° anticlockwise about (1, 4)
 vi 180° about (–1, 0).

Label your results i, ii etc.

7 Describe the rotation that takes each shape to its image.

Example

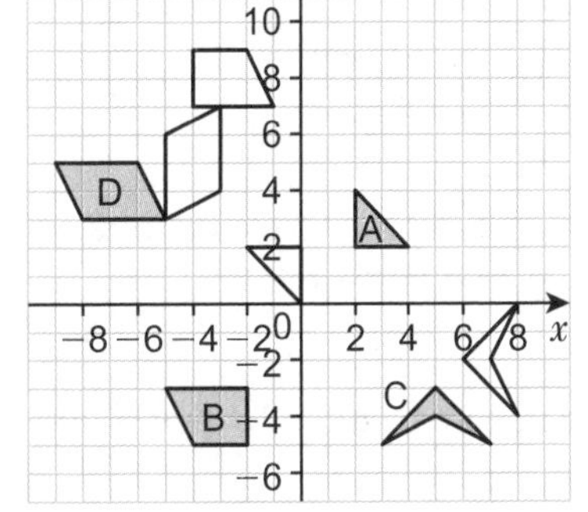

8 **P** Draw a coordinate grid from –5 to +5 on both axes.

a Plot the points (2, 1), (4, 2) and (1, 3). Join them and label the shape A.
b Reflect shape A in the x-axis. Label the image B.
c Reflect shape B in the y-axis. Label the image C.
d Describe the transformation that takes shape A to shape C.

9 **R** 'The rotation of a square can always be described as a reflection.'
Decide whether this statement is sometimes true, always true or never true.

10 **Exam-style question**

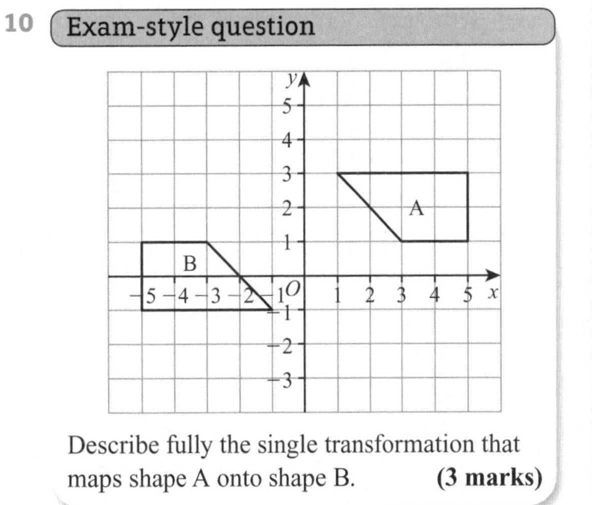

Describe fully the single transformation that maps shape A onto shape B. **(3 marks)**

8.3 Enlargement

1 Copy the diagram, drawing the x-axis from −2 to 8 and the y-axis from −2 to 5.

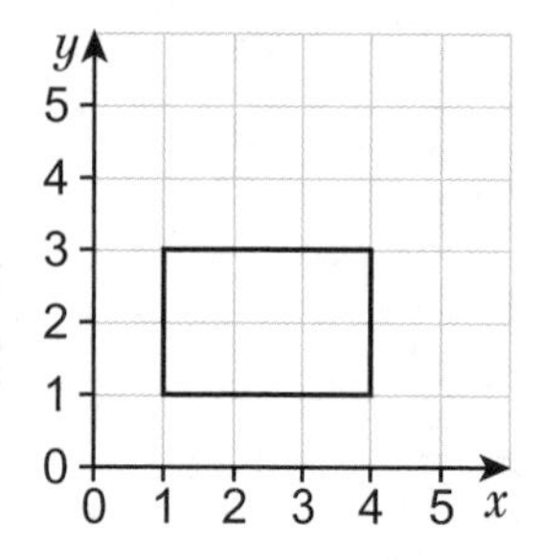

Enlarge the rectangle by scale factor 2, with these centres of enlargement.

a (1, 4) b (1, 1) c (3, 2)

2 Shape A has been enlarged to give shape B.

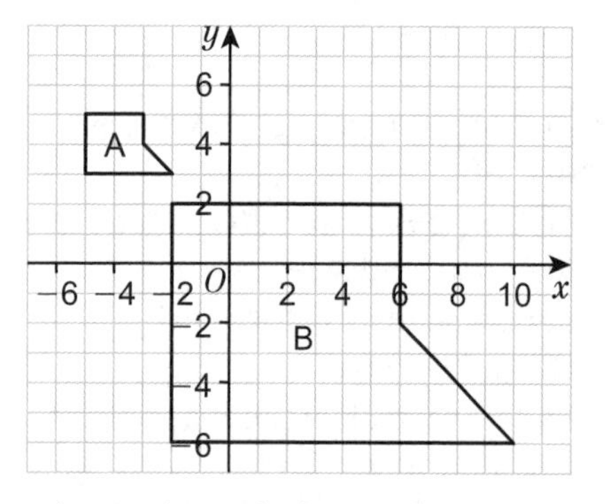

a What is the scale factor of the enlargement?

b Copy the diagram.
Join corresponding vertices on the object and the image with straight lines.
Extend the lines until they meet at the centre of enlargement.

c Write the coordinates of the centre of enlargement.

d Copy and complete to describe the enlargement from A to B.
Enlargement by scale factor ____ , centre (____ , ____).

3 **P** Draw a triangle A, with base 4 cm and height 3 cm.

a Work out the area of the triangle.

b Shape A is enlarged by scale factor 2 to make shape B.
Work out the area of shape B.

c Shape A is enlarged by scale factor 3 to make shape C.
Work out the area of shape C.

d Shape A is enlarged by scale factor 4 to make shape D.
Work out the area of shape D.

e Copy and complete this table.

Shape	Scale factor	$\frac{\text{Area of enlarged shape}}{\text{Area of shape A}}$
B	2	
C	3	
D	4	

4 Copy these diagrams.
Enlarge each shape by the scale factor given.

a scale factor $\frac{1}{2}$

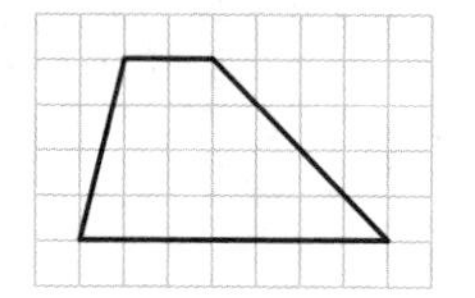

b scale factor $\frac{1}{3}$

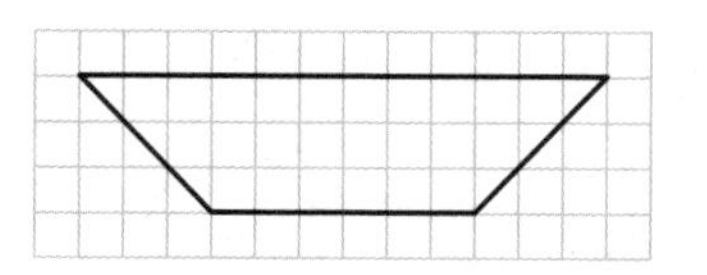

5 a Copy and enlarge each shape by the given scale factor about the centre of enlargement shown.

i scale factor $\frac{1}{3}$

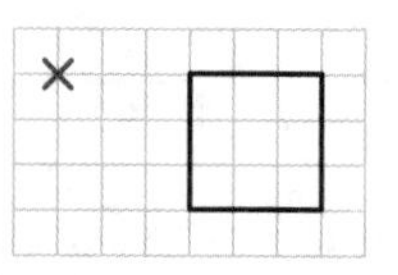

ii scale factor $\frac{1}{2}$

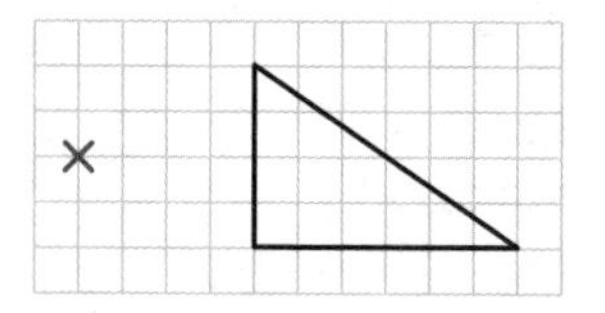

iii scale factor $\frac{1}{3}$

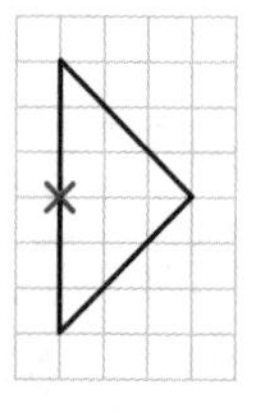

b **R** When a shape is enlarged by scale factor $\frac{1}{3}$, is its area enlarged by scale factor $\left(\frac{1}{3}\right)^2$? Explain.

6 **P** Describe the enlargement that maps shape A onto shape P.

a

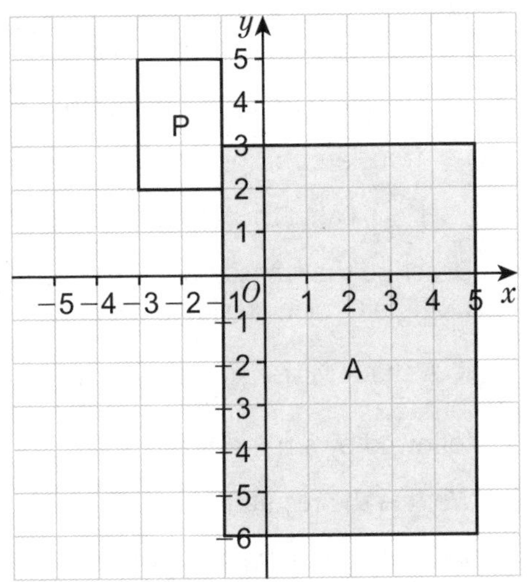

b

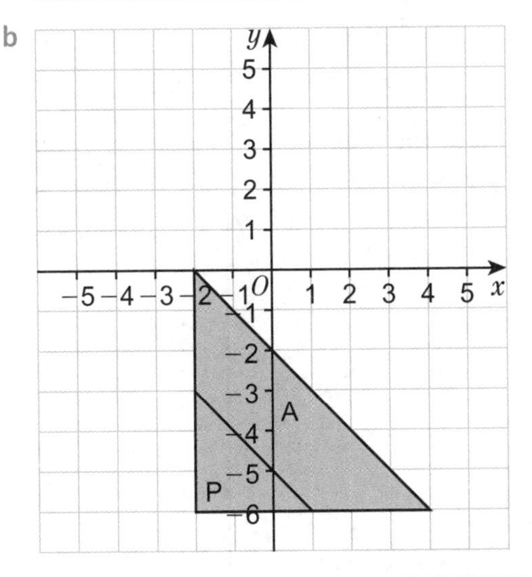

7 **P** Draw a coordinate grid from −12 to +12 on both axes.

Join the points (6, 3), (8, 3) and (6, 6) to make a triangle.

Enlarge the triangle

a by scale factor −2, centre of enlargement (5, 2)

b by scale factor −3, centre of enlargement (5, 4)

c by scale factor −1, centre of enlargement (6, 3).

Example

8 **P** Describe fully the transformation that maps shape S onto shape T.

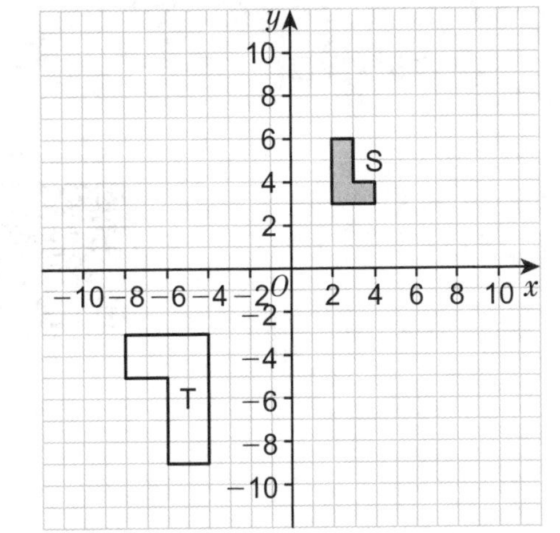

Q8 hint Draw lines from corresponding vertices to find the centre.

9 India said, 'An enlargement does not change the orientation of the shape.'

Is India correct?

Explain your answer.

10 **Exam-style question**

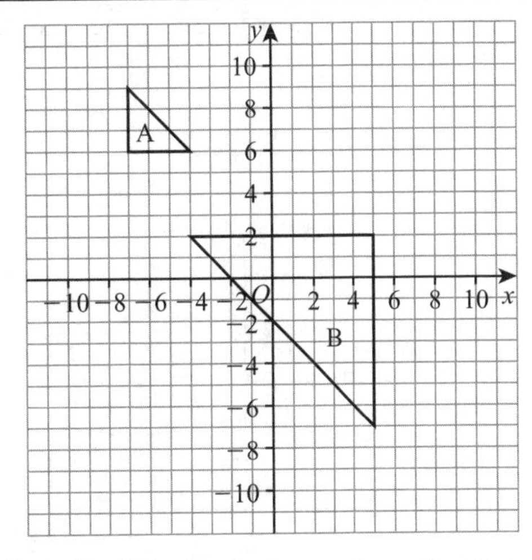

Describe fully the single transformation that maps shape A onto shape B. **(3 marks)**

Exam hint.
For 3 marks, give 3 pieces of information about the transformation.

8.4 Translations and combinations of transformations

1 Copy the diagram. Translate shape A by the vectors

Example

a $\begin{pmatrix}1\\3\end{pmatrix}$ to B b $\begin{pmatrix}2\\3\end{pmatrix}$ to C

c $\begin{pmatrix}0\\-7\end{pmatrix}$ to D d $\begin{pmatrix}-2\\0\end{pmatrix}$ to E

e $\begin{pmatrix}-5\\-3\end{pmatrix}$ to F

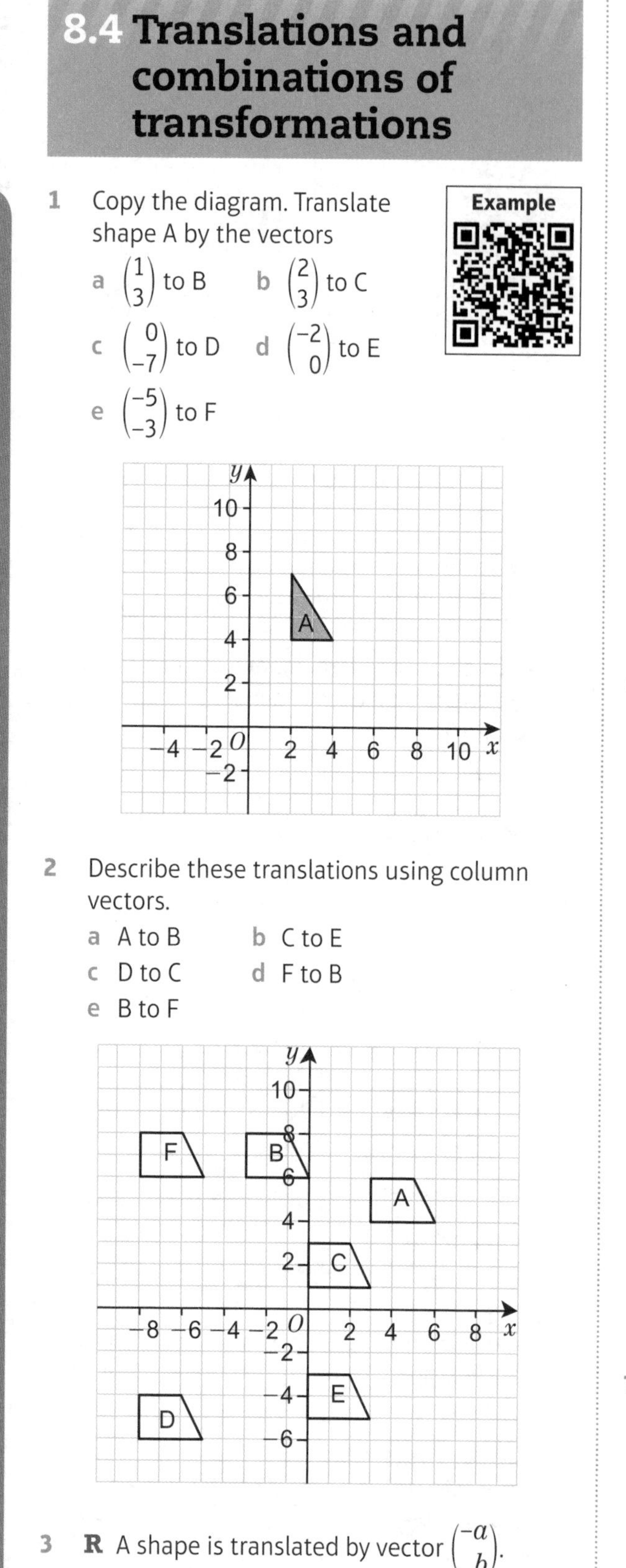

2 Describe these translations using column vectors.

a A to B b C to E

c D to C d F to B

e B to F

3 **R** A shape is translated by vector $\begin{pmatrix}-a\\b\end{pmatrix}$.

What vector would translate the shape back to its original position?

Explain your answer.

4 Draw a coordinate grid from −6 to +6 on both axes.

a Plot a triangle with vertices at (−2, 3), (0, 1) and (1, 4). Label the triangle P.

b i Translate triangle P by vector $\begin{pmatrix}2\\-7\end{pmatrix}$.

ii Translate this new triangle by vector $\begin{pmatrix}-4\\3\end{pmatrix}$. Label the image Q.

c Describe the translation of triangle P to triangle Q, using a single vector.

5 a **R** A shape is translated by vector $\begin{pmatrix}-2\\3\end{pmatrix}$ followed by a translation by vector $\begin{pmatrix}3\\1\end{pmatrix}$.

What is the resultant vector?

b The resultant of two vectors is $\begin{pmatrix}3\\-4\end{pmatrix}$.

The first vector is $\begin{pmatrix}-1\\2\end{pmatrix}$.

What is the second vector?

6 Copy this diagram and shape A only on a coordinate grid from −6 to +6.

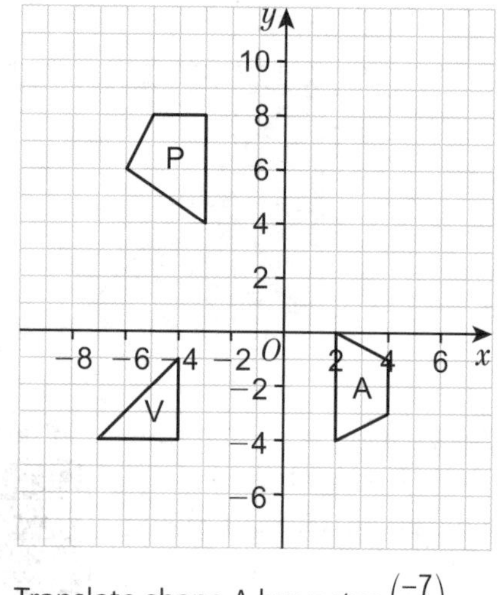

a Translate shape A by vector $\begin{pmatrix}-7\\3\end{pmatrix}$. Label the image B.

b Reflect shape B in the line $x = -2$. Label the image C.

7 Copy the diagram from **Q6** and shape P only.

a Translate shape P by vector $\begin{pmatrix}6\\-2\end{pmatrix}$. Label the image Q.

b Reflect shape Q in the line $y = 1$. Label the image R.

c Translate shape R by vector $\begin{pmatrix}-6\\-2\end{pmatrix}$. Label the image S.

d Describe the reflection that maps shape P onto shape S.

8 Copy the diagram from **Q6** and triangle V only.

a Rotate triangle V through 180° about the point (−1, 0). Label the image W.

b Reflect triangle W in the line $y = x$. Label the image X.

c Translate triangle X by vector $\begin{pmatrix} -9 \\ 0 \end{pmatrix}$. Label the image Y.

d Describe the single transformation that maps triangle V onto triangle Y.

9 **P / R** A tessellation is made by transforming shape A.

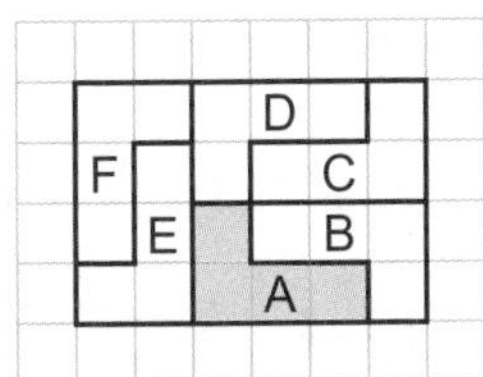

Copy the tessellation onto a coordinate grid with the vertices of A at (0, 0), (3, 0), (3, 1), (1, 1) (1, 2) and (0, 2).

Describe the transformation that would move shape

a A to B b B to C c C to D

d A to E e E to F.

f What transformation of shape A would make a shape with 4 times the area?

10 **Exam-style question**

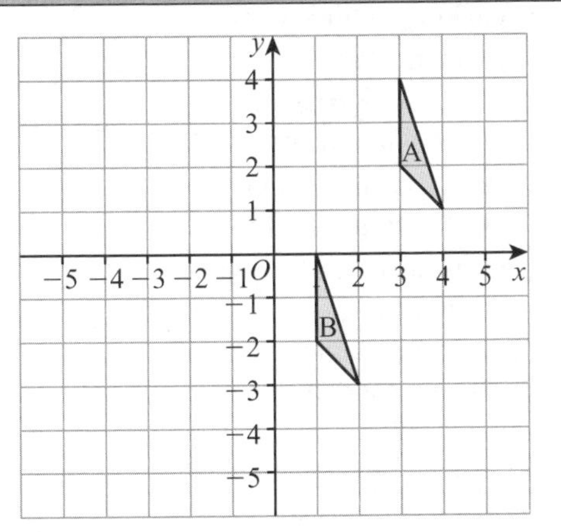

a Describe fully the single transformation that maps triangle A onto triangle B. **(2 marks)**

b Reflect triangle B in the y-axis and label the image C. Then translate shape C by vector $\begin{pmatrix} 4 \\ 0 \end{pmatrix}$ and label the image D. **(3 marks)**

c Describe the single transformation that maps triangle B onto triangle D. **(2 marks)**

11 Adam says, 'A shape and its transformed image always have the same area.'

Do you agree with this statement?

If not, give a counter example and explain your answer.

8.5 Bearings and scale drawings

1 **P** During a search and rescue mission the positions of a frigate, lifeboat and helicopter are marked in relation to the man overboard.

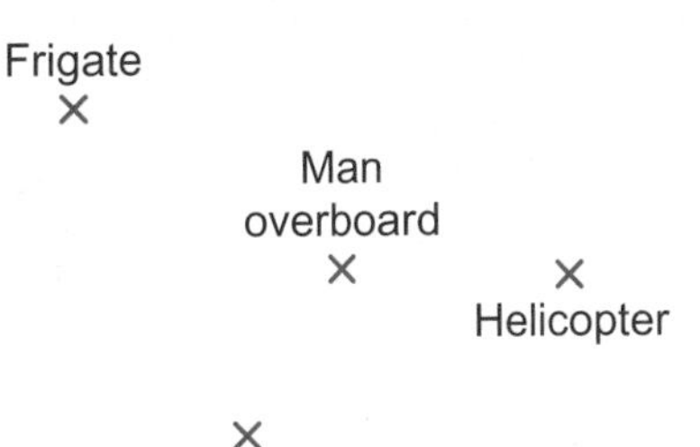

The real-life distance from the frigate to the helicopter is 16 km.

a What scale has been used on the map?

b From the map, estimate the distance from

i the lifeboat to the man overboard

ii the helicopter to the man overboard.

c The helicopter flies at a speed of 140 km/hour.

How long will it take it to fly from its current position to the man overboard?

Give your answer in minutes.

2 Adrian is using a map with a scale of 1 : 250 000.

He measures these distances on the map.

a 5 cm b 10 cm

c 3 cm d 0.5 cm

What are the distances in real life?

Write your answers in kilometres.

Example

3 The scale on a map is 1 : 50 000.

What is the distance on the map, in cm, for a real distance of

a 30 km b 5 km c 13 km?

4 The scale on a map is 1:10 000.

a On the map, the distance between the post office and the library Is 3.5 cm. Work out the real distance between them. Give your answer in metres.

b The real distance between the two churches is 1.5 km. Work out the distance between the churches on the map. Give your answer in cm.

5 a The scale of the map is 1:1 000 000. Calculate the distance in km between

i St Peter Port and St Helier

ii St Helier and Carteret.

b Which town is 57 km from Cherbourg?

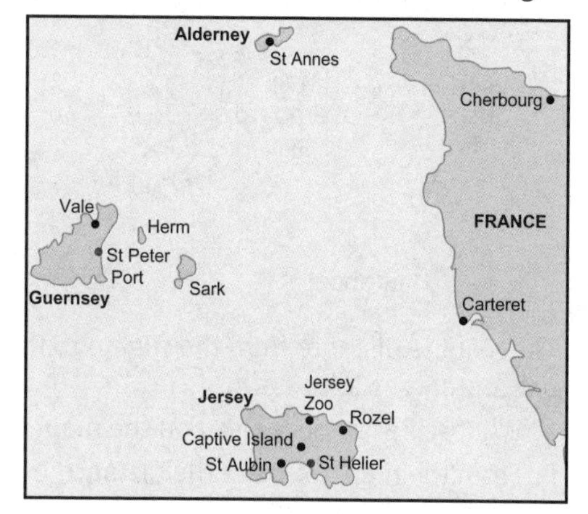

6 The diagram shows the relative positions of towns A, B and C.

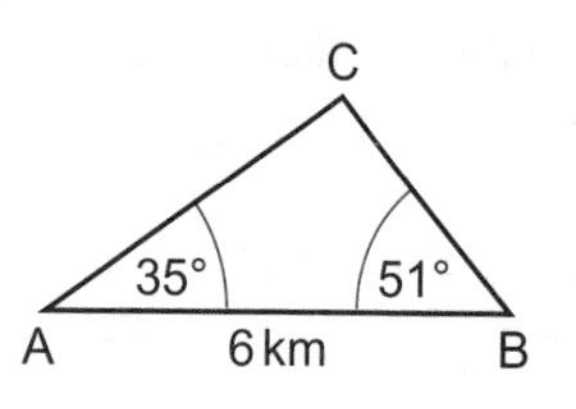

a Make an accurate scale drawing using a scale of 1:50 000.

b Work out the real distances AC and CB.

7 **P** The distance between Heathrow and Gatwick is 35 km.

The bearing of Gatwick form Heathrow is 155°.

Make an accurate scale map of the locations of the two airports, using a scale of 1 cm to 5 km.

8 **P** A ship is 120 km east of a port. The ship then sails on a bearing of 310° for 90 km.

a Make an accurate scale drawing. Use a scale of 1 cm to 20 km.

b What is the bearing of the port from the ship?

9 **P** A man walks 7 km on a bearing of 120°. He then turns and walks 8 km on a bearing of 320°.

a Use a scale of 2 cm to 1 km to draw an accurate scale drawing of the walk.

b How far is the man from his starting point?

c What is the bearing he should walk on to return to his starting point?

10 **P** The bearing from Manchester Airport to Amsterdam Airport is 105°.

Calculate the bearing of Manchester Airport from Amsterdam Airport.

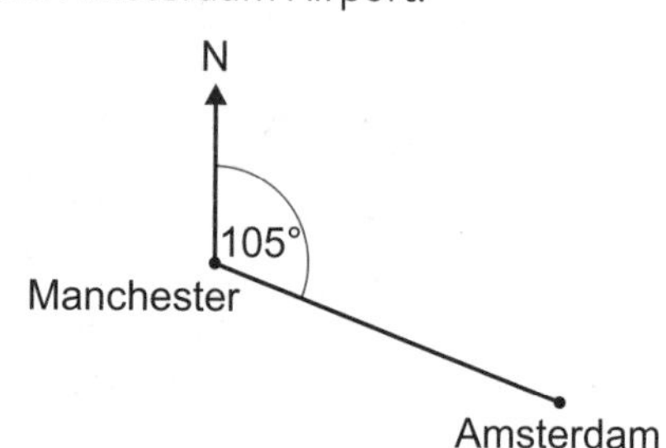

11 **P** a The bearing of B from A is 125°. Work out the bearing of A from B.

b The bearing of C from D is 322°. Work out the bearing of D from C.

12 

Exam-style question

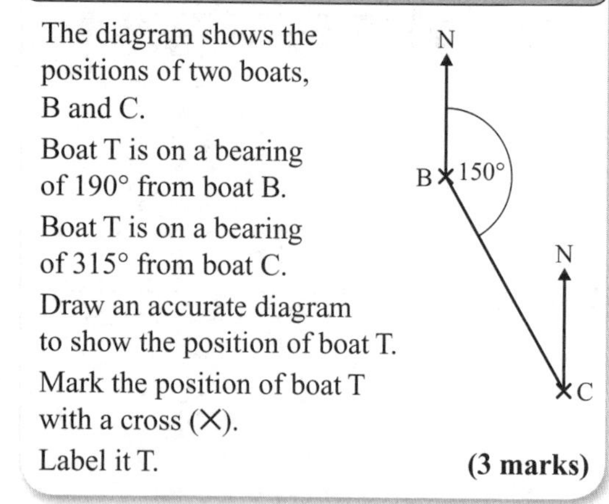

The diagram shows the positions of two boats, B and C.

Boat T is on a bearing of 190° from boat B.

Boat T is on a bearing of 315° from boat C.

Draw an accurate diagram to show the position of boat T.

Mark the position of boat T with a cross (X).

Label it T. **(3 marks)**

8.6 Constructions 1

1 Construct an accurate drawing of this triangle.

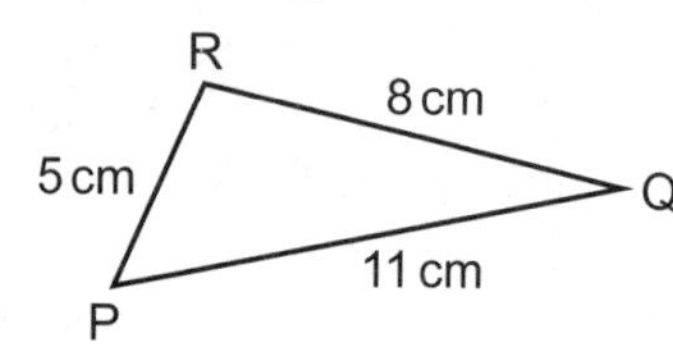

2 Construct each triangle PQR.
   a PQ = 4 cm, QR = 7 cm, PR = 5 cm
   b PQ = 9 cm, QR = 6 cm, PR = 4 cm
   c PQ = 9.5 cm, QR = 6.5 cm, PR = 4.5 cm

> **Q2 hint** Sketch each diagram first and label the lengths.

3 Construct an equilateral triangle with sides 7.5 cm.
   Check the angles using a protractor.

4 **R** Explain why it is impossible to construct a triangle with sides 3 cm, 5.5 cm and 12 cm.

5 Construct an accurate scale drawing of this flower bed.
   Use a scale of 1 : 100.

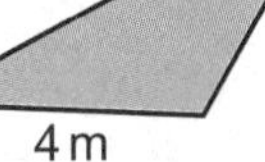

6 The diagram shows the profile of a triangular shed.
   Using a scale of 1 cm to 2 m, construct an accurate scale drawing of this profile.

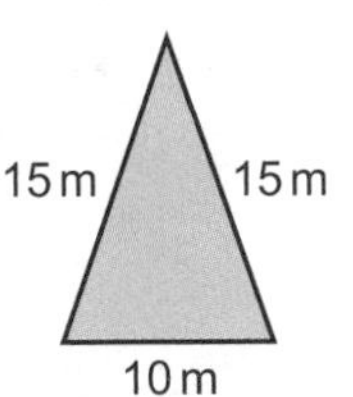

7 **P** This chocolate box is in the shape of triangular prism.

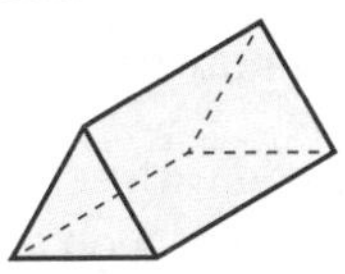

   The end faces are equilateral triangles with sides of length 5 cm.
   The whole box has length 7 cm.
   Construct an accurate net for the box.

8 a Draw a line segment PQ 10 cm long. Construct the perpendicular bisector of PQ.
   b Use a ruler and protractor to check that it bisects your line at right angles.
   c Mark any point R on your perpendicular bisector.
   Measure its distance from P and from Q.

Example

9 **P** Two schools A and B are 0.8 km apart.
   a Using a scale of 1 : 10 000, draw an accurate scale drawing of the schools.
   b A pelican crossing is equidistant from the two schools.
   Construct a line to show where the pelican crossing could be.

10 Follow these instructions to draw the perpendicular from point L to the line MN.
   a Open your compasses and draw an arc with centre L. Label the two points where it intersects with the line MN as Q and R.

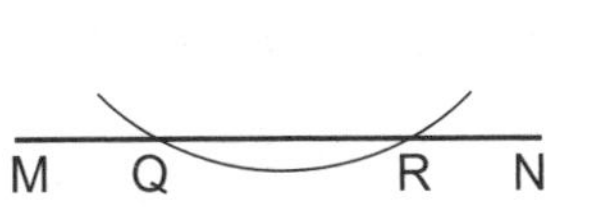

   b From each point Q and R, keeping your compasses the same, draw an arc underneath the line so they cross.
   c Join this point with point L.

11 **P** A walker is in the middle of a field and wants to walk the shortest possible distance to the marked footpath.
   The scale is 1 : 1000

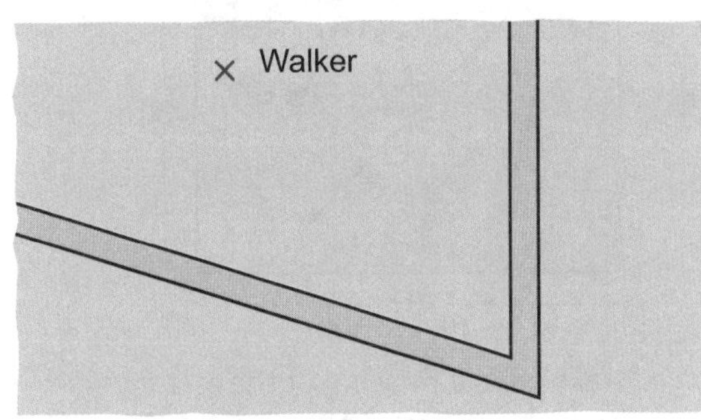

   a Trace the diagram and construct the shortest path for the walker to get to each of the two straight parts of the footpath.
   b Work out the difference in the distances.
   c The speed of the walker is 1.5 m/s.
   How long would it take to walk the shorter distance?

8.7 Constructions 2

1 For each angle
   i trace the angle
   ii construct the angle bisector using a ruler and compasses
   iii check your two smaller angles using a protractor.

Example

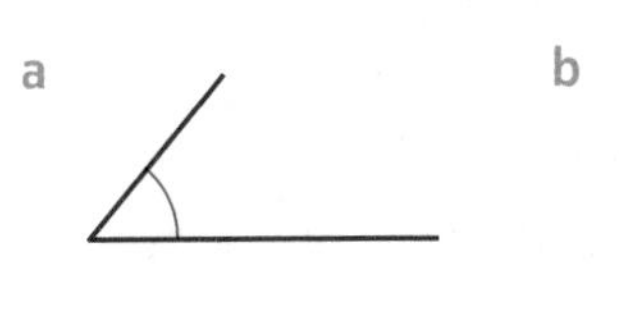

2 **P** Use a ruler and compasses to construct these angles.

a 90° b 135°

> **Q2a hint** What angle will you get when you bisect a straight line?

3 **P** Use a ruler and compasses to construct these angles.

a 22.5° b 112.5°

4 **P** Use a ruler and compasses to construct a 210° angle.

5 **P** A rectangular piece of land is being split into two plots A and B.

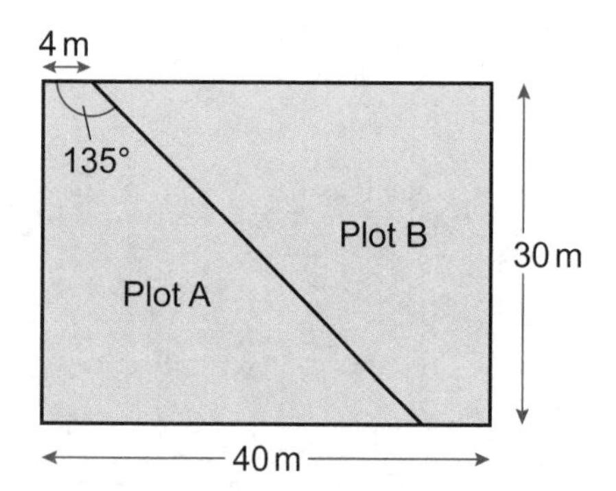

a Make a scale drawing of the rectangular plot using a scale of 1 cm to 4 m.

b Use a ruler and protractor to construct the diagonal line.

c Calculate the area of plot A.

6 **P** A piece of cloth in the shape of an equilateral triangle (ABC) is cut into two equal right-angled triangles to make two sails for a boat. Angle CBD is bisected and the lower part of the sail is dyed red.

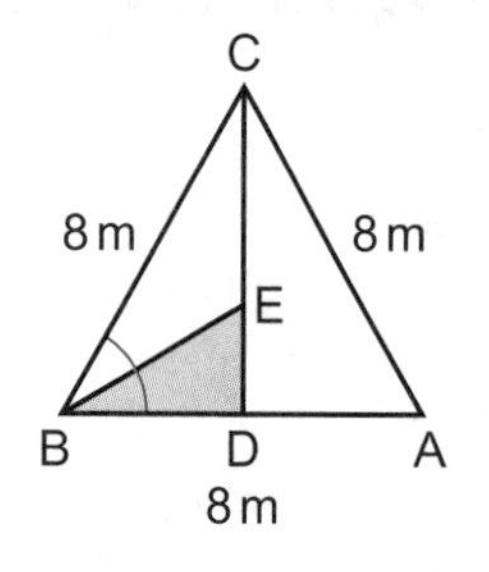

a Construct a scale drawing.
Use a scale of 1 cm to 1 m.

b Measure the length of the line DE.

c What percentage of sail BCD is dyed red?

7 **P** a Use a ruler, protractor and compasses to construct the triangle PQR.

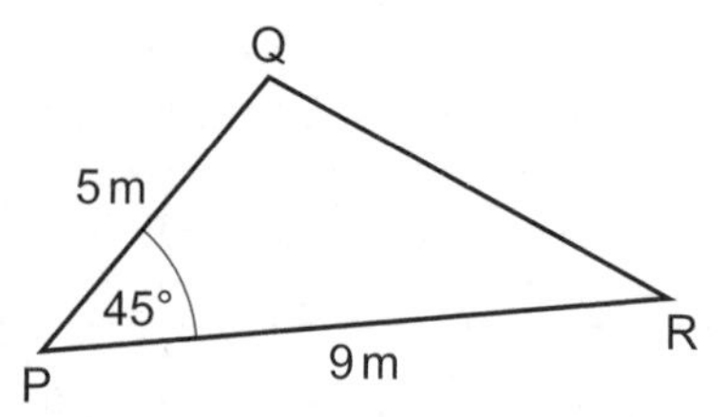

b Construct a line that is perpendicular to PR and passes through Q. Label the point where it meets PR as S.

c Calculate the perimeter of triangle PQS.

8 **P** The net of a regular tetrahedron is made from equilateral triangles. Construct the net using a ruler and compasses.

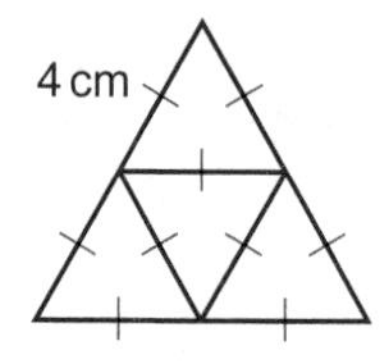

9 a Draw any triangle ABC.

b Construct the bisector of each angle.

c Label the point where the bisectors cross as point D.

d Construct the perpendicular from D to each of the sides of the triangle.

e Measure the distances from D to where each perpendicular meets the sides. What do you notice?

10 **R** a Draw a circle with centre O and radius 6 cm. Mark a point A on its circumference.

b Keep the compasses the same size as the radius and draw an arc from point A. Label the point where the arc cuts the circle B.

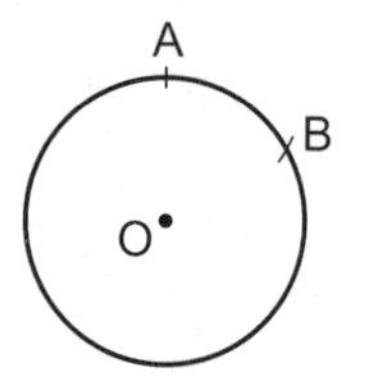

c Keeping the compasses the same, repeat from point B. Repeat until you have six points on the circumference.

d Join alternate points on the circumference.

e Name the shape you have constructed.

11 **P** Draw a regular hexagon in a circle of radius 6 cm.

> **Q11 hint** What angle will you need to construct from the centre to two consecutive points on the circumference?

8.8 Loci

1 A dog is tied to a post by a 6 m length of rope. Sketch the locus of where the dog can roam.

2 **P** Draw a line 10 cm long.
Draw the locus of all points which are 5 cm from the line.

3 **P** A rectangular bowling green is marked out to measure 20 m by 3 m.

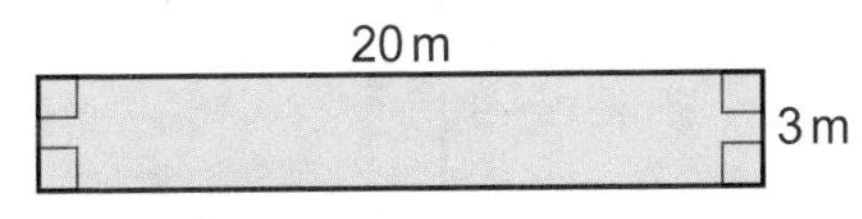

A fence is put up 1 m outside the perimeter.

a Draw a plan of the fenced area using a scale of 1 cm to 2 m.

b Construct the locus of the fence.

> **Q3 hint** Think carefully about what happens at the corners.

4 **R / P** a Draw two points 8 cm apart and label them A and B.

b Construct the perpendicular bisector of the line AB.

c Mark a point which is
  i 5 cm from A and 5 cm from B
  ii 7 cm from A and 7 cm from B.

d What do you notice about these points?

5 **P** A cinema is being built equidistant from two towns P and Q. P and Q are 40 km apart.
Using a scale of 1:1 000 000, construct the locus of points where the cinema could be built.

6 A gardener wants to plant a row of bulbs in a flowerbed equidistant from the edge AB and the edge AD.

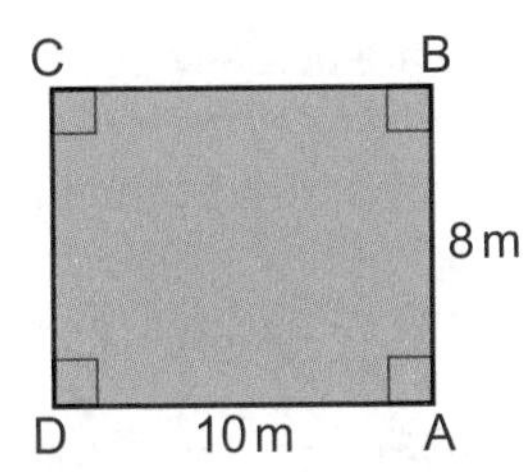

a Copy the diagram using a scale of 1 cm to 1 m.

b Construct the locus of points where the bulbs will be planted.

7 **P** This triangle is rotated anticlockwise about C.

a Copy the diagram.

b The triangle is then rotated 90° clockwise about B′. Add this to the diagram.

c Draw the locus of vertex A.

d Draw the locus of vertex C.

8 **P** Two lighthouses are 100 km apart.
Lighthouse A is on a bearing of 100° from lighthouse B.
Each lighthouse can be seen at a distance of 80 km.
Draw an accurate scale drawing of the lighthouses using a scale of 1 cm to 20 km.
Shade the region where both lighthouses can be seen.

Example

9 **Exam-style question**

Here is a scale drawing of a rectangular garden ABCD.

B C

A D

Scale: 1 cm represents 2 metres.

Jane wants to plant a tree in the garden

- at least 5 m from point C
- nearer to AB than to AD
- and less than 3 m from DC.

On a copy of the diagram, shade the region where Jane can plant the tree. **(4 marks)**

March 2013, Q15, 1MA0/1H

10 **P** A dog is tied to a rope at the corner of a barn which measures 10 m by 7 m.

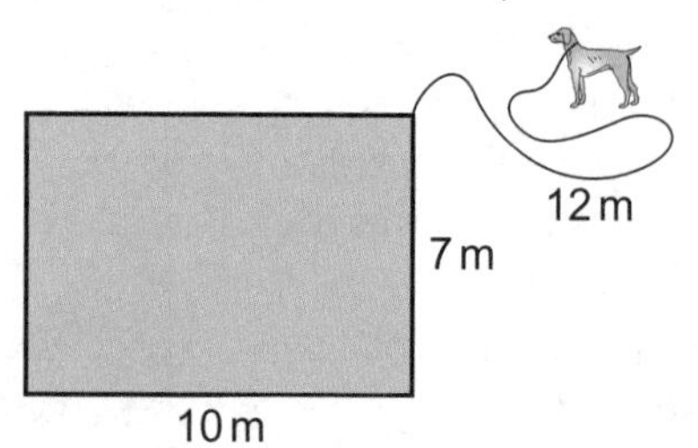

Make an accurate scale drawing of this barn. Use a scale of 1 cm to 2 m.

Accurately shade the region where the dog can roam.

11 **P** ABCD is a rectangle with sides 7 cm and 3 cm.

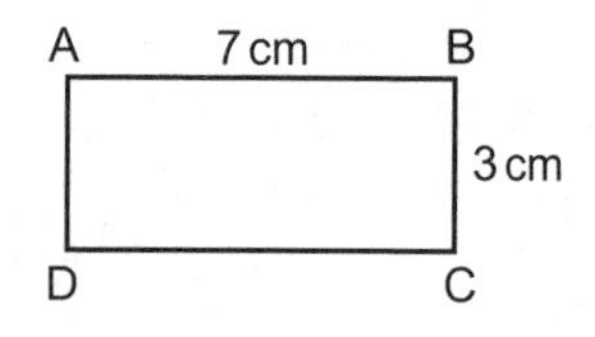

Copy the diagram. Shade the region that is closer to AB than BC and more than 2 cm from A.

12 A graph $x^2 + y^2 = 40\,000$ shows the region covered by a radio mast, where x and y are in metres.

- **a** Draw a coordinate grid from –300 to +300 on both axes. Plot the graph.
- **b** What area does the radio mast cover? Give your answer to the nearest thousand square metres.

8 Problem-solving

Solve problems using these strategies where appropriate:

- **Use pictures or lists**
- **Use smaller numbers**
- **Use bar models**
- **Use x for the unknown**
- **Use a flow diagram.**

1 Kelly needs to draw some construction drawings for a model she is making. Use the picture to draw the plan, side elevation and front elevation for Kelly's model.

2 **R** Look at these designs.

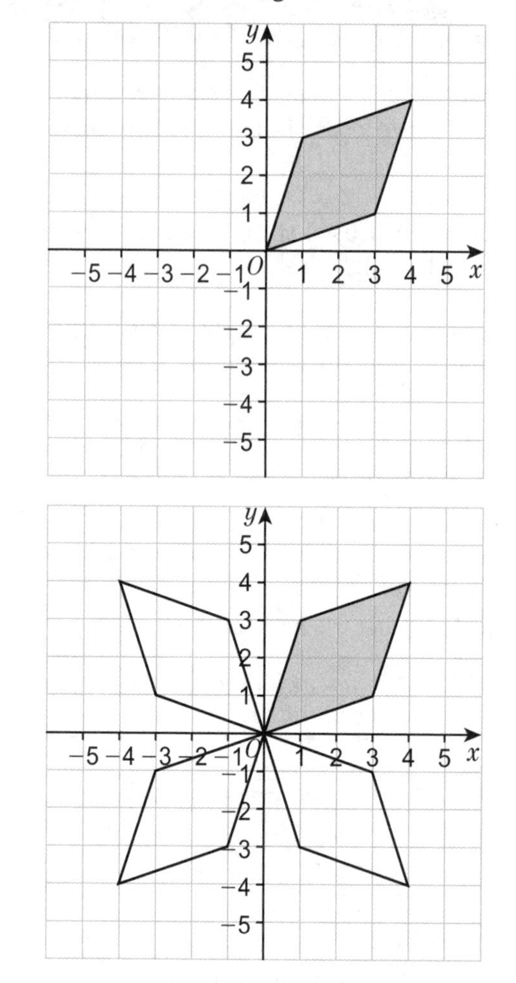

Sue-Ellen is designing a patchwork quilt. She started with the shape on the first grid and then used three transformations to create the pattern on the second grid.

What could the three transformations be?

3 An ancient fort has 9 sides.

- **a** What is the sum of the interior angles of the floor plan of the fort?
- **b** If all 9 sides of the fort were of equal length, what would each interior angle be?
- **c** The angles are actually one of three sizes. Three of the angles measure $x°$, three measure $2x°$, and the remaining three measure $3x°$.
  What are the three sizes of angles?

4 Michaela bought a car for £5700 plus VAT at 20%. She will pay the cost of the car in 12 equal monthly payments.

What is the amount of each monthly payment?

5 **R** Pete and Andy are hiking to the nearest village. Their map has a scale of 1 : 100 000. The path they need is 9 cm long on the map. Pete says that it will only take a few hours to get to the village as it is only 9 km away. Andy says it is too far to hike to as it is 90 km away.

a Who is correct? Explain your answer.

b There is a sign for a village 12 km away. What distance would this be on the map?

c What would the scale be if 8 cm represents 80 km?

6 This grid shows the position of a set of patio furniture.

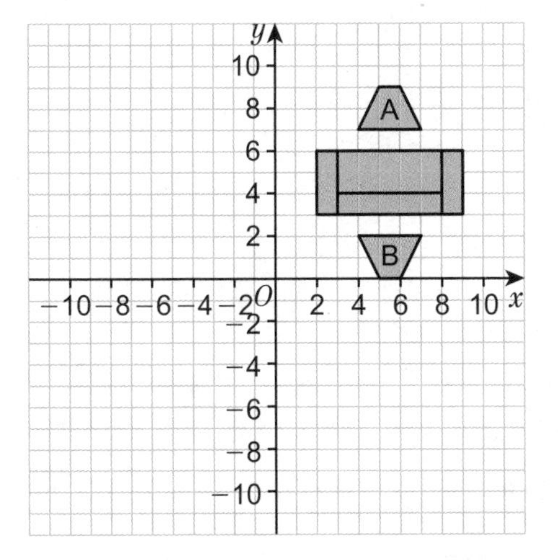

The grid shows the new positions of the furniture after it was blown by strong winds.

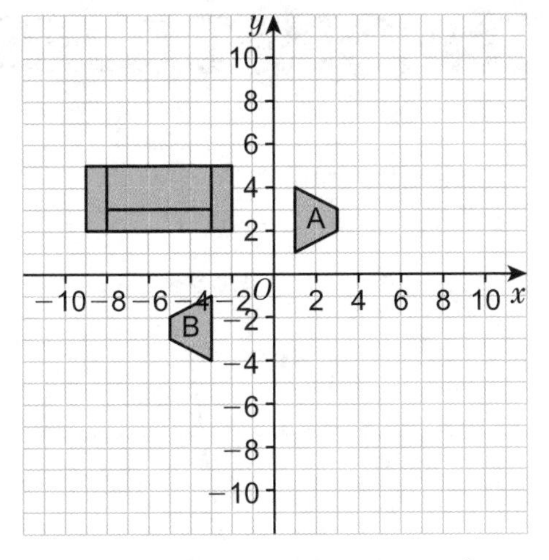

Fully describe the transformation of

a the table

b chair A

c chair B.

7 The manual for a photocopier needs a diagram to show how to fix a paper jam. Lever A needs to be rotated 120° clockwise and lever B needs to be rotated 90° anticlockwise.

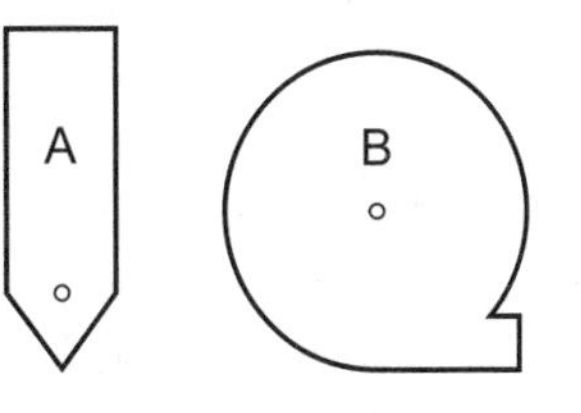

The centre of rotation is marked on each lever (o).

Copy the drawings onto a grid and then use dotted lines to draw the images of where the levers need to rotate to.

8 A farmer has invested in five wind turbines to supply energy to his farm and village. He draws a grid to show where four of the turbines are to be placed.

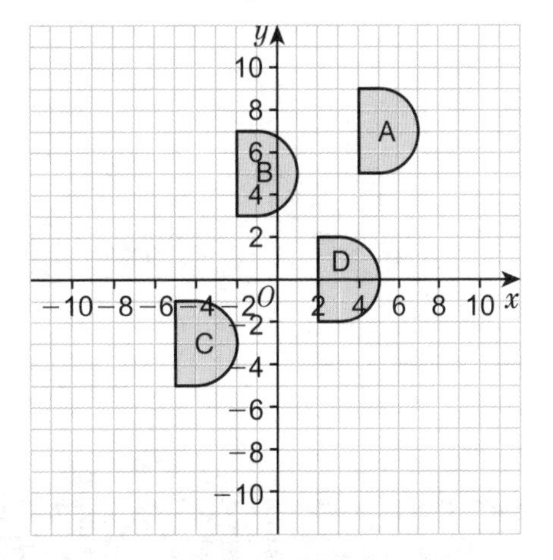

Use column vectors to describe the translation that would move turbine

a A to B b B to C c C to D

The farmer places turbine E by translating it by vector $\begin{pmatrix}-2\\-7\end{pmatrix}$ from turbine D.

d Copy the grid and draw turbine E in its correct position.

9 **R** There are some red counters and some blue counters in a bag.

7 red counters are added and the number of red and blue counters is now the same.

Then 22 blue counters are added and the ratio of red : blue becomes 1 : 3.

How many red counters were originally in the bag?

10 **Exam-style question**

In this question, use a ruler and a pair of compasses.
Leave in all your construction lines.
The scale drawing shows part of a school building and the playground outside.
The head teacher wants to paint a mini-maze on the playground.
It needs to be

- at least 4 m from the fence KL
- at least 3.6 m from the tree T
- nearer the maypole M than to the goal post G.

Construct and shade the region where the mini-maze can go.

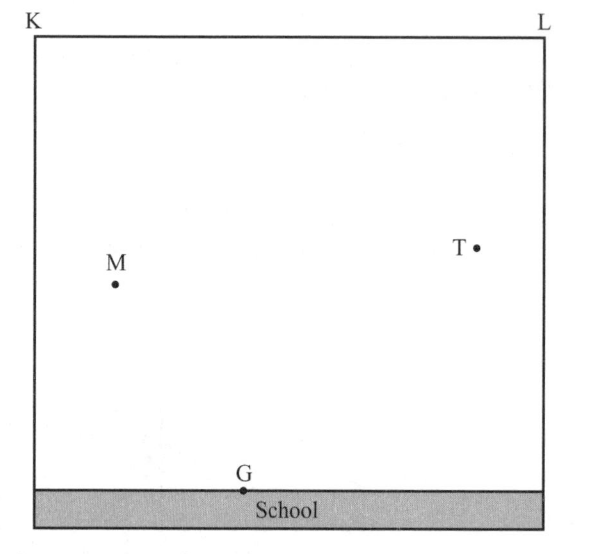

Scale: 1 cm represents 2 metres. **(5 marks)**

9 EQUATIONS AND INEQUALITIES

9.1 Solving quadratic equations 1

1 Find the solutions to these quadratic equations.

a $5x^2 = 45$

b $3x^2 - 3 = 72$

c $4x^2 - 144 = 0$

2 Solve

a $x^2 + 10x + 21 = 0$

b $a^2 + 3a - 18 = 0$

c $p^2 - 8p + 12 = 0$

d $y^2 - y - 20 = 0$

3 Find the roots of these functions.

a $x^2 - 3x$ b $x^2 - 36$ c $100 - z^2$

4 a Use this graph to solve the equation $x^2 - 2x - 8 = 0$

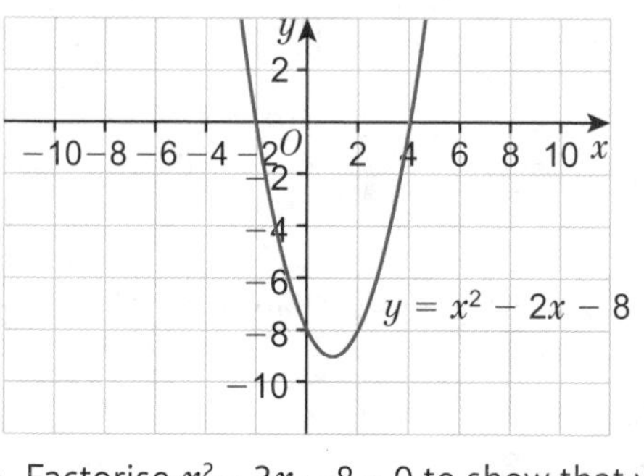

b Factorise $x^2 - 2x - 8 = 0$ to show that you get the same solutions.

5 Solve

a $x^2 + 7x = -12$ b $x^2 - x = 2$

c $x^2 = 7x - 10$ d $x^2 = 8x$

Q5a hint Rearrange into the form $x^2 + \square x + \square = 0$

6 **P** Write any function that will give the roots $x = 2$ and $x = -5$.

9.2 Solving quadratic equations 2

1 Write and solve an equation to find x.

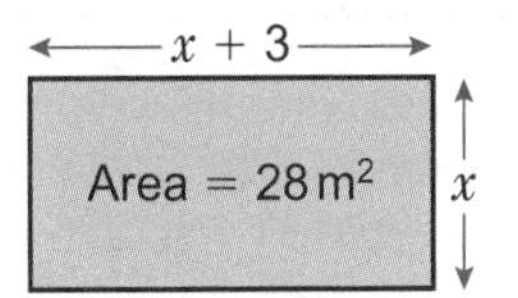

2 **P** A tile design is available in two sizes.
A small tile has dimensions $l \times l$.
A large tile has dimensions $2l \times (l + 2)$.
The area of a large tile is $160\,\text{cm}^2$.
What are the dimensions of the small tile?

3 Copy and complete to factorise the expression.
$2x^2 + 5x - 7 = (2x \_\_\_\_)(x \_\_\_\_)$

4 Factorise these expressions.

a $3x^2 + 12x + 9$

b $2x^2 + 7x + 3$

c $6x^2 - 10x - 4$

d $5x^2 + 13x - 6$

e $3x^2 - 2x - 8$

5 Solve

a $(3x - 9)(x + 5) = 0$

b $(2y + 7)(3y - 21) = 0$

c $(4t - 5)(2t - 5) = 0$

d $(3d + 7)(3d - 10) = 0$

Q5a hint Either $3x - 9 = 0$ or $x + 5 = 0$

6 Solve

a $2x^2 + 7x + 3 = 0$

b $3x^2 + 7x + 4 = 0$

c $4x^2 + 5x - 21 = 0$

d $4x^2 - 16x + 15 = 0$

7 **P** A photo frame has a space for a photo that is 7 inches by 5 inches. It has a wooden border that is the same width all the way round. The area of the wooden frame is 45 square inches.

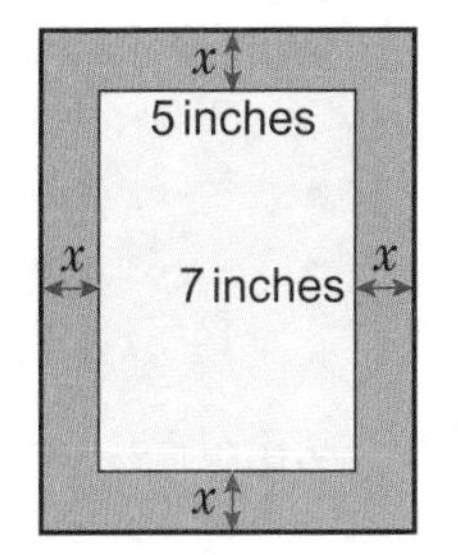

a Write an equation for the area of the wooden frame.

b Solve your equation to find x.

8 **Exam-style question**

Solve, by factorising, the equation
$6x^2 - 19x + 15 = 0$ **(3 marks)**

9 Solve, giving your solutions in surd form

a $x^2 + 8x + 5 = 0$

b $x^2 + 5x + 2 = 0$

c $x^2 + 4x - 3 = 0$

d $x^2 + 6x - 4 = 0$

e $2x^2 + 10x + 7 = 0$

10 Solve, giving your solutions to 2 decimal places

a $x^2 + 4x - 6 = 0$ b $2x^2 + 3x - 8 = 0$

c $3x^2 - 4x - 11 = 0$ d $5x^2 + 10x - 9 = 0$

11 Solve $3x^2 - 5x - 8$

a by factorising

b by using the quadratic formula.

12 **Exam-style question**

Solve $3x^2 + 9x - 10 = 0$
Give your solutions correct to 2 decimal places. **(3 marks)**

9.3 Completing the square

1 Write a quadratic expression for the area of the large square.

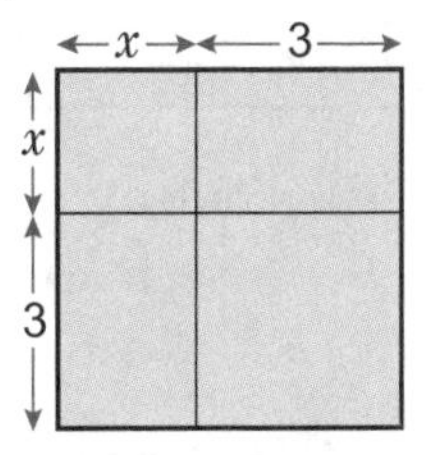

Q1 hint Write an expression with two brackets. Expand to get a quadratic expression.

2 Write these expressions in the form $(x + 3)^2 + \square$ or $(x + 3)^2 - \square$

a $x^2 + 6x + 10$

b $x^2 + 6x + 7$

c $x^2 + 6x - 1$

3 Write these as perfect squares.

a $x^2 + 4x + 4$

b $x^2 + 10x + 25$

c $x^2 + 16x + 64$

d $x^2 + 20x + 100$

4 Write these in the form $(x + p)^2 + q$

Example

a $x^2 + 4x + 7$

b $x^2 + 10x + 20$

c $x^2 + 14x$

d $x^2 + 8x + 20$

e $x^2 - 6x + 10$

5 Copy and complete to solve the quadratic equation, giving your answer in surd form.

$$x^2 + 6x + 6 = 0$$
$$(x + \square)^2 - \square + 6 = 0$$
$$(x + \square)^2 = \square$$
$$(x + \square) = \pm\sqrt{\square}$$
$$x = \square - \sqrt{\square} \text{ or } x = \square + \sqrt{\square}$$

6 Solve these quadratic equations, giving your answers in surd form.

a $x^2 + 4x + 2 = 0$

b $x^2 + 10x + 18 = 0$

c $x^2 + 2x - 4 = 0$

7 Copy and complete to write the expression $2x^2 + 8x + 35$ in the form $p(x + q)^2 + r$

$$2x^2 + 8x + 35 = 2(\square + \square) + 35$$
$$= 2[(x + \square)^2 - \square] + 35$$
$$= 2(x + \square)^2 - 8 + 35$$
$$= 2(x + \square)^2 + \square$$

Q7 hint Factorise the x^2 and x terms. Then complete the square for the expression inside the brackets. Simplify so that you have $p(x + q)^2 + r$

8 Write these in the form $a(x + p)^2 + q$

a $3x^2 + 6x - 5$

b $4x^2 - 16x + 11$

c $2x^2 + 12x - 7$

d $3x^2 - 18x + 10$

9 Solve these equations by completing the square. Give your answers in surd form.

a $2x^2 + 8x - 14 = 0$

b $3x^2 - 24x + 30 = 0$

10 Copy and complete to solve $3x^2 + 12x - 18 = 0$. Give your answer correct to 2 decimal places.

$$3x^2 + 12x - 18 = 0$$
$$3(x^2 + \square x - \square) = 0$$
$$x^2 + \square x - \square = 0$$
$$(x + \square)^2 - \square^2 - \square = 0$$
$$(x + \square)^2 - \square = 0$$
$$(x + \square)^2 = \square$$
$$x + \square = \pm\sqrt{\square}$$
$$x = \square + \sqrt{\square} \text{ or } x = \square - \sqrt{\square}$$

Q10 hint Begin by dividing every term by the coefficient of x^2, 3. Then complete the square for the first two terms. Rearrange terms and then take the square root of both sides.

11 Solve these quadratic equations, giving your answers correct to 2 decimal places.

a $2x^2 + 12x - 4 = 0$

b $4x^2 - 32x - 9 = 0$

c $3x^2 - 6x - 8 = 0$

d $6x^2 + 24x - 15 = 0$

e $4x^2 + 20x - 7 = 0$

12 **Exam-style question**

Solve $5x^2 + 30x + 8 = 0$ by completing the square.

Give your answer correct to 2 decimal places. **(3 marks)**

9.4 Solving simple simultaneous equations

1 Solve the simultaneous equations

a $y = 4$
$3x + y = 13$

b $y = -2$
$4x + y = 10$

c $y = 3$
$5x + 2y = 31$

d $y = 3x$
$2x + 3y = 22$

e $y = x - 1$
$x + y = 9$

f $y = x + 2$
$x + 2y = 13$

g $4x - y = 0$
$3x + 2y = 11$

h $y - 2x = 0$
$3x + y = 20$

2 **P** Two shirts and one pair of trousers cost £40. A pair of trousers costs £4 more than a shirt.

a How much is a shirt?

b How much is a pair of trousers?

3 **P** Alfie buys 3 fish and 2 portions of chips for £12. Lauren buys 4 fish and 4 portions of chips for £18.

How much does one portion of chips cost?

4 Solve these simultaneous equations.

Example

a $4x + y = 16$
$3x - y = 5$

b $3x + y = 11$
$x + y = 7$

c $6x - 2y = 20$
$5x - 2y = 16$

5 Emily solves simultaneous equations in this way.

$2x - y = 7$
$x + y = 5$
$y = 5 - x$
$2x - (5 - x) = 7$
$2x - 5 + x = 7$
$3x - 5 = 7$
$3x = 12$
$x = 4$
$4 + y = 5$
So $y = 1$
$(2 \times 4) - 1 = 7$ ✓

Solve the simultaneous equations in **Q4** part **a** using Emily's method.

Q5 hint Write the second equation with y as the subject. Substitute $y = \square$ into the first equation then collect like terms to find x. Substitute $x = \square$ into the second equation and check your solutions work by substituting them into the first equation.

6 a Copy and complete to solve the simultaneous equations.

$$3x - y = 13 \quad ①$$
$$2x + 2y = 14 \quad ②$$
$$\times 2 \qquad \square x - 2y = \square \quad ③$$

$$2x + 2y = 14 \quad ②$$
$$\square x - 2y = \square \quad ③$$
$$\square x + 0 = \square$$
$$x = \square$$

b Substitute your value of x into equation ① to find y.

Q6a hint Multiply every term in equation ① by 2. Then add equations ② and ③.

7 Solve these simultaneous equations.

a $2x - y = 4$
$4x + 3y = 18$

b $2x - 2y = 6$
$x + 3y = 7$

c $x - 2y = -6$
$2x + 3y = 2$

d $3x - y = -10$
$2x + 3y = -3$

8 **Exam-style question**

Solve the simultaneous equations
$3x - y = -9$
$2x + 3y = 5$ **(3 marks)**

9 Solve the simultaneous equations
$4x + y = 0$
$x + 4y = 0$

10 **P** The sum of two numbers is 29 and their difference is 7.
Let the two numbers be x and y.
Write two equations and solve them to find the two numbers.

11 **P** A plumber charges a call-out fee of £x plus an hourly rate of £y per hour. Callum pays £90 to call out the plumber for 3 hours. Louise pays £110 for 4 hours.

a Work out the call-out charge.

b How much would you have to pay for 5 hours?

12 **Exam-style question**

The diagram shows a rectangle.
All sides are measured in centimetres.

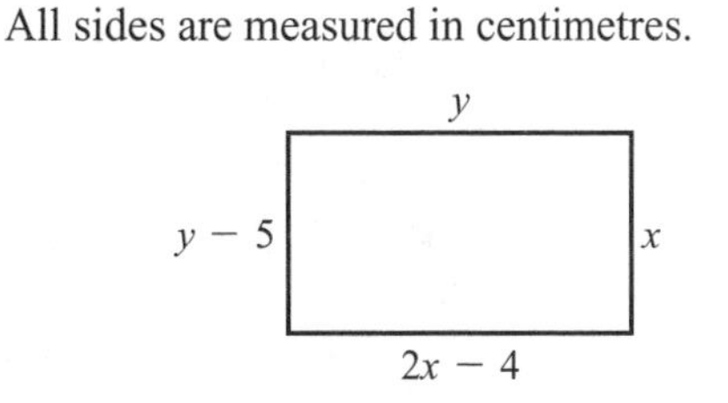

Amy says the perimeter of the rectangle is less than 50 cm.
Show that Amy is correct. **(5 marks)**

9.5 More simultaneous equations

1 a Write the equation of a line through (2, 7).

b Write the equation of a line through (3, 11).

c Solve your simultaneous equations from parts **a** and **b** to find m and c.

d Write the equation of the line through the points (2, 7) and (3, 11).

Q1a hint Substitute $x = 2$, $y = 7$ into $y = mx + c$

Q1d hint Substitute your values of m and c from part **c** into $y = mx + c$

2 Find the equation of the line through the points (–3, 11) and (2, 1).

3 Solve these simultaneous equations.

a $4x + 3y = 24$
$5x - 2y = 7$

b $7x + 2y = 15$
$4x + 4y = 20$

c $5x + 3y = 9$
$3x - 4y = 17$

d $3x - 4y = 4$
$2x + 2.5y = 13$

e $2x + 3y = 9$
$3x - 2y = -19$

4 **P** In a café, two sandwiches and two bananas cost £4, and two sandwiches and one banana cost £3.50.
Work out the cost of a sandwich and the cost of a banana.

5 **P** Henry buys two adult and three child cinema tickets for £41.20.
Rachel buys two adult and four child tickets for £48.40.
How much is an adult ticket and how much is a child ticket?

6 **P** Holly buys two pens and three pencils in a shop and pays £3.45.
At the same time Leah buys three pens and two pencils and pays £2.70.
What is the cost of a pen?
What is the cost of a pencil?

7 **Exam-style question**

Solve the simultaneous equations.
$5x + 4y = -4$
$3x + 8y = 6$ **(3 marks)**

8 **P** The maximum load of a service lift is 200 kg. Two possible maximum loads are 4 small crates and 2 large crates, or 1 small crate and 3 large crates.
What is the mass of a small crate?
What is the mass of a large crate?

9 **P** Hire charges for a taxi consist of £x fixed charge + y pence for each mile travelled.
A 5-mile journey costs £6.
A 10-mile journey costs £10.
How much would a 20-mile journey cost?

10 The diagram shows a rectangle with all measurements in centimetres.

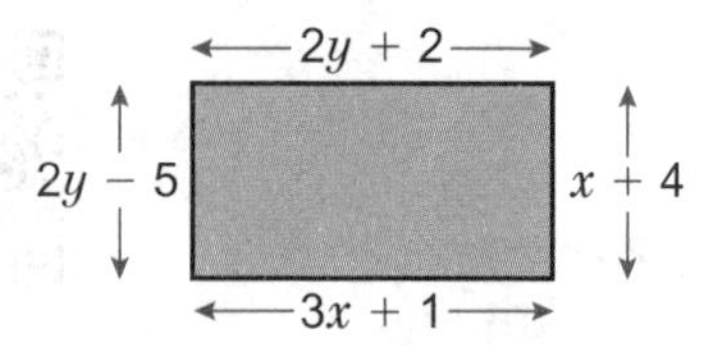

a Write down a pair of simultaneous equations in x and y.

b Solve the equations.

c Give the dimensions of the rectangle.

9.6 Solving linear and quadratic simultaneous equations

1 Solve these simultaneous equations.

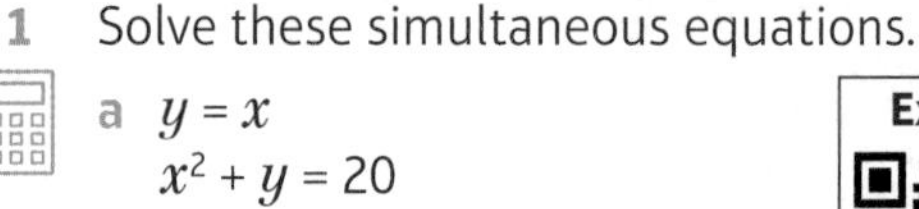

a $y = x$
$x^2 + y = 20$

b $x - y = 5$
$x^2 - 11 = y$

c $y + 2x = 7$
$2x^2 + 5x + 10 = y$

d $y = 3x + 2$
$y = 3x^2 + 11x + 7$

e $x^2 + y^2 = 16$
$y = 2x - 2$

2 Solve these simultaneous equations.
Give your answers correct to 2 decimal places.

a $y - 2x = 5$
$y = x^2 - 4x + 3$

b $3y + 2x = 11$
$y = x^2 + 2x - 5$

3 **R** The diagram shows a sketch of the curve $y = 3(x^2 - x)$. The curve crosses the straight line with equation $y = 2 - 2x$ at two points.

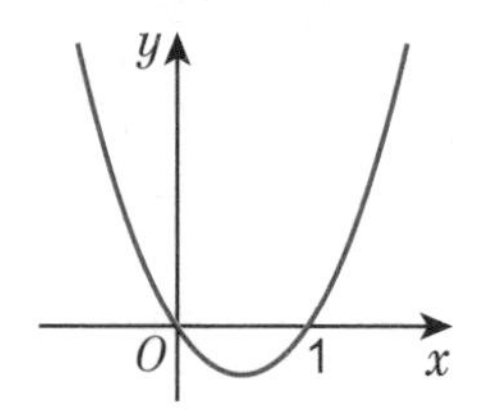

Find the coordinates of the points where they intersect.

4 Solve these simultaneous equations.

a $y = 2x^2 - 3x + 1$
$x + y = 5$

b $y = 3x^2 - 9x + 3$
$y = 2x + 7$

c $y = x^2 - 9x + 10$
$y + 3x = 2$

d $y = 2x^2 + 3x - 10$
$y - 2 = 5x$

e $y = x - 2$
$y^2 = 14 - x$

f $y = 2x + 3$
$y^2 = 8x + 33$

5 **R** The diagram shows a patio within a garden. The length of the patio is half the length of the garden.

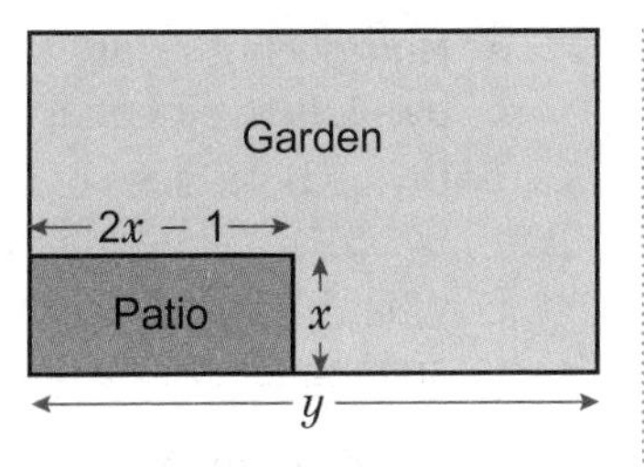

a Write an equation to represent the length of the garden (y).

b Write an equation to represent the area of the patio, which is 15 m^2.

c Use these two equations to find the value of x and hence the length of the garden (y).

6 A curve with equation $y = x^2 + 5x - 12$ crosses a straight line with equation $y = 3x - 4$ in two places.
Find the coordinates of the points where they intersect.

7 **Exam-style question**

C is the curve with equation $y = x^2 - x - 1$
L is the straight line with equation $y = 2x + 3$
L intersects C at two points, P and Q.
Calculate the exact length of PQ. **(6 marks)**

Exam hint
There are 2 marks each for finding P and Q, and 2 marks for finding the length of PQ.

8 **R** The diagram shows a circle of diameter 6 cm with centre at the origin.

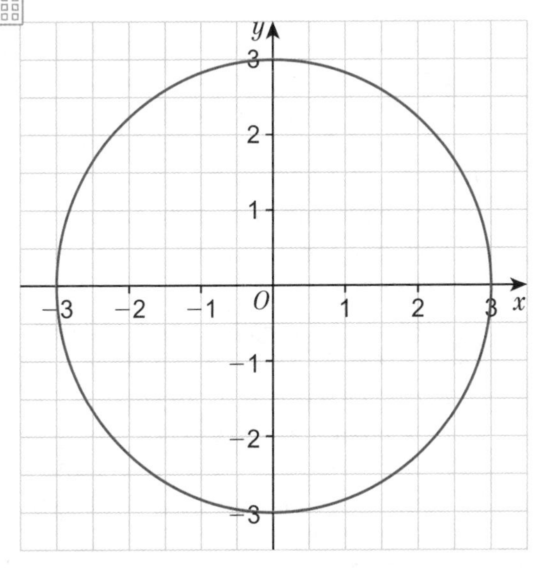

a Write the equation for the circle.

b Use an algebraic method to find the points where the line $y = 3x + 1$ crosses the circle.

9 Solve these simultaneous equations.
Give your answers correct to 3 significant figures.

a $x^2 + y^2 = 24$
$y = x + 2$

b $x^2 + y^2 = 30$
$y = 2x - 3$

c $x^2 + y^2 = 65$
$y = 5 - 2x$

9.7 Solving linear inequalities

1 Write down four integers that satisfy each inequality.

a $x + 3 > 5$

b $x - 2 < -1$

c $2x - 1 > 6$

d $4x + 3 > 1$

e $-4 < x \leqslant 5$

f $-4 \leqslant x < 0$

g $4 > x \geqslant -2$

2 Write the inequalities that these number lines represent for x.

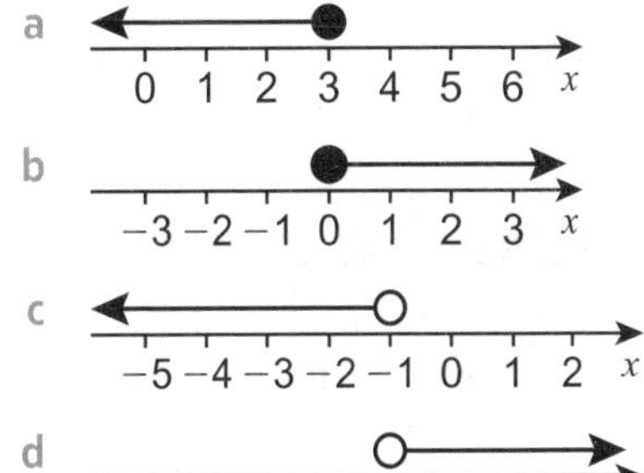

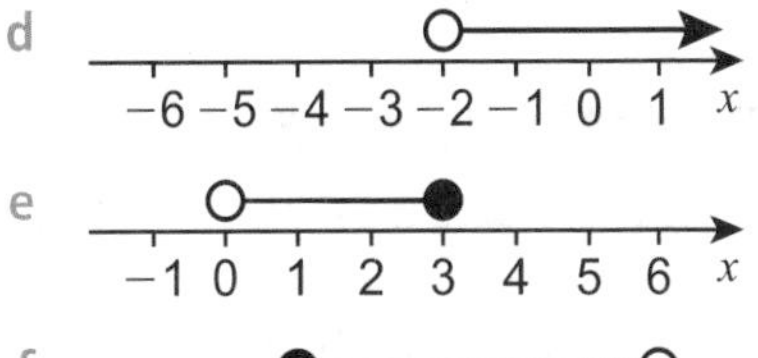

3 **Exam-style question**

$-2 \leqslant x < 5$ where x is an integer.

a Write down all the possible values of x. **(2 marks)**

b Write down the inequality represented on the number line.

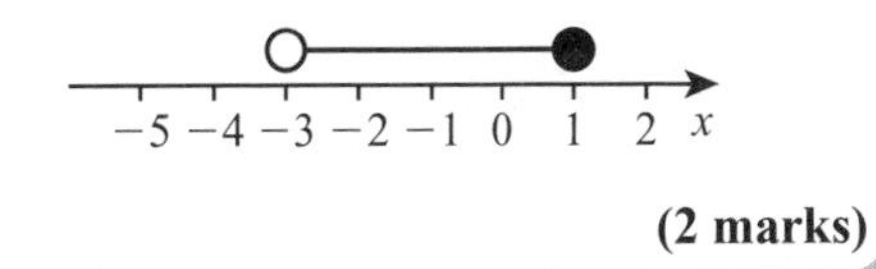

(2 marks)

4 Draw number lines to show these inequalities.

a $x < 3$

b $2 \leqslant x \leqslant 4$

c $-4 < x < 0$

d $-3 < x \leqslant 2$

5 $\{x : x > 2\}$ means the set of all x values such that x is greater than 2.

Write the meaning of these sets.

a $\{x : x > 4\}$

b $\{x : x \leqslant 2\}$

c $\{x : x < 0\}$

d $\{x : x > 0\}$

e $\{x : x \geqslant -3\}$

6 Show the sets in **Q5** on number lines.

7 Solve each inequality and show your answer on a number line.

Write the solutions using set notation.

Example

a $2x < 10$

b $4x \geqslant 8$

c $3x + 2 > 5$

d $3x - 4 \leqslant 8$

8 **Exam-style question**

$4x - 7 < 12$

Find the largest integer value of x. **(3 marks)**

Q8 hint Use the inequality symbol on each line of your working.

9 Solve these inequalities and write the solutions using set notation.

a $4(x + 1) < 12$

b $3(x - 2) \geqslant 2x - 2$

c $5x - 1 > 2(2x - 1)$

d $3(7 - x) < 4(3x - 6)$

10 Solve

a $-3 < 2x - 1 \leqslant 9$

b $-4 \leqslant 3x + 2 < 11$

c $-3 < \frac{3x}{2} < 6$

d $-1 \leqslant \frac{2x + 3}{5} < 1$

11 a Multiply both sides of the inequality $4 < 7$ by -1. Is the inequality still true?

b Divide both sides of the inequality $12 > 9$ by -3. Is the inequality still true?

c What happens when you multiply or divide an inequality by a negative number?

12 Find the possible integer values of x in these inequalities.

a $-7 < -x < -3$

b $-8 < -2x < 6$

c $6 \leqslant 10 - 2x \leqslant 16$

d $-4 \leqslant 8 - 3x < 14$

13 Solve these inequalities and write the solutions using set notation.

a $2(x + 3) \geqslant 4x + 2$

b $4x - 3 > 7 - x$

c $-7 \leqslant 3x - 4 < 5$

d $4 \leqslant 2(3 - x) \leqslant 18$

9 Problem-solving

Solve problems using these strategies where appropriate:

- **Use pictures or lists**
- **Use smaller numbers**
- **Use bar models**
- **Use x for the unknown**
- **Use a flow diagram.**

1 Three lookout towers of a castle form the corners of a triangle. The North Tower is 85 m from the East Tower. The East Tower is 65 m from the West Tower, and the West Tower is 70 m from the North Tower.

Use a scale drawing to show the shape and dimensions of the triangle formed by the three towers.

2 **Exam-style question**

$-2 \leqslant n < 5$ where n is an integer.

a Write down all possible values for n. **(2 marks)**

b Write down the inequality represented on the number line.

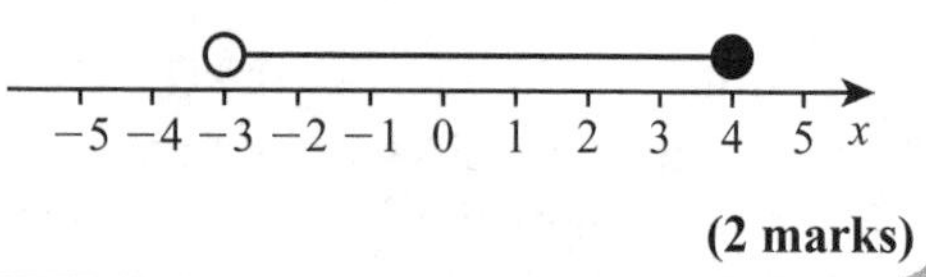

(2 marks)

3 **R** Marcia is decorating a gift box she has made in the form of a triangular prism. She has enough glitter to cover 400 cm².

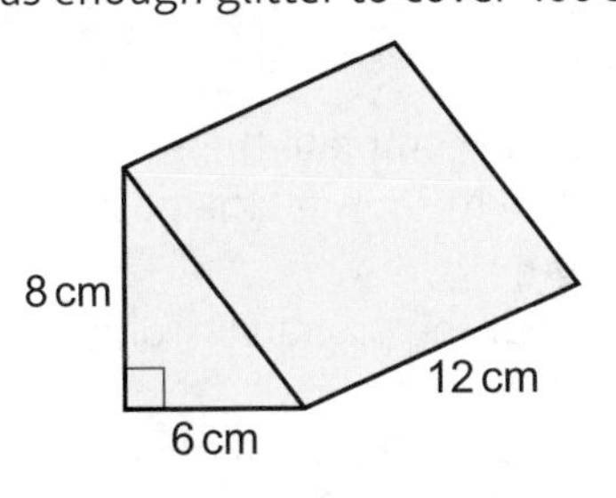

a Does Marcia have enough glitter to completely cover the box?

b What extra information did you need to find before you could answer part **a**? How did you find it?

c What is the volume of the box?

4 **R** Jessica is putting together a flat-pack bookcase. Some of the shelves are slightly longer than others even though they are all meant to be the same length. She works out that the lengths have been rounded to the nearest centimetre.

a If a shelf is meant to be 75 cm what are the possible values it could be?

b If one shelf is the upper bound value and one shelf is the lower bound value, what is the size of the gap that would be created?

c How would rounding to the nearest tenth of a centimetre affect the measurements of the shelves?

5 Jonathan has two bags of oranges and three bags of apples. They weigh 17 lb.

Corinne has three bags of oranges and four bags of apples. They weigh 24 lb.

How much does one bag of apples weigh?

6 Find the possible integer values of x in these inequalities.

a $-10 < 3x - 4 < 8$

b $-3 \leqslant 2x - 5 < 3$

7 Solve $2x^2 + 4x - 3 = 0$.

Give your answer correct to 2 decimal places.

8 **R** The outdoor area of a pre-school is 42 m².

The storage shed in the outdoor area has dimensions $b \times b$.

The dimensions of the outdoor area are $3b \times (b + 5)$.

What are the dimensions of the shed?

Q8 hint Draw a picture.

9 Solve these simultaneous equations.

$5x + y = 6$

$x^2 + y = 12$

10 **R** Three equations are written on the board.

$x^2 - 5x + 7 = 0$

$x^2 + y^2 = 34$

$y = 3x + 4$

a Clay has been asked to choose one of the equations to draw a circle.

Which equation does Clay use to draw the circle?

b What is the centre point of this circle?

c Clay is asked to choose one of the other equations to draw a straight line.

Which equation does he choose?

d Use the equations from parts **a** and **c** and solve them simultaneously to find the points where the line intersects the circle.

10 PROBABILITY

10.1 Combined events

1 T-shirts are made in five different colours (red, blue, navy blue, green and pink) and with four different logos (cat, dog, fox, owl).
   a How many possible combinations are there?
   b What is the probability that the combination chosen is red with an owl logo?
   c What is the probability that the combination chosen is not red with a fox logo?

2 At a school, students choose two of five different activities: drama, mountain biking, hockey, art and orchestra.
   a How many combinations of two activities are there?
   b What is the probability that a student will choose art?
   c What is the probability that a student will choose mountain biking?
   d What is the probability that a student chooses art and drama?

3 A coin is flipped and an ordinary six-sided dice is rolled.
   a Write a list of all the possible outcomes.
   b How many outcomes are there altogether?
   c Work out
      i P(head and 2)
      ii P(tail and odd number).

4 Amelia rolls a six-sided dice and spins a four-sided spinner, then multiplies the results together.

		Dice					
		1	2	3	4	5	6
Spinner	1	1	2				
	2	2	4				
	3	3					
	4						

   a Copy and complete the sample space diagram to show all the possible outcomes.
   b Work out the probability of getting
      i a product of 6
      ii a product that is a prime number
      iii a product of 20.

5 **Exam-style question**

Bethan rolls two dice.
She adds the scores on the dice together.
She records the possible scores in a table.

+	1	2	3	4	5	6
1	2	3	4	5	6	7
2	3	4	5	6	7	8
3	4	5	6	7	8	9
4	5	6	7	8	9	10
5	6	7	8			
6	7	8	9			

   a Complete the table of possible scores. **(1 mark)**
   b Write down all the ways in which Bethan can get a total score of 7
   One way has been done for you.
   (1, 6) **(2 marks)**

Both dice are fair.
   c Find the probability that Bethan's total score is prime. **(2 marks)**

6 Aiden spins a three-sided spinner labelled with the numbers 1, 3, 5 and a four-sided spinner labelled with the numbers 2, 4, 6, 8, then finds the sum of the two numbers from his spins.
   a Draw a sample space diagram to show all of the possible outcomes.
   b How many possible outcomes are there altogether?
   c Work out the probability of getting a total of
      i 5
      ii an odd number
      iii less than 12.
   d Which total are you most likely to get?

7 Two bags, A and B, contain coloured marbles.
Bag A has 2 green, 1 red and 1 blue.
Bag B has 3 red and 1 blue.
Logan takes a marble at random from each bag.
   a Draw a sample space diagram to show all the possible outcomes.
   b Work out the probability that the marbles will be
      i both green
      ii the same colour
      iii different colours.

8 Shen rolls a dice twice.
Work out the probability that he will roll the same number on both rolls.

> **Q8 hint** List all possible outcomes.

9 A group of six students, Mohammed, Susan, Jackson, Henry, Charles and Eloise, sit together in a science class. Two students from the group are picked at random to demonstrate an experiment.
Work out the probability that both the students are male.

10.2 Mutually exclusive events

1 A five-sided spinner lettered A, B, C, D, E is spun.
Which two of the following events are mutually exclusive?
   A Spinning a vowel
   B Spinning a letter in the word ADD
   C Spinning a consonant

2 A fair six-sided dice is rolled once.
Work out the probability of rolling
   a a prime number or an odd number
   b a factor of 6 or an even number
   c a square number or a prime number.

3 A standard pack of cards is shuffled and a card is picked at random.
Find the probability of picking
   a a king or a 2
   b a red queen or a club.

> **Q3 hint** A standard pack of 52 cards is equally split into four suits: hearts, diamonds, clubs and spades. For each suit there is an ace, 2, 3, 4, 5, 6, 7, 8, 9, 10, jack, queen and king.

4 The chance of a traffic light being red is 35%.
The chance of it being amber is 17%.
What is the probability that next time you are at the lights it will be green?

5 The lettered cards are shuffled. A card is picked at random.

Work out the probability of picking a vowel, a letter in the word BAD or a letter in the word CAT.

6 **P** The table gives the probability of getting each of 1, 2, 3, 4, 5 and 6 on a biased dice.

Number	1	2	3	4	5	6
Probability	0.5	$2a$	$4a$	$2a$	a	a

Work out the probability of getting
   a 2 or 4   b an odd number.

7 A fair four-sided spinner showing the numbers 1–4 is spun.
Work out the probability of getting
   a 4   b not 4.

Example

8 A letter is picked at random from the alphabet.
Find the probability of picking
   a a vowel   b not a vowel.

9 The probability that it will snow tomorrow is 0.37.
Work out the probability that it will not snow tomorrow.

10 **Exam-style question**

Here is a four-sided spinner.
The spinner is biased.
The table shows the probabilities that the spinner will land on 1 or on 3.

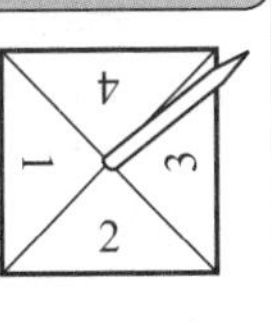

Number	1	2	3	4
Probability	0.2		0.1	

The probability that the spinner will land on 2 is the same as the probability that the spinner will land on 4.
Work out the probability that the spinner will land on 4. **(3 marks)**

March 2014, Q4, 1MA0/2H

11 **R** William has a box of biscuits. In the box there are chocolate, wafer and sugar-coated biscuits in the ratio 2 : 4 : 5.
William doesn't like the wafers.
Work out the probability that he will pick a biscuit at random that is not a wafer.

12 A and B are two mutually exclusive events.
P(A) = 0.3 and P(A or B) = 0.73
Work out the value of P(B).

> **Q12 hint** For two mutually exclusive events, P(A or B) = P(A) + P(B).

13 **P** C and D are two mutually exclusive events.
P(D) = 0.11 and P(C or D) = 0.7
Work out P(not C).

10.3 Experimental probability

1 Eli dropped a piece of buttered toast lots of times. It landed either butter up or butter down. He recorded his results in a frequency table.

Position	Frequency
Butter down	48
Butter up	52

a Work out the total frequency.
b Work out the experimental probability of the toast landing
i butter up ii butter down.
c He drops the toast 50 more times.
How many times do you expect it to land butter down?

2 **Exam-style question**

A car manufacturer wants to work out an estimate for the number of cars of each colour that will be bought next year.
The Managing Director says to record the colours of the next 100 cars sold.
The Assistant Director says to record the colours of the next 1000 cars sold.
Who is more likely to get the better estimate?
Give a reason for your answer. **(1 mark)**

> **Exam hint**
> You only get the mark for the job title and a reason. Just writing the correct name scores 0 marks.

3 Ahuva rolls a ten-sided dice numbered 1–10.

Example

a What is the theoretical probability that the dice will land on 7?
Ahuva rolls the dice 200 times.
b Estimate how many times the dice will land on 7.

4 There are toffees, chocolates and mints in a bag in the ratio 2 : 7 : 3.
a What is the probability of picking a mint?
A sweet is picked at random from the bag and then replaced. This is done 240 times.
b How many mints would you expect to be picked?

5 The probability of a person failing their driving test is 0.64.
Work out an estimate for how many of the next 200 people who take the test will pass.

6 There are 24 counters in a bag. There are red, blue, yellow and green counters.
A counter is picked from the bag and then replaced.
The table shows the results.

Colour	Frequency	Relative frequency
Red	35	
Blue	14	
Yellow	33	
Green	18	

a Copy and complete the table, calculating the relative frequency for each outcome.
b What is the experimental probability of picking a red?
c The experiment is repeated 400 times. How many times would you expect to get a red?
d Are there the same number of counters of each colour? Explain your answer.

7 The table shows the results of rolling a dice.

Number	1	2	3	4	5	6
Frequency	18	13	14	16	17	22

Is the dice fair? Give a reason for your answer.

8 Isaac rolls two dice and records the result.
He does this 180 times.
One possible outcome is (2, 2). Estimate the number of times he will get two 2s.

9 The probability of winning a game is $\frac{1}{10}$.
Ghayth says that if he plays 10 times he will win.
Is he right? Give a reason for your answer.

10 Ben flips two coins 100 times.
How many times would you expect him to get
a two heads b one head and one tail?

11 A vet estimates that the probability of an animal brought to the vet's being a dog is 0.47.
Of the next 150 animals brought in, 37 are dogs.
Is the vet's estimate a good one?
Explain your answer.

10.4 Independent events and tree diagrams

1 There were 30 customers in a restaurant one lunchtime. Of these, 17 chose the set menu.
13 of the customers who chose the set menu had coffee.
25 customers ordered coffee in total.
Copy and complete the frequency tree for the 30 customers.

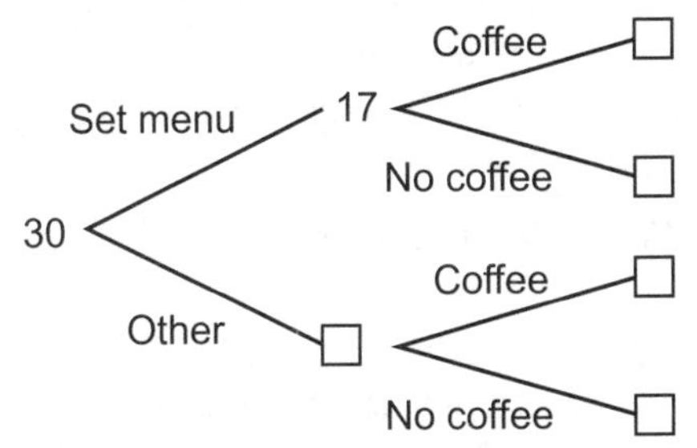

2 A driving test centre examines 50 students one day.
Of the 50 students, 27 are female.
A total of 32 students passed, 18 of whom were male.

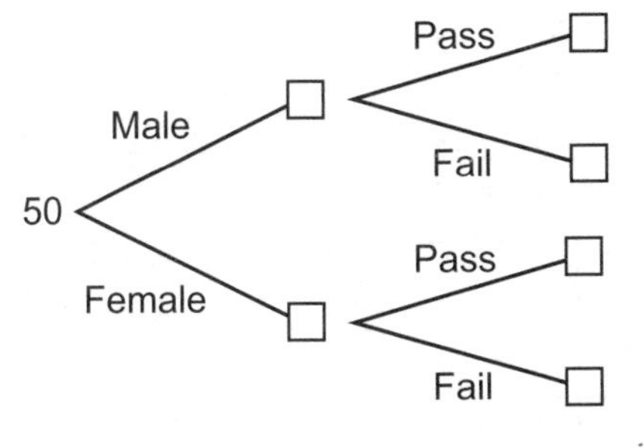

a Copy and complete the frequency tree.
b Work out the probability that a student picked at random is a male who fails.

3 60 customers bought a holiday at a travel agent's one day.
Of the 60 customers, 45 bought a holiday abroad.
A total of 35 of the customers paid in full.
Of these, 13 were going on holiday in the UK.

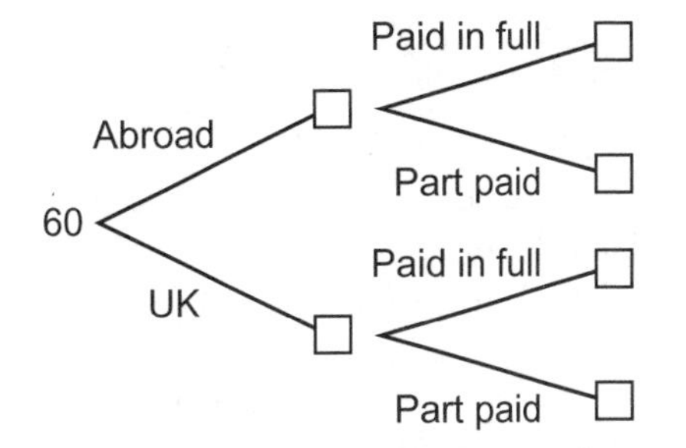

a Copy and complete the frequency tree.
b Work out the probability that a person who goes abroad pays in full.

4 Two teams, A and B, play each other at hockey and at football.
The probability that Team A wins the hockey is 0.5. The probability that Team B wins the football is 0.7.
Assuming that the two events are independent, work out the probability that Team A wins both games.

5 **R** The probability that it rains is 0.3.
The probability that Amy remembers her coat is 0.7.
Assuming that the two events are independent, work out the probability that
a it rains and Amy has her coat
b it doesn't rain and Amy doesn't have her coat
c it rains and Amy doesn't have her coat.

6 Two letters are picked one after the other from a set of alphabet cards.
(The same letter can be picked twice.)
Work out the probability that
a both letters are vowels
b both letters are consonants
c the first letter is a vowel and the second letter is a consonant
d the first letter is a vowel and the second is in the word CONSONANT
e the first letter is a consonant and the second is a vowel.

7 A fair coin is flipped. What is the probability of flipping three heads one after the other?

8 In a game, a set of cards is numbered 1–10.

If you pick a prime number you win a prize.

James plays the game twice.

a Copy and complete the tree diagram to show the probabilities

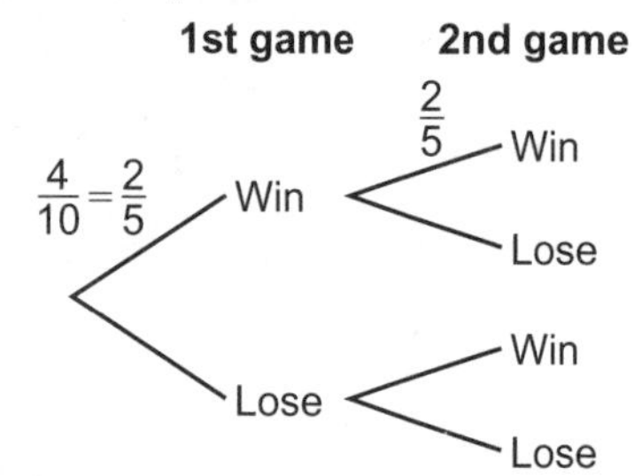

b What is the probability of winning

i two prizes ii nothing

iii one prize iv at least one prize?

9 Marganita has two bags of sweets.

Bag A contains 7 toffees and 3 mints.

Bag B contains 4 toffees and 6 mints.

She picks a sweet at random from each bag.

a Copy and complete the tree diagram to show the probabilities.

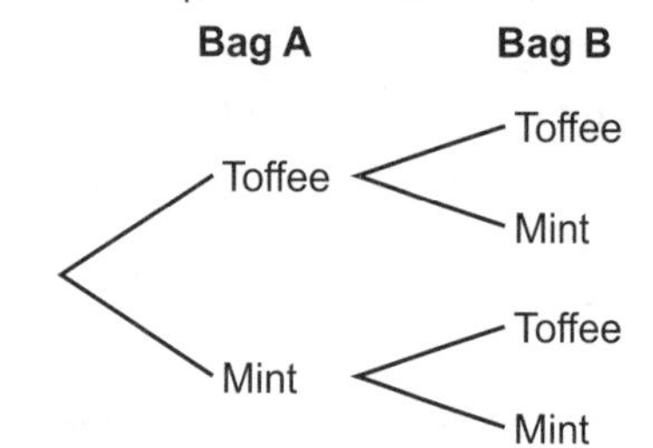

b Work out the probability of picking

i two sweets the same

ii one toffee and one mint

iii no mints

iv at least one toffee.

10 Terry plays two games on his phone. He has a 1 in 10 chance of winning the first game and a 1 in 5 chance of winning the second.

a Copy and complete the tree diagram to show the probabilities.

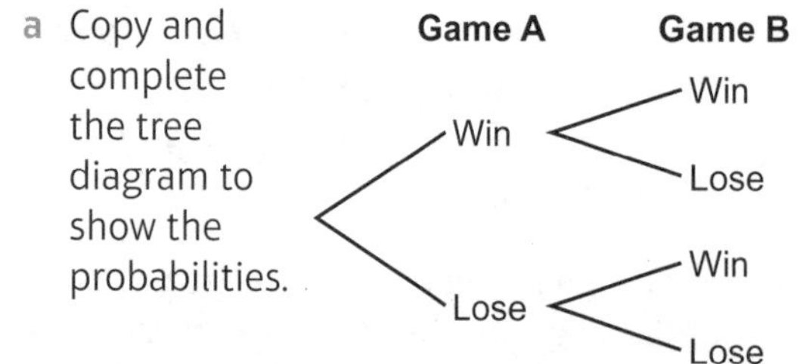

b Work out the probability that

i he wins both games

ii he wins the first game but not the second.

11 **Exam-style question**

A fair dice is rolled.

Hanifa notes whether it is a square number.

She then repeats this process.

a Complete the probability tree diagram.

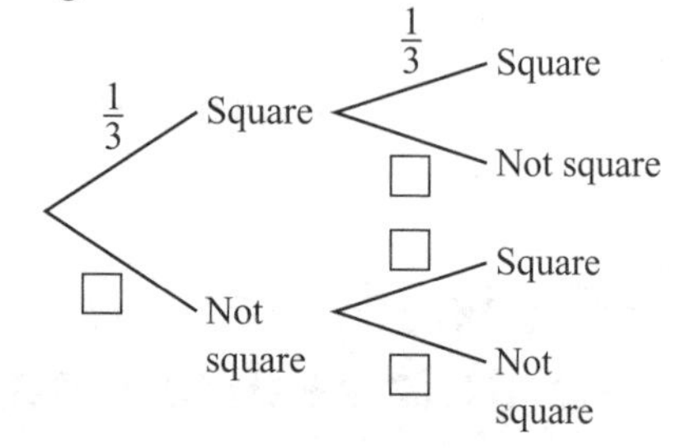

(2 marks)

b Work out the probability that Hanifa rolls at least one square number. **(3 marks)**

12 **R** Daniel spins two spinners.

On spinner 1, P(blue) = 0.1

On spinner 2, P(blue) = 0.72

a Draw a tree diagram to show all the possible outcomes.

b What is the probability of only one spinner landing on blue?

c What is the probability of both spinners landing on blue?

d If each spinner was spun 1000 times, how many times would you expect them both to land on blue?

10.5 Conditional probability

1 For each of the events, state if the events are independent or dependent.

a Picking a card from a pack, replacing it and then picking another one.

b Flipping a coin and rolling a dice.

c Picking two marbles from a bag, one after the other.

d Picking a counter from a bag, then rolling a dice.

e Picking a student from a class, then picking another student.

2 GCSE students in a school have to study one subject from each of two lists.

The table shows their choices.

		List A			
		History	Geography	French	Total
List B	DT	12	7	8	27
	ICT	8	17	12	37
	Art	3	6	7	16
	Total	23	30	27	

Work out the probability that a student chosen at random

a studies history

b studies DT

c studies ICT, given that he studies history

d studies French, given that she studies art.

3 Marion has a bag containing 6 chocolates and 6 mints.

She chooses a sweet at random and eats it.

She then chooses another sweet at random.

a Copy and complete the tree diagram to show all the probabilities.

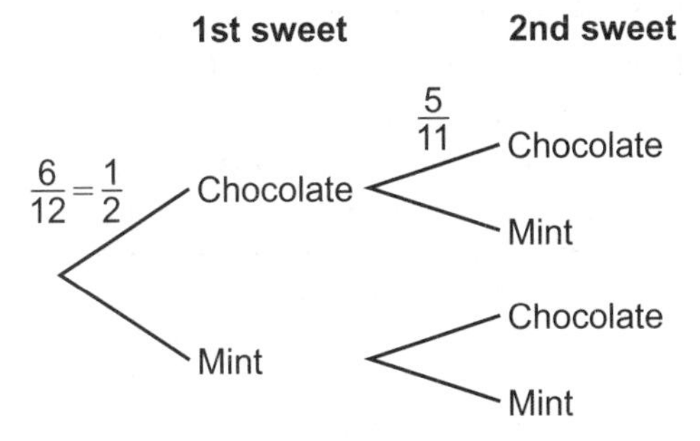

b Work out the probability that the sweets will be

i both chocolate ii one of each

iii both mint.

4 **P** In a survey, 45% of the people asked were male. 35% of the men and 40% of the women supported a football team.

One person is chosen at random.

Find the probability that this is a woman who does not support a football team.

5 **P** A train is either late or on time.

The probability it is late is 0.85. If the train is late, the probability Mr Murphy is late is 0.7.

If the train is on time, the probability he is late is 0.1.

Work out the probability that Mr Murphy arrives at work on time.

6 There are 3 red, 4 blue and 5 green marbles in a bag. Husni picks a marble then John picks a marble.

a Copy and complete the diagram to show all the probabilities.

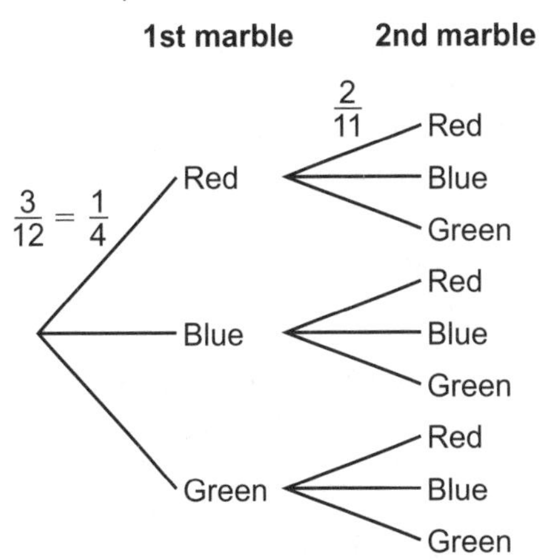

b What is the probability that they both pick the same colour?

7 **P** There are 3 tins of red paint and 11 tins of green paint.

James and Melek both take a tin at random.

Work out the probability that they do not pick the same colour.

8 **P** A box contains 3 new batteries, 5 partly used batteries and 4 dead batteries.

Melissa takes two batteries at random.

Work out the probability that she picks new batteries.

9 **Exam-style question**

Here are seven tiles.

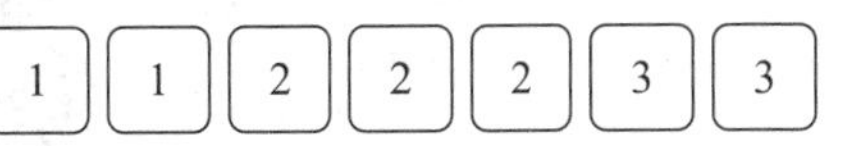

Jim takes at random a tile.

He does not replace the tile.

Jim then takes at random a second tile.

a Calculate the probability that both the tiles Jim takes have the number 1 on them. **(2 marks)**

b Calculate the probability that the number on the second tile Jim takes is greater than the number on the first tile he takes. **(3 marks)**

November 2012, Q21, 1MA0/2H

Q9 hint Draw a tree diagram to help you.

10.6 Venn diagrams and set notation

1 A = {square numbers $<$ 10}
B = {positive odd numbers $<$ 10}
a List the numbers in each set.
A = {1, …} B = {…}
b Write 'true' or 'false' for each statement.
i $5 \in A$
ii $9 \in B$
iii $11 \in B$

2 The Venn diagram shows two sets, P and Q, and the set of all numbers being considered, ξ.

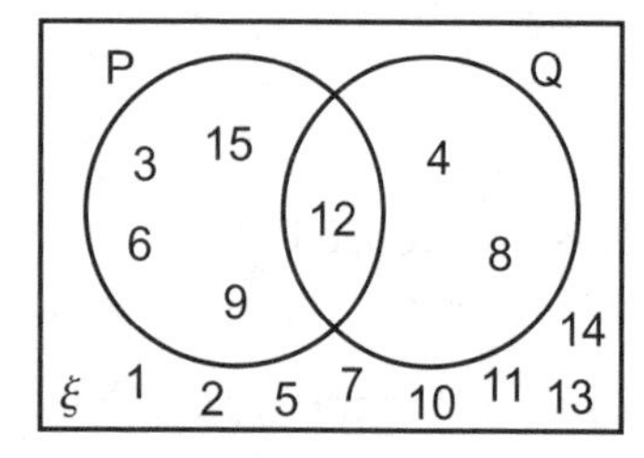

Write all the elements of each set inside curly brackets { }.
a P b Q c ξ
d Which set is {multiples of 3 $\leqslant$ 15}?
e Write descriptions of the other two sets.

3 For the Venn diagram in **Q2**, write these sets.
a $P \cup Q$ b $P \cap Q$
c P' d Q'
e $P' \cap Q$ f $Q' \cap P$

4 The Venn diagram shows two events when a 12-sided dice is rolled: square numbers and factors of 6.
X = {number is square}
Y = {number is a factor of 6}

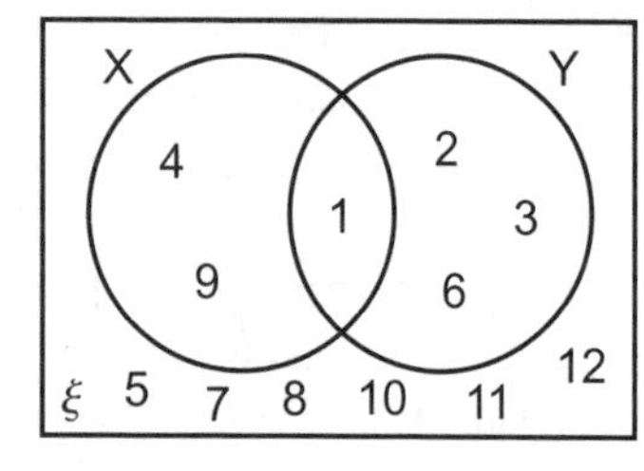

Work out
a P(X) b P(Y)
c $P(X \cap Y)$ d $P(X \cup Y)$
e $P(X')$ f $P(Y')$
g $P(X \cap Y')$ h $P(X' \cup Y')$.

5 **R** Dan asked the 30 students in his class if they were studying French (F) or Spanish (S). 15 were studying both and a total of 21 were studying French.
They were all studying at least one language.
a Draw a Venn diagram to show Dan's data.
b Work out
i P(S)
ii $P(F \cap S)$
iii $P(F \cup S)$
iv $P(F' \cap S)$.

6 100 customers were surveyed to find out about their shopping habits for groceries (G) and clothing (C). Of those surveyed:
64 bought their groceries online
13 bought both groceries and clothing online
8 bought neither online.
a Draw a Venn diagram to show the data.
b Work out the probability that a customer from the sample buys clothing online.
c Work out the probability that a customer from the sample buys groceries online, given that they buy clothing online.

7 At a school, the students can play football (F), hockey (H) or tennis (T).
Nahal carried out a survey to find out which sport students in his year played.
He recorded his results in a Venn diagram.

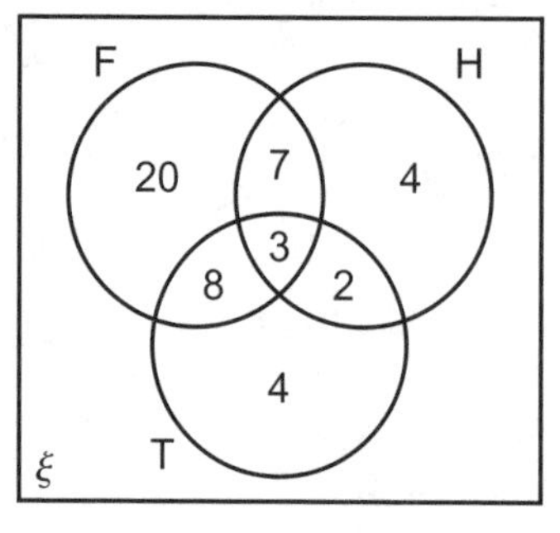

a How many students took part in the survey?
One of the students is chosen at random.
b Work out
i P(H)
ii $P(F \cap T \mid F)$
iii $P(T \cap H \mid H)$.

8 The Venn diagram shows the instruments played by members of an orchestra: violin (V), flute (F) and piano (P).

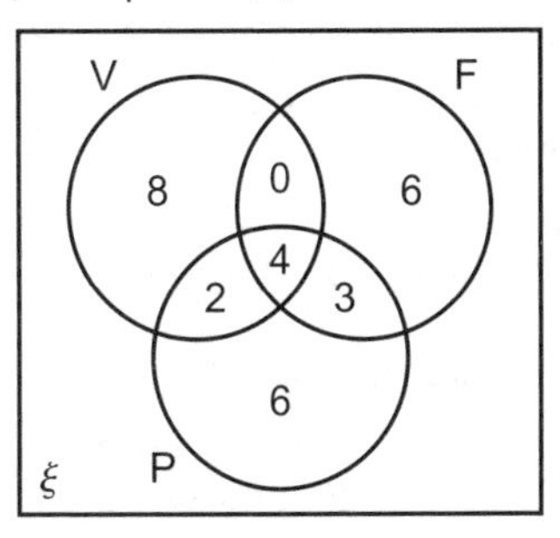

a How many people play all three instruments?

b How many people are in the orchestra?

c Work out

i $P(V \cap P \cap F)$

ii $P(V \cap F)$

iii $P(V \cup P \mid P)$.

9 **Exam-style question**

There are 100 students studying maths.

All 100 study at least one of three courses, Pure, Mechanics and Statistics.

18 study all three.

24 study Pure and Mechanics.

31 study Pure and Statistics.

22 study Mechanics and Statistics.

57 study Pure.

37 study Mechanics.

a Draw a Venn diagram to show this information. **(3 marks)**

One of the students is chosen at random.

b Work out the probability that this student studies Pure but not Mechanics. **(2 marks)**

Given that the student studies Statistics,

c work out the probability that this student also studies Mechanics. **(2 marks)**

10 Problem-solving

Solve problems using these strategies where appropriate:

- **Use pictures or lists**
- **Use smaller numbers**
- **Use bar models**
- **Use x for the unknown**
- **Use a flow diagram.**

1 **R** Maya is playing a board game with Ellen. She has two dice and needs to roll two 6s. Ellen says that two 6s is the hardest roll to get.

a Draw the sample space diagram and find the probability of rolling two 6s.

b What is the probability of rolling

i two 4s

ii two 5s

iii a 3 and a 2?

c Was Ellen correct? Explain your answer.

2 Frankie is trying to find a matching pair of socks. A blue, a red, a black and two green socks have been washed and are in a clean pile. A blue, two white, a red and a black sock are still in the dryer. Frankie takes one sock from the dryer and one from the pile, both of them at random. What is the probability that she picks out a matching pair of socks? Draw a sample space diagram to show all the possible outcomes.

3 **R** Kai wants to make a set of nesting gift boxes. He uses a grid to draw bases for the the different sizes of box.

Kai draws a 3 cm × 4 cm rectangle using the points (0, 0), (4, 0), (4, 3) and (0, 3). Using (0, 0) as his centre of enlargement, he draws the bases of the next two boxes. Kai uses scale factors of 1.5 and 2.

a Use a grid of 0 to 12 on both axes. Draw the first box and the two enlargements.

b Kai wants to draw the bases for two more sizes. The fourth box has a width of 7.5 cm and the fifth has a width of 9 cm. What scale factors (from the first box) does Kai use for the fourth and fifth boxes?

c Draw the bases of the fourth and fifth boxes on the grid.

4 **R** A biased dice is twice as likely to land on a 6 as on a 5 or a 2.

It is three times as likely to land on a 1 as on a 2, but a 4 is only half as likely as a 1.

Landing on a 3 is equally as likely as landing on a 4.

Work out the probability for landing on each number. Write your answers as percentages.

Q4 hint Use a flow diagram to work out the probability of each event in turn.

5 **R** Blake is making a five-sided spinner labelled 1–5 for a board game he has made. He wants to check that the spinner is fair.

a How can Blake test the spinner?

Blake spins the spinner 200 times.
The table shows his results.

Number	1	2	3	4	5
Frequency	42	36	45	38	39

b Is the spinner fair?
Explain your answer.

6 **R** Jackson buys four apple and berry smoothies and six cartons of orange juice.
He pays £29.60.
April buys five apple and berry smoothies and three cartons of orange juice.
She pays £24.40.
How much does one apple and berry smoothie cost? How much does one carton of orange juice cost?

7 A glass hemisphere dome and base are being made for a garden fountain. The diameter of the dome is 1.2 m. What are the surface area and volume? Give your answers in terms of π.

8 **Exam-style question**

There are 30 dogs at the animal shelter.
Each dog is at least one colour from black, brown and white.
3 of the dogs are black, brown and white.
15 of the dogs are black and white.
5 of the dogs are brown and white.
8 of the dogs are black and brown.
22 of the dogs are black.
15 of the dogs are brown.

a Draw a Venn diagram to show this information. **(3 marks)**

One of the dogs is chosen at random.

b Work out the probability that this dog is brown but not white. **(2 marks)**

Given that the dog is black,

c work out the probability that this dog is also brown. **(2 marks)**

9 **R** A nursery school teacher has a box of stickers. There are 10 smiley face stickers, 5 bear stickers and 5 rainbow stickers.
Two children take a sticker at random.

a Draw a diagram to show all the probabilities.

b What is the probability that the two children will take different stickers?

10 Mike and Syeona are rolling two ordinary dice. They make a Venn diagram to show all the possible outcomes of the totals, and whether they are multiples of 2, multiples of 3 or a prime number.

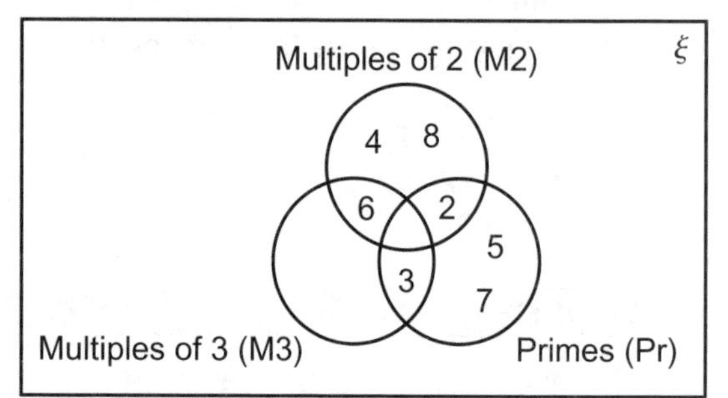

a Copy and complete the Venn diagram.

b Mike uses the diagram to find

i $P(M2 \cap M3)$

ii $P(M3')$

iii $P(Pr \cup M2 \mid Pr)$

c Syeona writes the following in set notation for Mike to find.
'Totals that are either multiples of 2 or 3, given that the total is not a multiple of both 2 and 3.'
Write the set notation and find the probability.

11 MULTIPLICATIVE REASONING

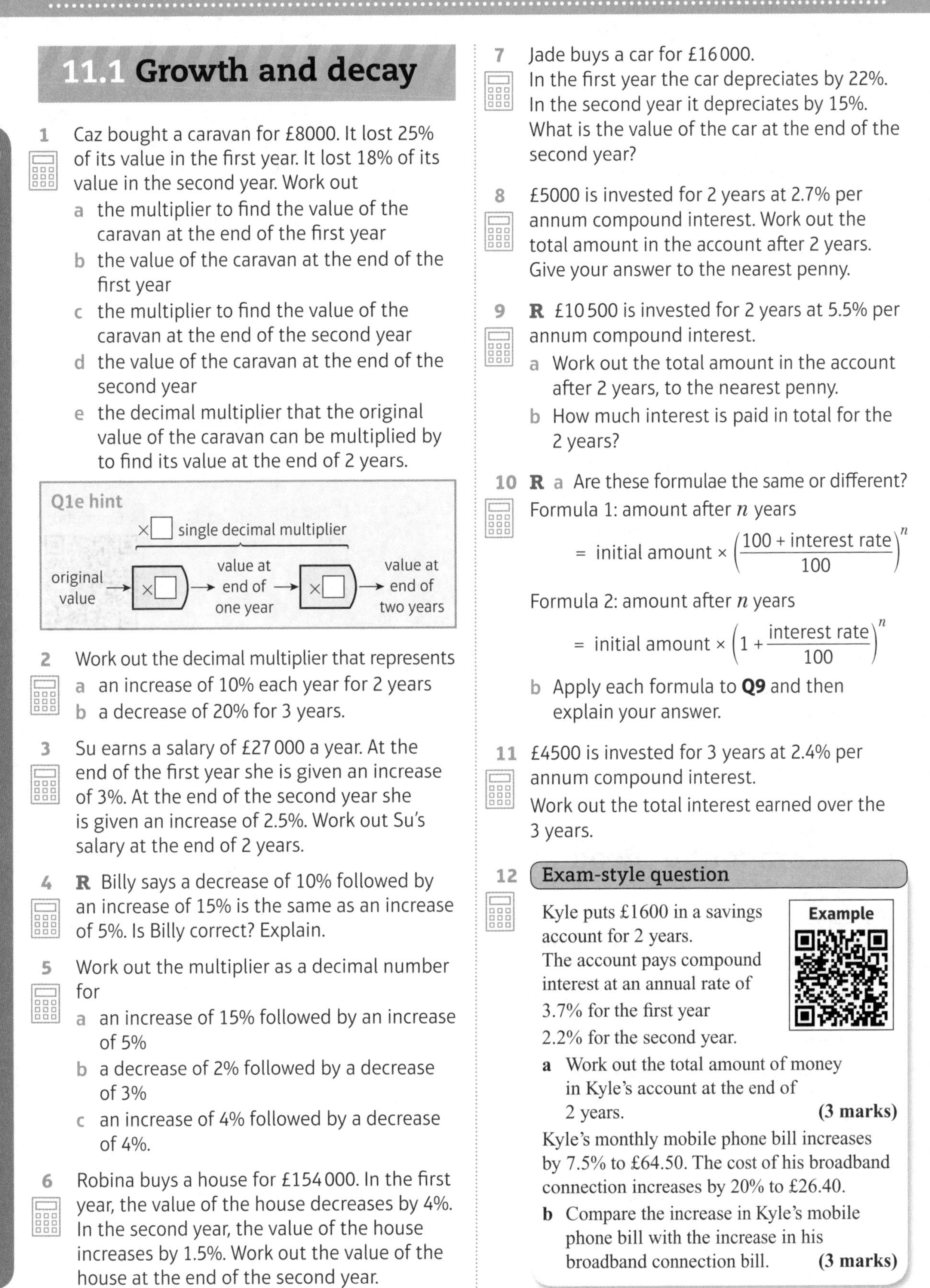

11.1 Growth and decay

1 Caz bought a caravan for £8000. It lost 25% of its value in the first year. It lost 18% of its value in the second year. Work out
 a the multiplier to find the value of the caravan at the end of the first year
 b the value of the caravan at the end of the first year
 c the multiplier to find the value of the caravan at the end of the second year
 d the value of the caravan at the end of the second year
 e the decimal multiplier that the original value of the caravan can be multiplied by to find its value at the end of 2 years.

Q1e hint

2 Work out the decimal multiplier that represents
 a an increase of 10% each year for 2 years
 b a decrease of 20% for 3 years.

3 Su earns a salary of £27 000 a year. At the end of the first year she is given an increase of 3%. At the end of the second year she is given an increase of 2.5%. Work out Su's salary at the end of 2 years.

4 **R** Billy says a decrease of 10% followed by an increase of 15% is the same as an increase of 5%. Is Billy correct? Explain.

5 Work out the multiplier as a decimal number for
 a an increase of 15% followed by an increase of 5%
 b a decrease of 2% followed by a decrease of 3%
 c an increase of 4% followed by a decrease of 4%.

6 Robina buys a house for £154 000. In the first year, the value of the house decreases by 4%. In the second year, the value of the house increases by 1.5%. Work out the value of the house at the end of the second year.

7 Jade buys a car for £16 000.
In the first year the car depreciates by 22%.
In the second year it depreciates by 15%.
What is the value of the car at the end of the second year?

8 £5000 is invested for 2 years at 2.7% per annum compound interest. Work out the total amount in the account after 2 years. Give your answer to the nearest penny.

9 **R** £10 500 is invested for 2 years at 5.5% per annum compound interest.
 a Work out the total amount in the account after 2 years, to the nearest penny.
 b How much interest is paid in total for the 2 years?

10 **R** a Are these formulae the same or different?

Formula 1: amount after n years

$$= \text{initial amount} \times \left(\frac{100 + \text{interest rate}}{100}\right)^n$$

Formula 2: amount after n years

$$= \text{initial amount} \times \left(1 + \frac{\text{interest rate}}{100}\right)^n$$

 b Apply each formula to **Q9** and then explain your answer.

11 £4500 is invested for 3 years at 2.4% per annum compound interest.
Work out the total interest earned over the 3 years.

12 **Exam-style question**

Kyle puts £1600 in a savings account for 2 years.
The account pays compound interest at an annual rate of
3.7% for the first year
2.2% for the second year.

Example

 a Work out the total amount of money in Kyle's account at the end of 2 years. **(3 marks)**

Kyle's monthly mobile phone bill increases by 7.5% to £64.50. The cost of his broadband connection increases by 20% to £26.40.

 b Compare the increase in Kyle's mobile phone bill with the increase in his broadband connection bill. **(3 marks)**

13 **P** Lee invests £75 000 in a savings account for 2 years. The account pays 4.47% compound interest per annum.
Lee has to pay 20% tax on the interest earned each year. The tax is taken from the account at the end of each year.
How much is in the account at the end of 2 years?

14 **R** Theo invested £5700 in a savings account. He is paid 2.75% per annum compound interest.
How many years before he has £6183.30 in the savings account?

15 The number of cells in a biological sample increase by 7% each hour.
A scientist estimated there were 1.2 million cells present at time $t = 0$.

a How many complete hours later will the number of cells reach 1.68 million?

b Estimate the number of cells 24 hours after time $t = 0$, to 3 significant figures.

16 In 2014 the rate of increase in the population of the UK was 0.3%. The population of the UK in 2014 was 63.7 million.
If the population keeps growing at the same rate, how big is the population likely to be in 2025?

17 A new virus has spread among a population at a rate of 17.5% per day. In a city there are now 371 cases.
How many people had the virus 7 days earlier?

11.2 Compound measures

1 **P** Abdul works for 40 hours from Monday to Friday. His rate of pay for working Monday to Friday is £7.90 an hour.
Abdul is paid time and a half for each hour he works on a Saturday and double time for each hour he works on a Sunday.

a How much is Abdul paid for a week when he works 40 hours from Monday to Friday, plus 5 hours on Saturday and 4 hours on Sunday?

b In one week Abdul works 40 hours from Monday to Friday and some hours on Sunday. He is paid £371.30 for the week.
How many hours did Abdul work on Sunday?

2 Water flows from a hosepipe at a rate of 1200 litres per hour.

a Work out how much water flows from the hosepipe in
i 10 minutes ii 35 minutes.

Two identical hosepipes are used at the same time to fill a garden pool. They each have the same flow rate as in part **a**. The pool has a capacity of 216 000 litres. Initially the pool is empty.

b Work out how many days it takes to fill the pool.

3 **R** Dave drives his motorbike for 180 miles and uses 25 litres of petrol.

a Work out the average rate of petrol usage. State the units with your answer.

b Estimate the amount of petrol Dave would use if he travels 275 miles on the motorbike. Give your answer to an appropriate degree of accuracy.

4 Convert these speeds from m/h to km/h.

Example

a 7500 m/h
b 970 m/h
c 36 500 m/h
d 243 600 m/h

5 **R** Convert these speeds from metres per second (m/s) to metres per hour (m/h).

a 3 m/s b 15 m/s
c 100 m/s d 0.5 m/s
e Would more or fewer metres be travelled in 1 hour than in 1 second?

6 Copy and complete the table.

Metres per second	Kilometres per hour
10	
25	
	126
	180

7 The average speed of a small plane is 360 km/h.
What is this speed in metres per second?

8 **P / R** The new fast train in the UK has an expected top speed of 272 km/h. Usain Bolt's fastest recorded running speed is 10.44 m/s (run on 16 August 2009).
How much faster than Bolt is the train expected to be?

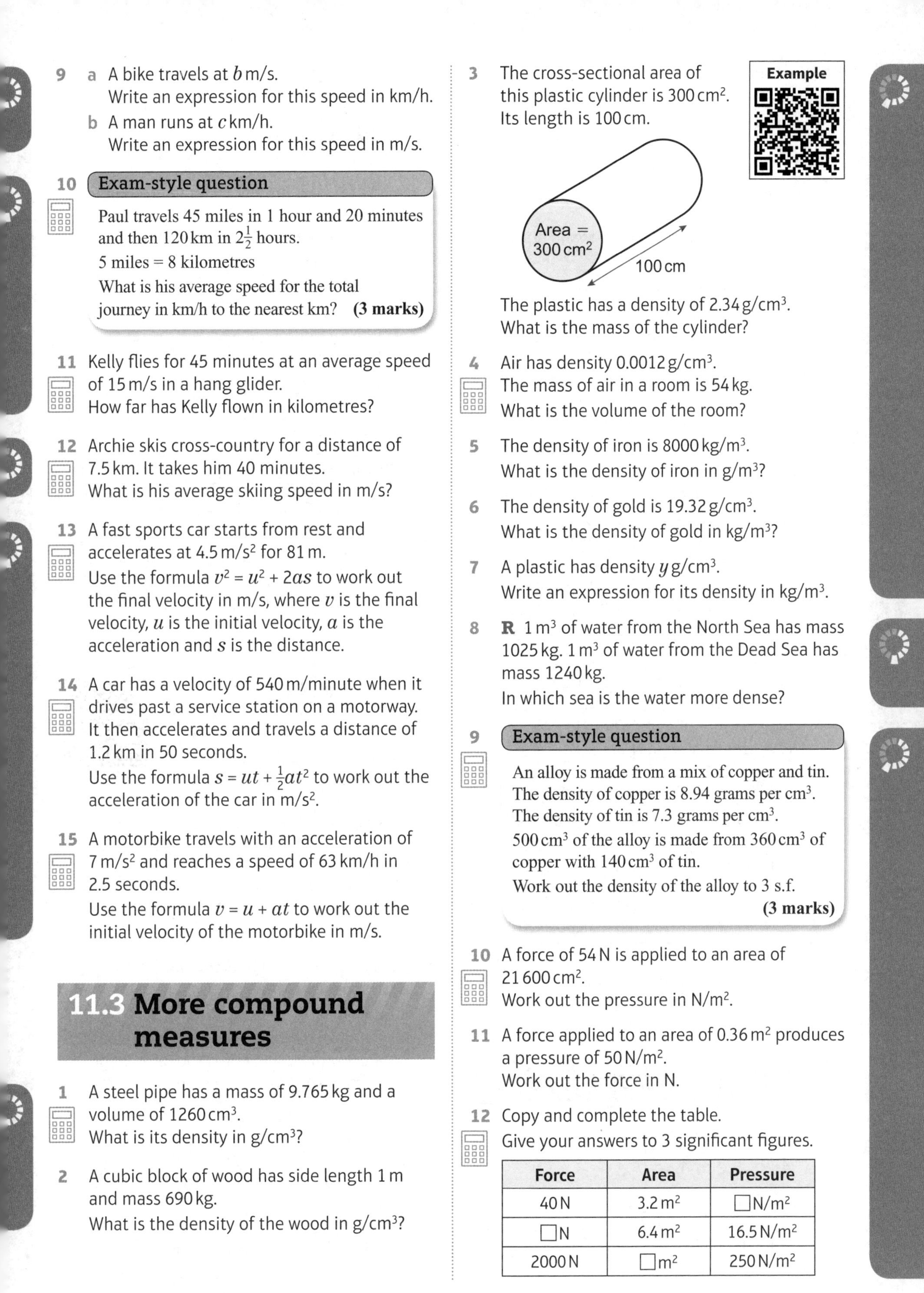

9 a A bike travels at b m/s.
Write an expression for this speed in km/h.
b A man runs at c km/h.
Write an expression for this speed in m/s.

10 **Exam-style question**

Paul travels 45 miles in 1 hour and 20 minutes and then 120 km in $2\frac{1}{2}$ hours.
5 miles = 8 kilometres
What is his average speed for the total journey in km/h to the nearest km? **(3 marks)**

11 Kelly flies for 45 minutes at an average speed of 15 m/s in a hang glider.
How far has Kelly flown in kilometres?

12 Archie skis cross-country for a distance of 7.5 km. It takes him 40 minutes.
What is his average skiing speed in m/s?

13 A fast sports car starts from rest and accelerates at 4.5 m/s² for 81 m.
Use the formula $v^2 = u^2 + 2as$ to work out the final velocity in m/s, where v is the final velocity, u is the initial velocity, a is the acceleration and s is the distance.

14 A car has a velocity of 540 m/minute when it drives past a service station on a motorway. It then accelerates and travels a distance of 1.2 km in 50 seconds.
Use the formula $s = ut + \frac{1}{2}at^2$ to work out the acceleration of the car in m/s².

15 A motorbike travels with an acceleration of 7 m/s² and reaches a speed of 63 km/h in 2.5 seconds.
Use the formula $v = u + at$ to work out the initial velocity of the motorbike in m/s.

11.3 More compound measures

1 A steel pipe has a mass of 9.765 kg and a volume of 1260 cm³.
What is its density in g/cm³?

2 A cubic block of wood has side length 1 m and mass 690 kg.
What is the density of the wood in g/cm³?

3 The cross-sectional area of this plastic cylinder is 300 cm². Its length is 100 cm.

Example

The plastic has a density of 2.34 g/cm³.
What is the mass of the cylinder?

4 Air has density 0.0012 g/cm³.
The mass of air in a room is 54 kg.
What is the volume of the room?

5 The density of iron is 8000 kg/m³.
What is the density of iron in g/m³?

6 The density of gold is 19.32 g/cm³.
What is the density of gold in kg/m³?

7 A plastic has density y g/cm³.
Write an expression for its density in kg/m³.

8 **R** 1 m³ of water from the North Sea has mass 1025 kg. 1 m³ of water from the Dead Sea has mass 1240 kg.
In which sea is the water more dense?

9 **Exam-style question**

An alloy is made from a mix of copper and tin.
The density of copper is 8.94 grams per cm³.
The density of tin is 7.3 grams per cm³.
500 cm³ of the alloy is made from 360 cm³ of copper with 140 cm³ of tin.
Work out the density of the alloy to 3 s.f. **(3 marks)**

10 A force of 54 N is applied to an area of 21 600 cm².
Work out the pressure in N/m².

11 A force applied to an area of 0.36 m² produces a pressure of 50 N/m².
Work out the force in N.

12 Copy and complete the table.
Give your answers to 3 significant figures.

Force	Area	Pressure
40 N	3.2 m²	□ N/m²
□ N	6.4 m²	16.5 N/m²
2000 N	□ m²	250 N/m²

13 A cylindrical plant pot has a circular base with diameter 0.3 m.
The plant pot exerts a force of 60 N on the ground.
Work out the pressure in N/cm^2.
Give your answer to 3 significant figures.

14 A cylinder with a movable piston in an engine contains hot gas.
The pressure of the gas is 350 N/m^2.
The area of the piston is 0.05 m^2.
What is the force exerted by the piston?

15 **R** Claire sits on a table with four identical legs. Each table leg has a flat rectangular base measuring 2 cm by 3 cm.
Claire has a mass of 68 kg and the table has a mass of 4 kg.

a Use $F = mg$ to work out the combined weight of Claire and the table, where g is the acceleration due to gravity.
Use $g = 9.8$ m/s^2.

b Work out the pressure on the floor exerted by the table only in N/m^2, when only the four table legs are in contact with the floor.

c The area of Claire's feet is 330 cm^2.
Work out the pressure on the floor when Claire is standing up in bare feet.
Give your answer in N/cm^2.
Use $g = 9.8$ m/s^2 and give your answer to 3 s.f.

d Is the pressure on the floor greater when Claire is sitting on the table or when she is standing up in bare feet?

Q15a hint Weight is a force on an object due to gravity and is measured in newtons (N).

16 a Convert 2500 N/m^2 to N/cm^2.
b Convert x N/cm^2 to N/m^2.

11.4 Ratio and proportion

1 Copy and complete these.

a $A:B = 5:2$ so $A = \frac{\square}{\square}B$

b $X:Y = 3:7$, so $X = \frac{\square}{\square}Y$

c $10P = Q$ so $Q:P = \square:\square$

2 The table shows some masses in both pounds and kilograms.

Kilograms	10	20	30	40
Pounds	22	44	66	88

a What is the ratio of kilograms : pounds in the form $1:n$?
b Plot a line graph for these values.
c Are pounds and kilograms in direct proportion? Explain your answer.
d What is the gradient of the line?
e Write a formula that shows the relationship between kilograms and pounds.

3 The table shows the distance (s) in kilometres travelled by a car over a period of time (t) in hours.

Distance, s (km)	15	30	45	60
Time, t (hours)	2	4	6	8

a Is s in direct proportion to t? Explain.
b What is the relationship between distance (s) and time (t)?
c Work out the distance travelled after 1.5 hours.

4 The cost of buying

Example

- 150 US dollars is £100
- 300 US dollars is £200.

a Show that the amount in US dollars, $\$D$, is directly proportional to the amount in sterling, £P.

b What is the relationship between D and P?
c How much is £570 in US dollars?

5 It takes 4 cleaners 3 hours to clean a house.
How long would it take 9 cleaners?
Give your answer in hours and minutes.

6 It takes 8 women 3 hours to dig a ditch.
How long will it take

a 2 women　　b 5 women?

c For both parts **a** and **b**, multiply the exact answer in hours (H) by the number of women (N). What do you notice?

7 P and Q are in inverse proportion.
Work out the values of A, B, C and D.

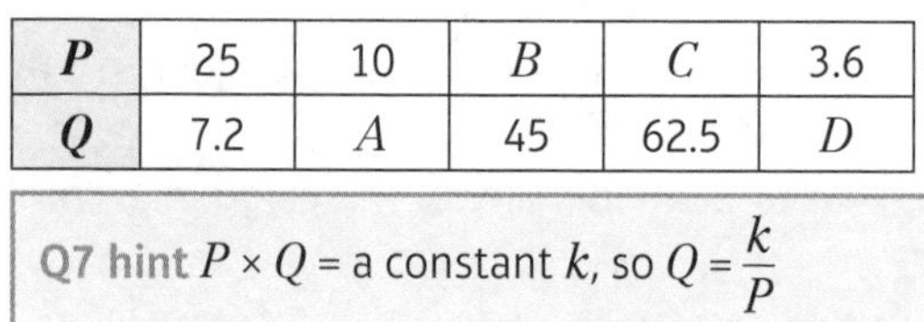

P	25	10	B	C	3.6
Q	7.2	A	45	62.5	D

Q7 hint $P \times Q$ = a constant k, so $Q = \frac{k}{P}$

8 Do these equations represent direct proportion, inverse proportion or neither?

a $y = \frac{2}{x}$ b $y = 5x$ c $y = 10x - 5$

d $\frac{y}{x} = 3$ e $xy = 6$

9 **P** The frequency, f Hz, of a sound wave is inversely proportional to the wavelength, λ m, of the sound wave.

When the frequency is 200 Hz, the wavelength is 1.7 m.

Find the wavelength when the frequency is 680 Hz.

10 **P** Boyle's law states that the pressure of a gas and its volume are inversely proportional, providing the temperature and the amount of gas stay constant.

A sample of gas occupies 0.0015 m^3 when the pressure is 100 000 N/m^2.

What is the volume of the sample when the pressure increases to 200 000 N/m^2?

11 v is inversely proportional to u.
$v = 35$ when $u = 0.2$.

a Find a formula for v in terms of u.

b Calculate the value of v when $u = 140$.

12 a Copy and complete the table for $q = \frac{12.5}{p}$

p	1	2	5	10
q				

b Use the table to sketch a graph of q against p.

c Work out the value of q when $p = 50$. Does it fit the shape of your graph?

d Work out the value of q when $p = 0.05$. Is your answer consistent with your sketch?

13 **Exam-style question**

The time, t minutes, it takes to pull an object up a ramp with an electric motor is directly proportional to the mass, m kg, of the object.

When $m = 1$ kg, $t = 2.5$ minutes.

a Find m when $t = 1.25$ minutes. **(3 marks)**

The time, t minutes, it takes to pull an object up a ramp with an electric motor is inversely proportional to the power, P watts, of the electric motor.

When $P = 1800$ W, $t = 3$ minutes.

b Find the value of t when $P = 1600$ W. **(3 marks)**

11 Problem-solving

Solve problems using these strategies where appropriate:

- **Use pictures or lists**
- **Use smaller numbers**
- **Use bar models**
- **Use x for the unknown**
- **Use a flow diagram**
- **Use arrow diagrams.**

Example

1 **R** Water flows from a hose at 0.42 litres per second. If the hose is turned off after 10 minutes, will it have filled more than half or less than half of a 500-litre pond? Explain how you found your answer.

2 **R** Four child tickets to a show cost £14. Five adult tickets to the show cost £31.25.

a What is the cost of tickets for 9 children and 13 adults?

b Two adults take some children to the show. The total price of the tickets is £26.50. How many children did they take to the show?

3 **R** Therese wrote down these values:
−5, −4, −3, −2, −1, 0, 1, 2

Write down **two** inequalities that these values would satisfy. Use the letter n to represent an integer in your inequality.

4 Malcolm ran for $4\frac{3}{4}$ miles.
Pritpal ran for $5\frac{2}{5}$ miles.

a What is the total distance run by Malcolm and Pritpal?

b How much further did Pritpal run than Malcolm?

Give your answers as mixed numbers where appropriate.

5 The amount, M, Sandy earns is directly proportional to the number of hours, H, she works. When $H = 21$, $M = £472.50$.
Find the value of M when $H = 35$.

6 A survey was taken of people coming out of a grocery shop. 35% of the people were men. 70% of these men bought milk. 15% of the women did not buy milk.

One of these customers was chosen at random to take part in a longer survey.

Find the probability that this customer was a woman who did buy milk.

7 A plastic block in the shape of a cuboid has a density of $3.9\,g/cm^3$.
It has a length of 24 cm.
The width of the block is $\frac{1}{3}$ of its height.
Its height is $\frac{1}{4}$ of its length.
What is the mass of the block?

8 **R** A company needs to complete a construction project in 8 days.
It will take 18 workers 12 days to complete the project.
What is the minimum number of workers they will need, to be able to complete the project on time?

> **Q8 hint** More than 18 workers will be needed to complete the project in less than 12 days.

9 **R** Yogesh has £9866.37 in his savings account. He has received 3% per annum compound interest since opening the account 2 years ago. He has not deposited any extra money since opening the account.

a How much did Yogesh originally deposit when he opened the account?

b Tara says that if Yogesh had originally deposited £500 more, he would now have £500 more in his savings account.
Is Tara correct? Explain how you know.

10 **Exam-style question**

h is inversely proportional to the square of r.
When $r = 5$, $h = 3.4$.
Find the value of h when $r = 8$. **(3 marks)**

June 2013, Q22, 1MA0/2H

12 SIMILARITY AND CONGRUENCE

12.1 Congruence

1 Here is a pair of congruent triangles.

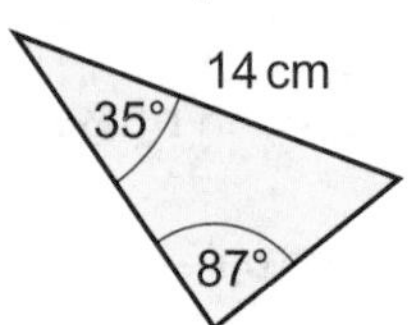

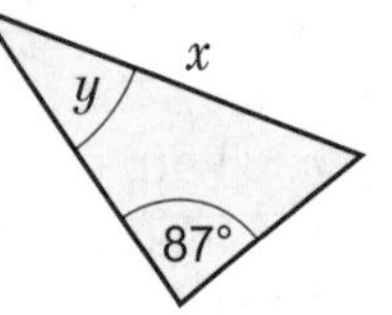

Write down

a the size of angle y b the length of side x.

2 Each pair of triangles is congruent.
Explain why.

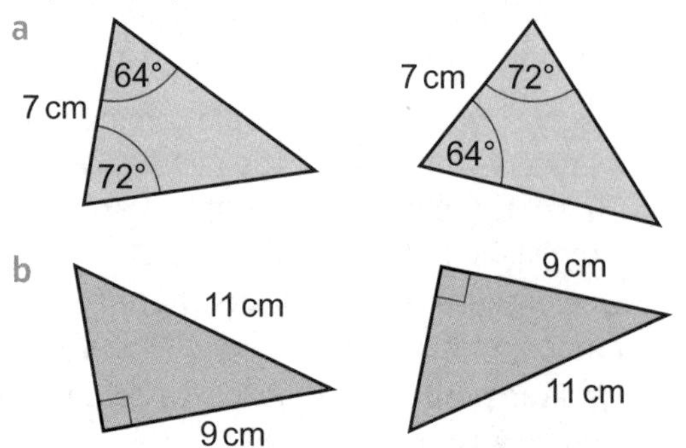

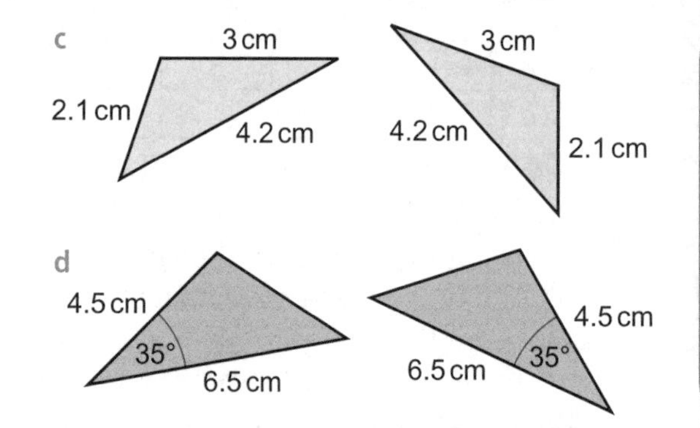

3 State whether each pair of triangles described below is congruent or not.
If the triangles are congruent, give the reason and write the corresponding vertices in pairs.

a ABC where AB = 5 cm, BC = 9 cm, angle B = 37°
PQR where PQ = 50 mm, QR = 95 mm, angle Q = 37°

b ABC where AB = 6 cm, angle B = 51°, angle C = 97°
PQR where PQ = 6 cm, angle Q = 97°, angle R = 51°

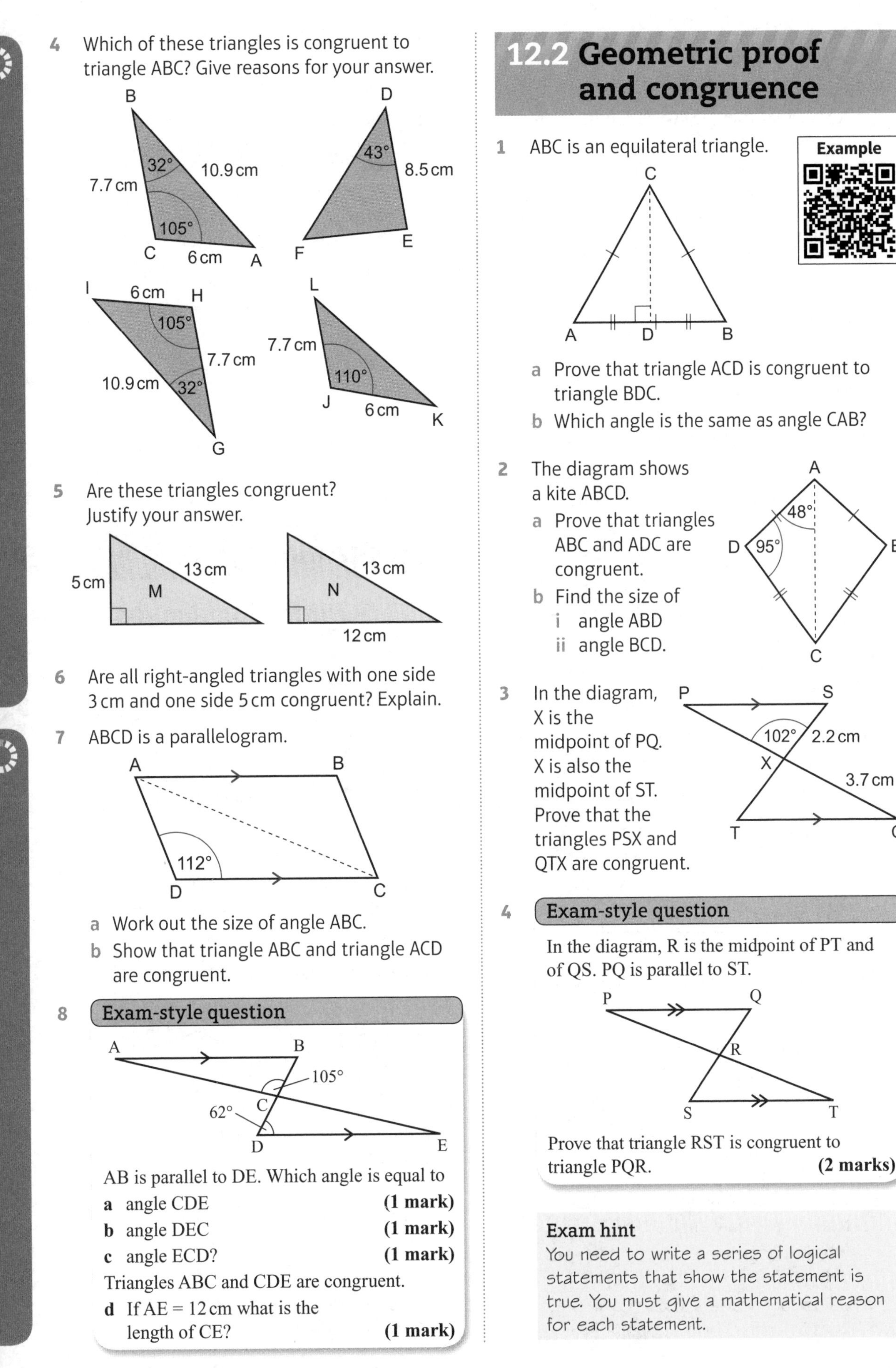

4 Which of these triangles is congruent to triangle ABC? Give reasons for your answer.

5 Are these triangles congruent? Justify your answer.

6 Are all right-angled triangles with one side 3 cm and one side 5 cm congruent? Explain.

7 ABCD is a parallelogram.

a Work out the size of angle ABC.

b Show that triangle ABC and triangle ACD are congruent.

8 **Exam-style question**

AB is parallel to DE. Which angle is equal to

a angle CDE **(1 mark)**

b angle DEC **(1 mark)**

c angle ECD? **(1 mark)**

Triangles ABC and CDE are congruent.

d If AE = 12 cm what is the length of CE? **(1 mark)**

12.2 Geometric proof and congruence

1 ABC is an equilateral triangle.

Example

a Prove that triangle ACD is congruent to triangle BDC.

b Which angle is the same as angle CAB?

2 The diagram shows a kite ABCD.

a Prove that triangles ABC and ADC are congruent.

b Find the size of

i angle ABD

ii angle BCD.

3 In the diagram, X is the midpoint of PQ. X is also the midpoint of ST. Prove that the triangles PSX and QTX are congruent.

4 **Exam-style question**

In the diagram, R is the midpoint of PT and of QS. PQ is parallel to ST.

Prove that triangle RST is congruent to triangle PQR. **(2 marks)**

Exam hint

You need to write a series of logical statements that show the statement is true. You must give a mathematical reason for each statement.

5 **P** FGH is an isosceles triangle with FG = GH. Point E lies on FH.
EG is perpendicular to FH.
 a Prove that triangle FGE is congruent to triangle GHE.
 b Given that angle GFE = 70° work out the size of angle FGE.

Q5 hint Sketch the triangle.

6 ABCD is a rectangle. Use congruent triangles to show that the area of ADC = the area of ABC.

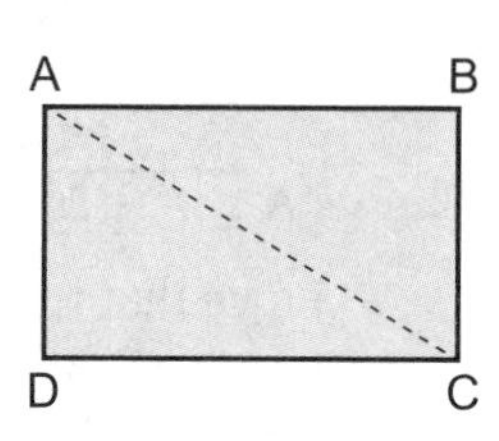

7 **P** PQRS is a parallelogram.

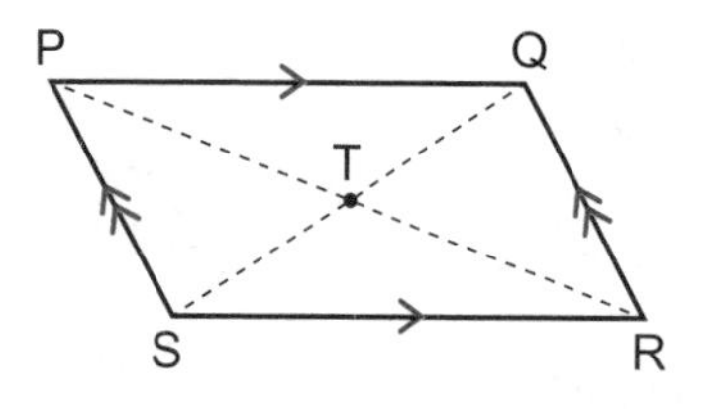

 a Prove that triangles PQT and SRT are congruent.
 b Hence, prove that the diagonals of a parallelogram bisect one another.

8 **P** ABCD is an isosceles trapezium.
Two lines are drawn from A and B that meet DC at right angles.

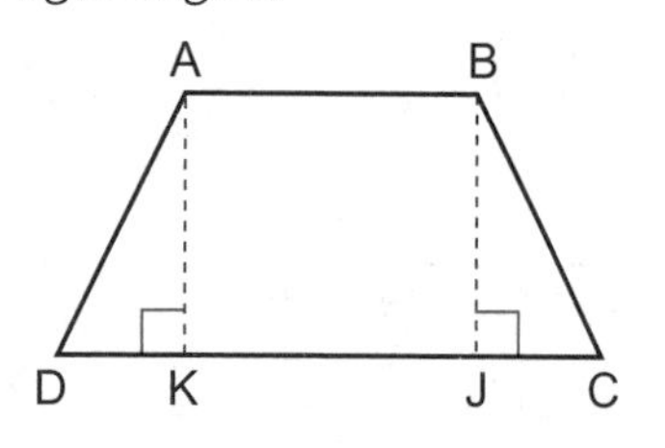

Prove that triangle ADK is congruent to triangle BCJ.

9 ABCD is a rectangle. AC and BD are the diagonals of the rectangle, which cross at point E.

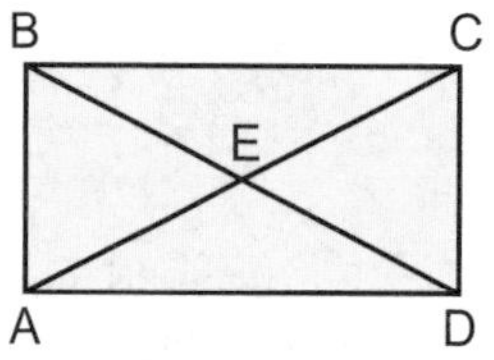

 a Draw the rectangle showing both diagonals.
 b Mark on all equal angles.
 c Which triangles are congruent in your diagram?
 d Using your answers to parts **a**–**c**, show that lines AC and BD bisect at point E.

10 **P** KLMN is a square. The diagonals KM and LN are drawn and meet at P.
Prove that the diagonals KM and LN are perpendicular bisectors.

11 **Exam-style question**

ABC is a right-angled isosceles triangle with AB = BC.
D is the midpoint of AC.

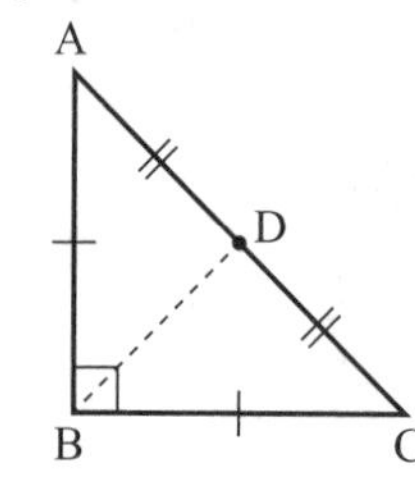

Prove that triangle ABD is congruent to triangle CBD. **(2 marks)**

12.3 Similarity

1 Write the pairs of corresponding sides in triangles X and Y.

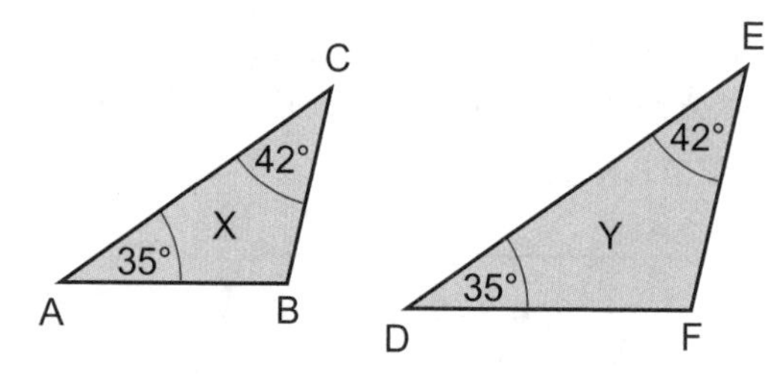

2 In **Q1** triangle X has side AB = 5 cm.
Triangle Y has side DF = 7.5 cm.
Write the ratio of corresponding sides as a fraction.

3 Here are two right-angled triangles.

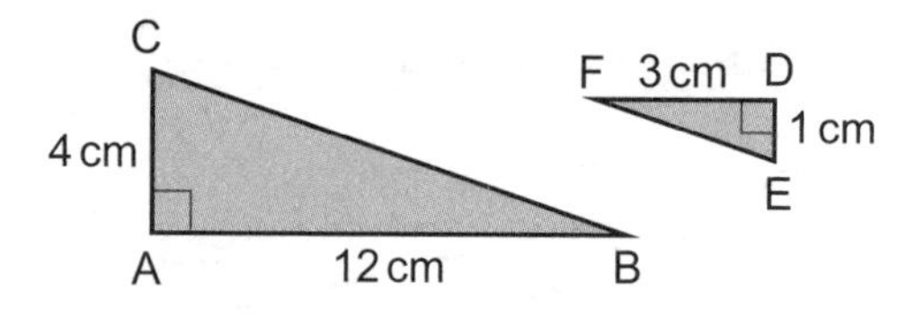

 a Which side in triangle ABC corresponds to
  i DF
  ii FE?
 b Work out the ratio $\frac{DF}{AB}$
 c Work out the ratio $\frac{DE}{AC}$
 d Use your answers to parts **a** and **b** to show that the triangles are similar.

4 The diagram shows parallelograms ABCD and EFGH.

Angles ABC and EFG are the same size.

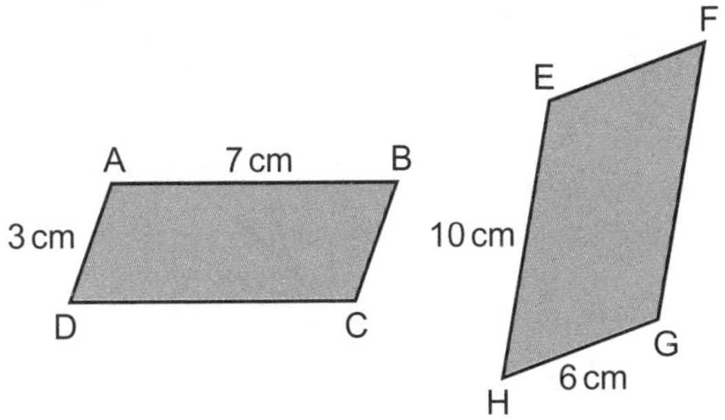

A 7 cm B, 3 cm, D C; E F, 10 cm, H 6 cm G

a Write the ratio $\frac{AD}{HG}$

b Write the ratios of the other corresponding sides.

c Are the parallelograms similar? Explain your answer.

5 State which of the pairs of shapes are similar.

a A 3 cm B, D C; E 6.2 cm F, H G

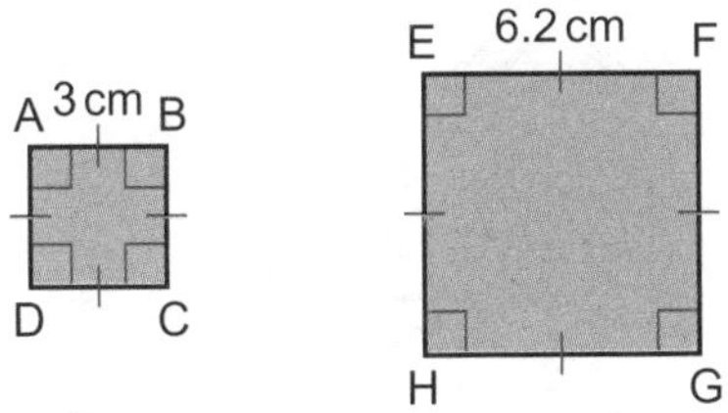

b Triangle ABC: AC = 12 cm, CB = 13 cm, AB = 5 cm. Triangle DEF: DF = 10 cm, FE = 10.4 cm, DE = 3 cm

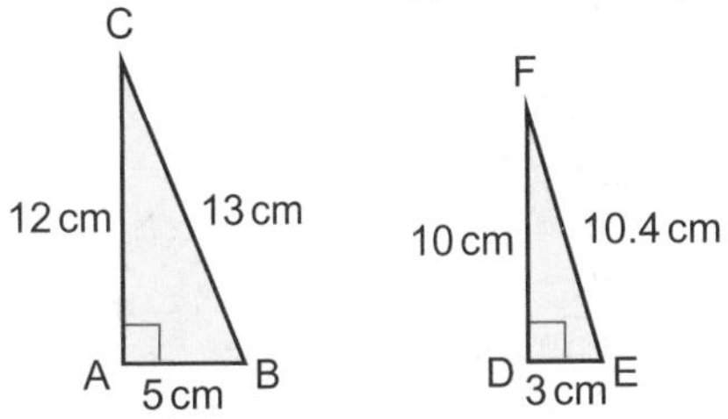

c Kite: 3 cm, 7 cm; kite: 1.5 cm, 3 cm

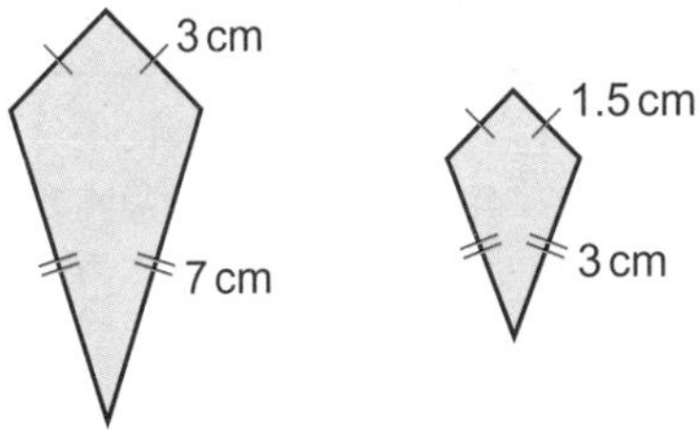

6 Show that quadrilateral ABCD is similar to quadrilateral EFGH.

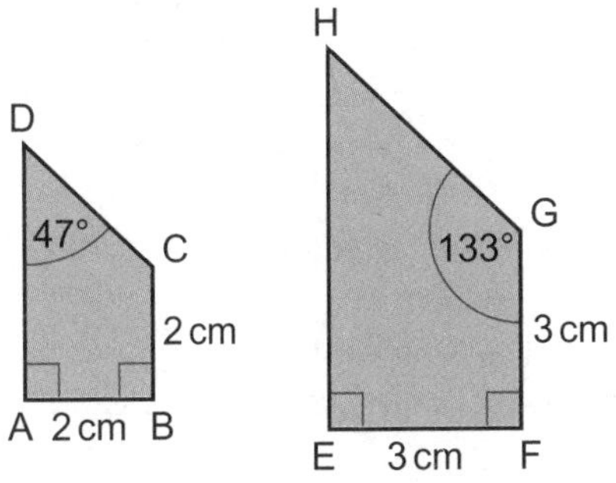

ABCD: 47°, 2 cm, A 2 cm B; EFGH: 133°, 3 cm, E 3 cm F

7 These two rectangles are similar.

Find the missing side length x in the larger rectangle.

Example

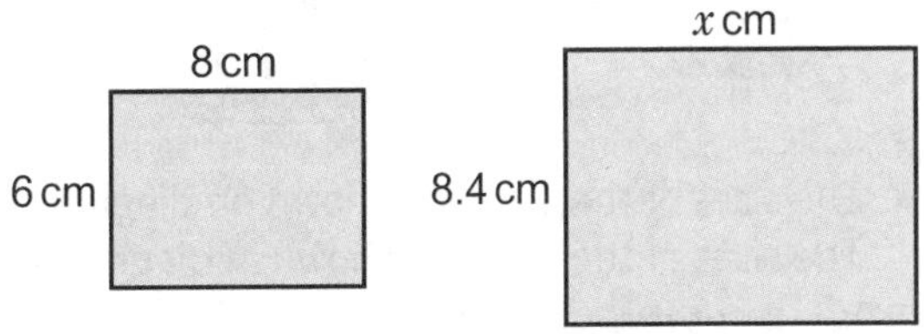

8 cm, 6 cm; x cm, 8.4 cm

8 **Exam-style question**

A scale diagram of a room has length 12 cm and width 8 cm.

Diagram **NOT** drawn accurately

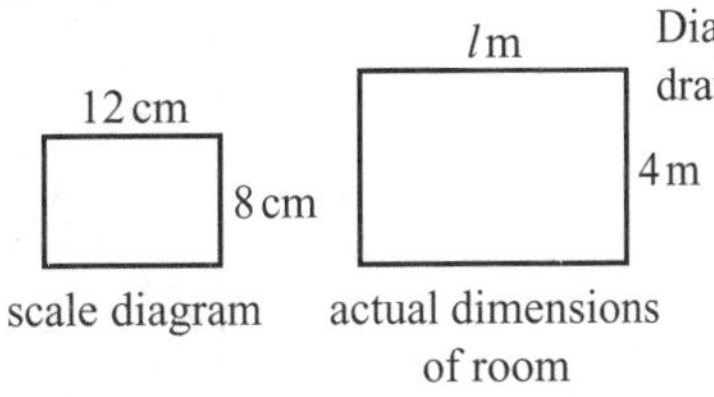

12 cm, 8 cm — scale diagram; l m, 4 m — actual dimensions of room

Work out the actual area of the room.

(3 marks)

9 **P** A scale drawing of a shed is 5 cm wide by 8 cm long.

The real shed is 7.5 m wide.

What is the length of the shed?

10 a Show that triangles A and B are similar.

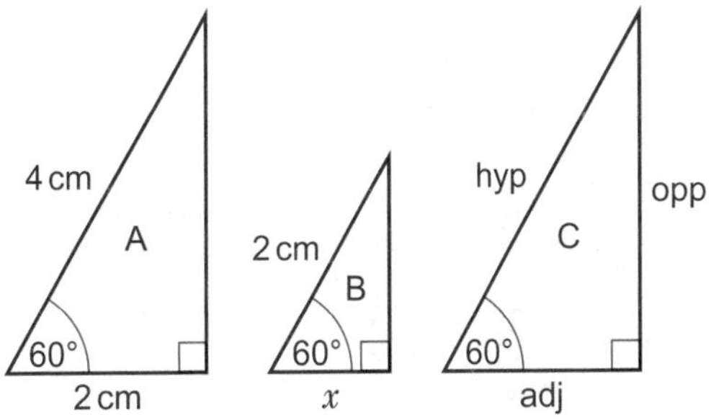

A: 4 cm, 60°, 2 cm; B: 2 cm, 60°, x; C: hyp, opp, 60°, adj

b Work out length x in triangle B.

c Show that triangle C is similar to A and B. Explain.

d Write down the value of $\frac{\text{adjacent}}{\text{hypotenuse}}$ for these triangles.

e What is another name for the ratio $\frac{\text{adjacent}}{\text{hypotenuse}}$?

11 **P** A chocolate bar has length 12 cm and width 4.5 cm. A miniature of the chocolate bar is similar to the large one but has length 4 cm. Find the width of the miniature bar.

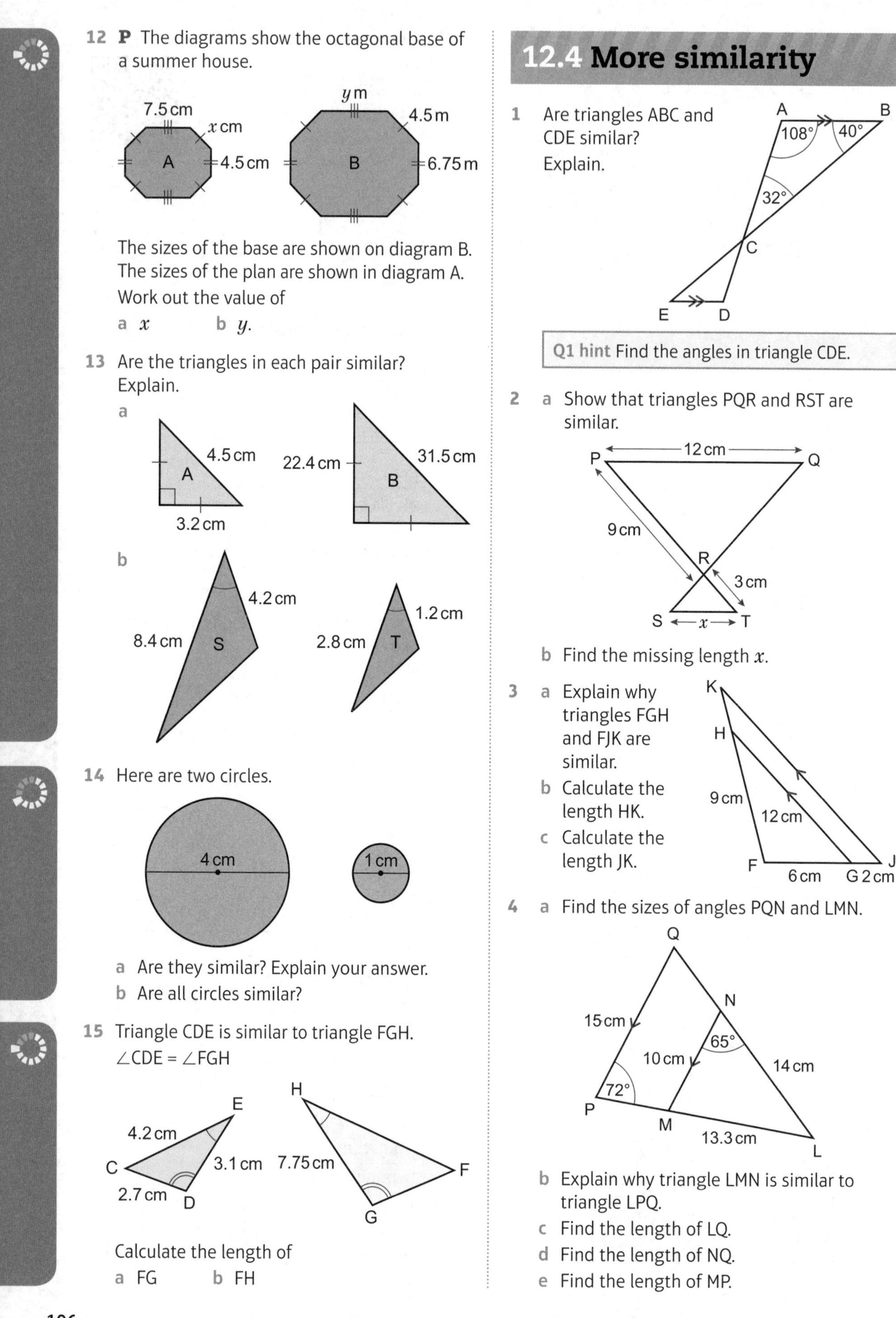

12 **P** The diagrams show the octagonal base of a summer house.

The sizes of the base are shown on diagram B.
The sizes of the plan are shown in diagram A.
Work out the value of

a x b y.

13 Are the triangles in each pair similar? Explain.

a

b

14 Here are two circles.

a Are they similar? Explain your answer.

b Are all circles similar?

15 Triangle CDE is similar to triangle FGH.
$\angle CDE = \angle FGH$

Calculate the length of

a FG b FH

12.4 More similarity

1 Are triangles ABC and CDE similar? Explain.

Q1 hint Find the angles in triangle CDE.

2 a Show that triangles PQR and RST are similar.

b Find the missing length x.

3 a Explain why triangles FGH and FJK are similar.

b Calculate the length HK.

c Calculate the length JK.

4 a Find the sizes of angles PQN and LMN.

b Explain why triangle LMN is similar to triangle LPQ.

c Find the length of LQ.

d Find the length of NQ.

e Find the length of MP.

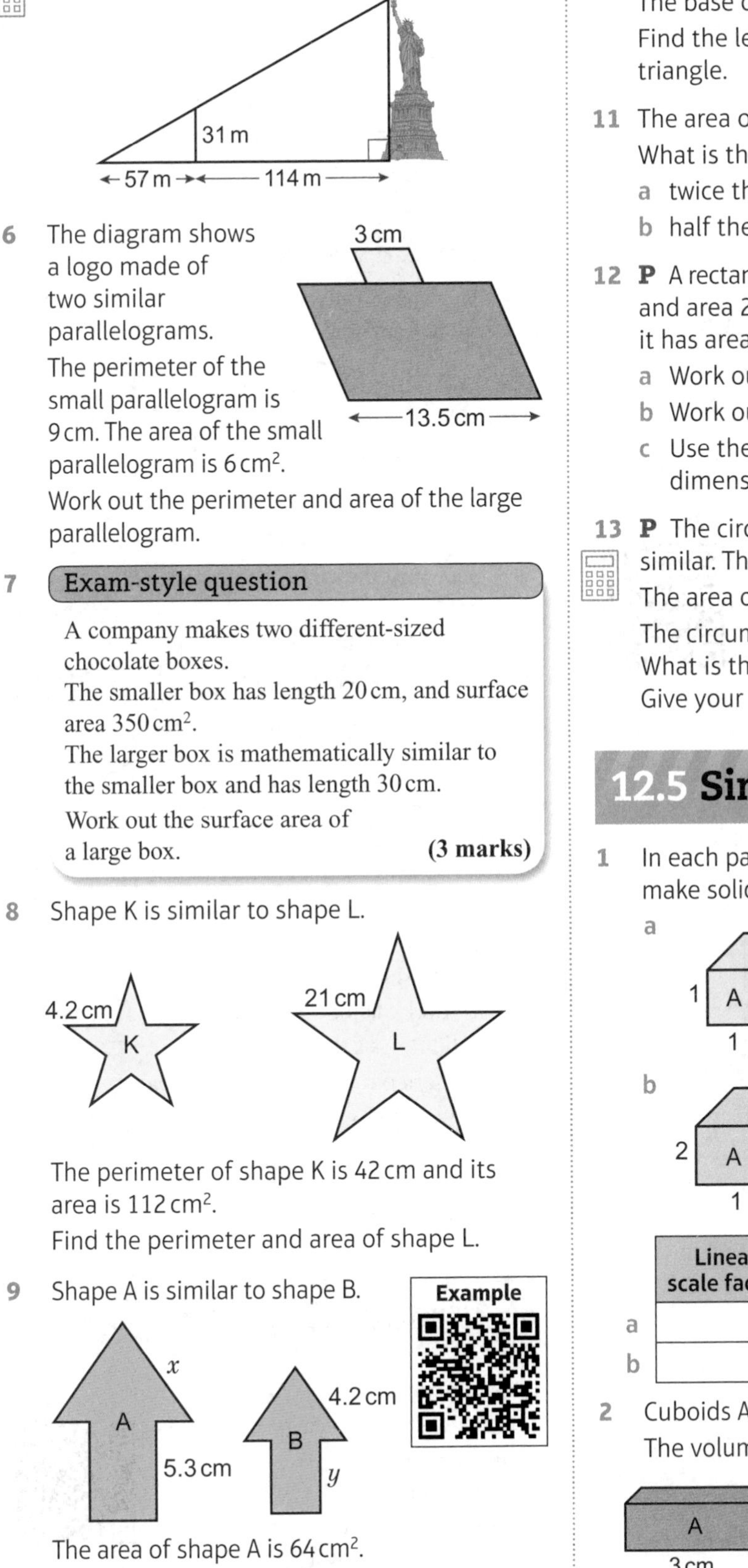

5 Calculate the height of the Statue of Liberty using similar triangles.

6 The diagram shows a logo made of two similar parallelograms.
The perimeter of the small parallelogram is 9 cm. The area of the small parallelogram is 6 cm².
Work out the perimeter and area of the large parallelogram.

7 **Exam-style question**

A company makes two different-sized chocolate boxes.
The smaller box has length 20 cm, and surface area 350 cm².
The larger box is mathematically similar to the smaller box and has length 30 cm.
Work out the surface area of a large box. **(3 marks)**

8 Shape K is similar to shape L.

The perimeter of shape K is 42 cm and its area is 112 cm².
Find the perimeter and area of shape L.

9 Shape A is similar to shape B.

The area of shape A is 64 cm².
The area of shape B is 10.24 cm².
Calculate
a length x b length y.

10 Two similar triangles have areas of 45 cm² and 405 cm² respectively.
The base of the smaller triangle is 3.5 cm.
Find the length of the base of the larger triangle.

11 The area of the front face of a coin is 12 cm².
What is the area of a similar coin with
a twice the diameter of the original coin
b half the diameter of the original coin?

12 **P** A rectangular photograph has width 6 inches and area 24 square inches. An enlargement of it has area 486 square inches.
a Work out the area scale factor.
b Work out the length scale factor.
c Use the length scale factor to work out the dimensions of the enlarged photograph.

13 **P** The circular tops of two tins of beans are similar. The area of the larger top is 75 cm².
The area of the smaller top is $8\frac{1}{3}$ cm².
The circumference of the larger top is 31 cm.
What is the circumference of the smaller top?
Give your answer to 1 decimal place.

12.5 Similarity in 3D solids

1 In each pair of diagrams, solid A is enlarged to make solid B. Copy and complete the table.

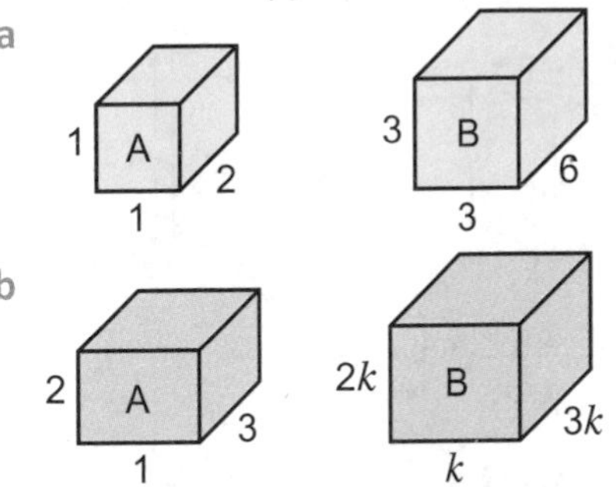

	Linear scale factor	Volume of A	Volume of B	Volume scale factor
a				
b				

2 Cuboids A and B are similar.
The volume of cuboid A is 6 cm³.

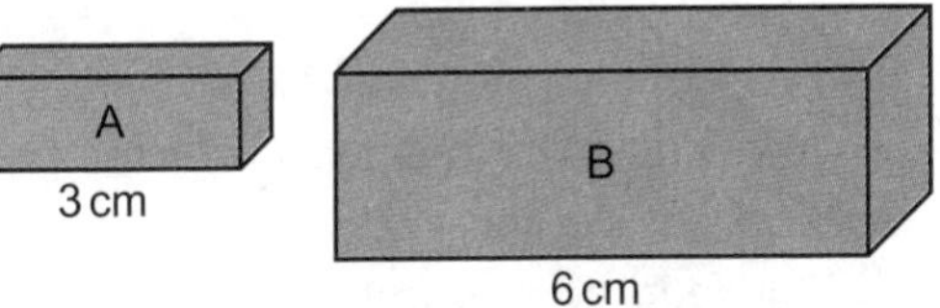

Calculate the volume of cuboid B.

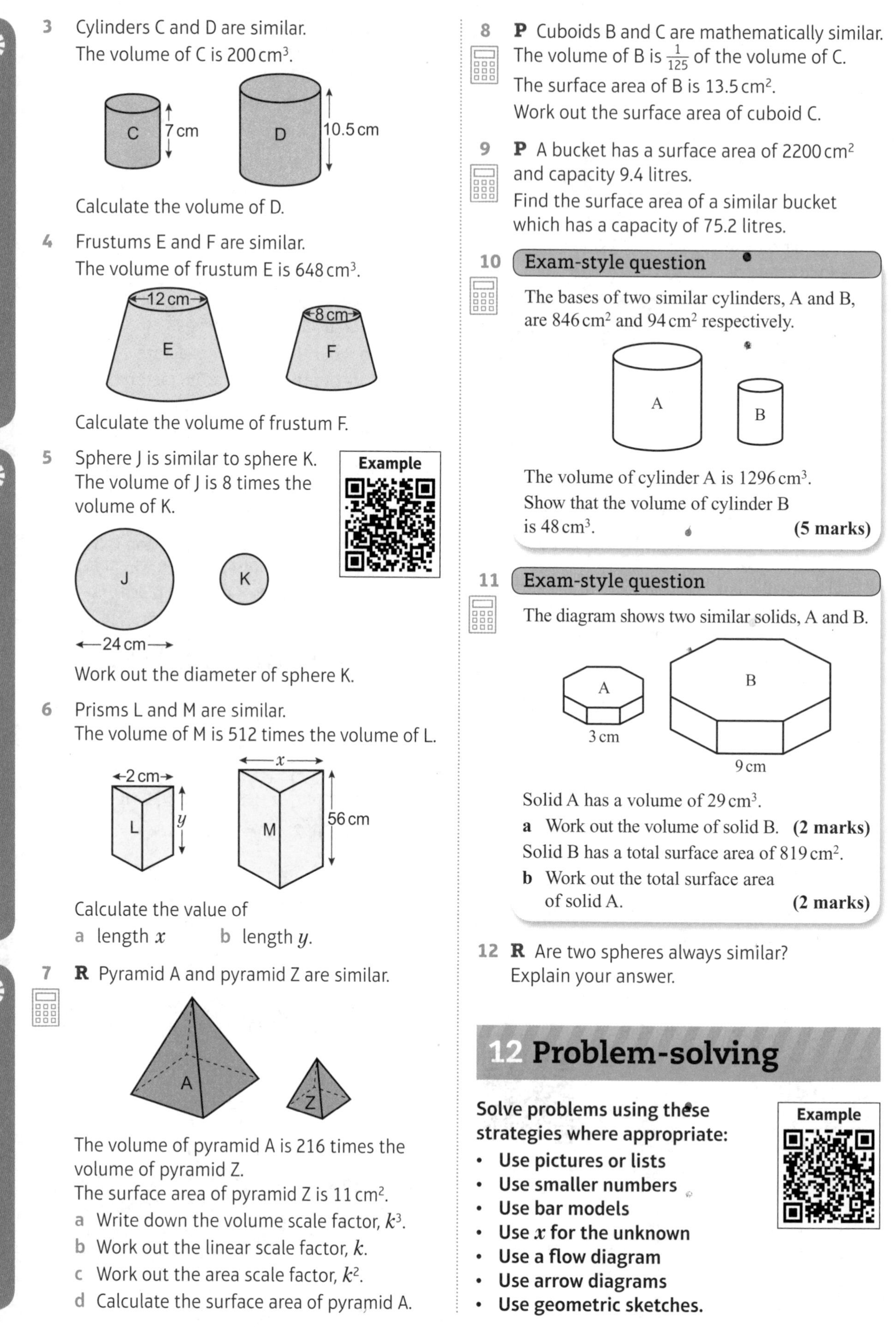

3 Cylinders C and D are similar.
The volume of C is 200 cm^3.

Calculate the volume of D.

4 Frustums E and F are similar.
The volume of frustum E is 648 cm^3.

Calculate the volume of frustum F.

5 Sphere J is similar to sphere K.
The volume of J is 8 times the volume of K.

Example

Work out the diameter of sphere K.

6 Prisms L and M are similar.
The volume of M is 512 times the volume of L.

Calculate the value of
a length x b length y.

7 **R** Pyramid A and pyramid Z are similar.

The volume of pyramid A is 216 times the volume of pyramid Z.
The surface area of pyramid Z is 11 cm^2.

a Write down the volume scale factor, k^3.
b Work out the linear scale factor, k.
c Work out the area scale factor, k^2.
d Calculate the surface area of pyramid A.

8 **P** Cuboids B and C are mathematically similar.
The volume of B is $\frac{1}{125}$ of the volume of C.
The surface area of B is 13.5 cm^2.
Work out the surface area of cuboid C.

9 **P** A bucket has a surface area of 2200 cm^2 and capacity 9.4 litres.
Find the surface area of a similar bucket which has a capacity of 75.2 litres.

10 **Exam-style question**

The bases of two similar cylinders, A and B, are 846 cm^2 and 94 cm^2 respectively.

The volume of cylinder A is 1296 cm^3.
Show that the volume of cylinder B is 48 cm^3. **(5 marks)**

11 **Exam-style question**

The diagram shows two similar solids, A and B.

Solid A has a volume of 29 cm^3.

a Work out the volume of solid B. **(2 marks)**

Solid B has a total surface area of 819 cm^2.

b Work out the total surface area of solid A. **(2 marks)**

12 **R** Are two spheres always similar?
Explain your answer.

12 Problem-solving

Solve problems using these strategies where appropriate:

- **Use pictures or lists**
- **Use smaller numbers**
- **Use bar models**
- **Use x for the unknown**
- **Use a flow diagram**
- **Use arrow diagrams**
- **Use geometric sketches.**

Example

1 **R** Jason draws a right-angled triangle with hypotenuse 45 cm and one side 27 cm long. Yasmin draws a triangle where one side is 36 cm long, the hypotenuse is 45 cm long and one angle is 90°.
Are the two shapes congruent?
Explain how you know.

2 At a bottle stall at a summer fair, tickets from 1 to 100 are placed in a bag.
Customers can pull out 5 tickets for £1.
To win a bottle from the stall, the number on a ticket must end in either 5 or 0.
The first customer picks out 5 tickets.

a What is the probability that the first ticket will be a winner?

b What is the probability that the first two tickets will not be winners?

The first customer finds one winning ticket out of the five they picked out.

c What is the probability that the next customer will find a winning ticket on their first turn?

3 A rectangular park has a pathway which goes diagonally across it.
The park is 21 m wide and 28 m long.
There is enough tarmac to make a path 32 m long.
Will there be enough tarmac for the path?

4 **R** Rectangles A and B are similar.
Rectangle C is an enlargement of rectangle B by the same scale factor as rectangle B is an enlargement of rectangle A.

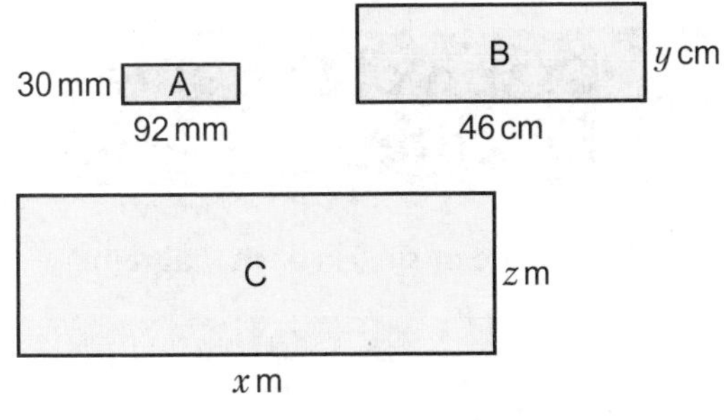

a Work out the value of x.

b Work out the value of y.

c Work out the value of z.

Q4 hint Find the scale factor of the enlargement.

5 Martha has 60 m of fencing and two plots of land. The two plots are similar. The perimeter of Plot B is 38 m. Martha puts fencing around both plots of land. What length of fencing will she have left?

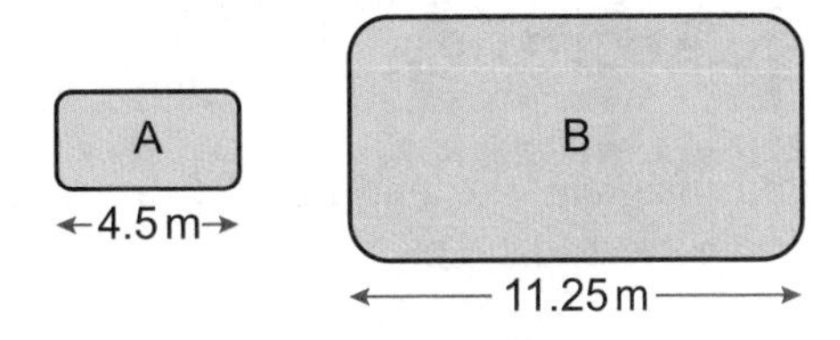

6 What are the roots of these quadratic functions?

a $y^2 + 11y + 24 = 0$ b $t^2 - t - 12 = 0$

c $b^2 + 6b + 8 = 0$

7 **R** Jermain has two similar boxes, A and B. Box A is shown below. Box B has a height of 9 cm. What is the volume of box B?

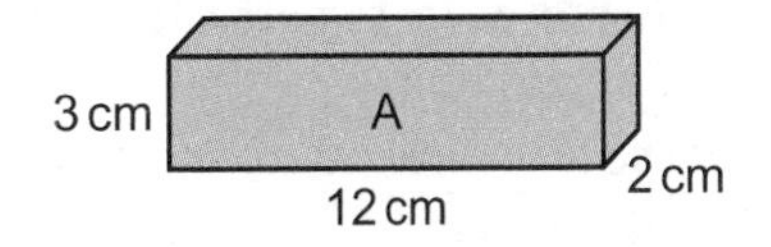

8 A delivery company charges £80 per cubic metre for express delivery. Mason pays to send two containers. On Monday, he sends a container with a volume of 15 m³. On Thursday, he sends a container that is mathematically similar and twice as wide.

a How much did Mason pay to send the container on Monday?

b How much did Mason pay to send the similar container on Thursday?

9 **R** XYZ is a triangle.
XY = YZ = XZ
Point O is the midpoint of XY.

a What type of triangle is XYZ?
Explain your answer.

b Jamie draws the line OZ to make triangles OXZ and OYZ.
Prove these two triangles are congruent.

10 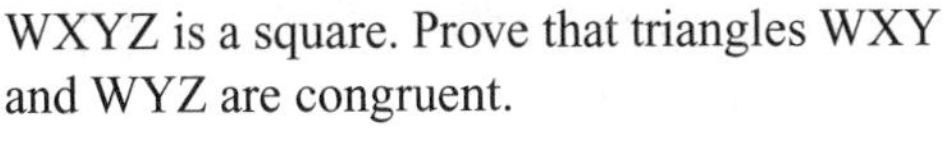

WXYZ is a square. Prove that triangles WXY and WYZ are congruent.

(2 marks)

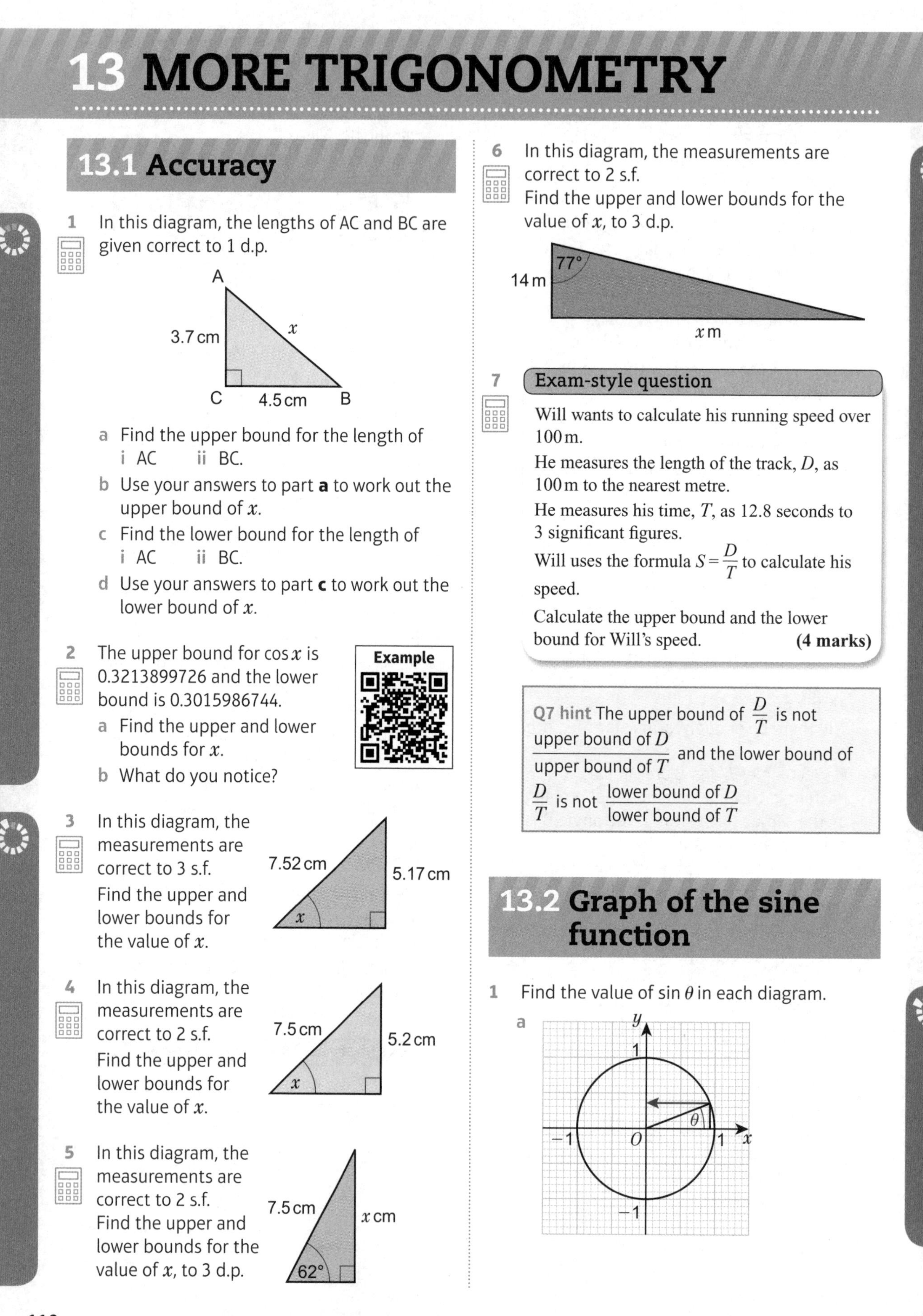

13 MORE TRIGONOMETRY

13.1 Accuracy

1 In this diagram, the lengths of AC and BC are given correct to 1 d.p.

a Find the upper bound for the length of
i AC   ii BC.

b Use your answers to part **a** to work out the upper bound of x.

c Find the lower bound for the length of
i AC   ii BC.

d Use your answers to part **c** to work out the lower bound of x.

2 The upper bound for $\cos x$ is 0.3213899726 and the lower bound is 0.3015986744.

a Find the upper and lower bounds for x.

b What do you notice?

3 In this diagram, the measurements are correct to 3 s.f.
Find the upper and lower bounds for the value of x.

4 In this diagram, the measurements are correct to 2 s.f.
Find the upper and lower bounds for the value of x.

5 In this diagram, the measurements are correct to 2 s.f.
Find the upper and lower bounds for the value of x, to 3 d.p.

6 In this diagram, the measurements are correct to 2 s.f.
Find the upper and lower bounds for the value of x, to 3 d.p.

7 **Exam-style question**

Will wants to calculate his running speed over 100 m.

He measures the length of the track, D, as 100 m to the nearest metre.

He measures his time, T, as 12.8 seconds to 3 significant figures.

Will uses the formula $S = \frac{D}{T}$ to calculate his speed.

Calculate the upper bound and the lower bound for Will's speed. **(4 marks)**

Q7 hint The upper bound of $\frac{D}{T}$ is not $\frac{\text{upper bound of } D}{\text{upper bound of } T}$ and the lower bound of $\frac{D}{T}$ is not $\frac{\text{lower bound of } D}{\text{lower bound of } T}$

13.2 Graph of the sine function

1 Find the value of $\sin \theta$ in each diagram.

a

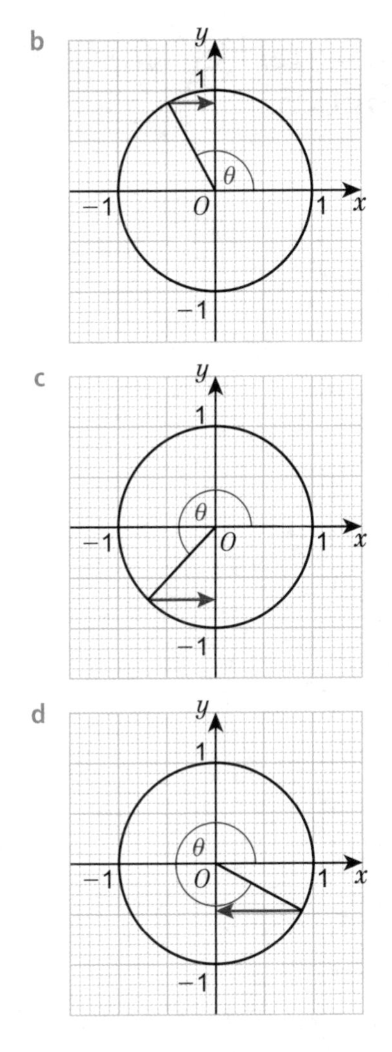

2 a What is the largest value that $\sin\theta$ can take?

b What is the smallest value that $\sin\theta$ can take?

c Find two values of θ so that $\sin\theta = 0$.

3 $\sin 60° = \frac{\sqrt{3}}{2}$

Use the diagram to find an obtuse angle θ such that $\sin\theta = \frac{\sqrt{3}}{2}$.

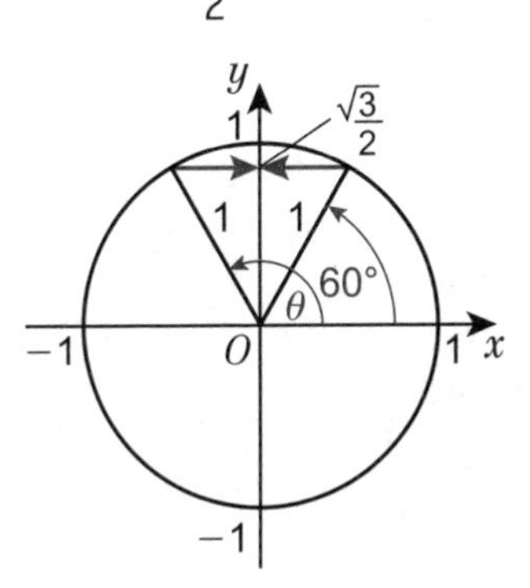

4 As θ increases from 0° to 90°, $\sin\theta$ increases from 0 to 1.

Copy and complete these statements in the same way.

a As θ increases from 90° to 180°, $\sin\theta$ ....

b As θ increases from 180° to 270°, $\sin\theta$ ....

c As θ increases from 270° to 360°, $\sin\theta$ ....

5 Here is the graph of $y = \sin x$ for $0° \leqslant x \leqslant 180°$.

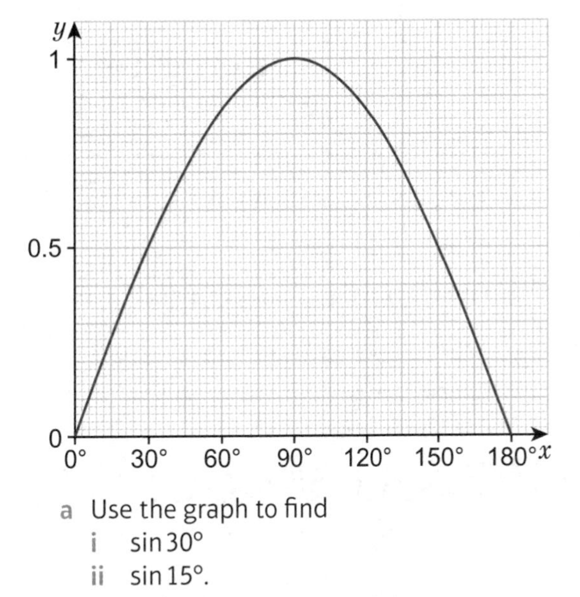

a Use the graph to find

i $\sin 30°$

ii $\sin 15°$.

b Describe the symmetry of the curve.

c Use the graph to check your answer to **Q3**.

d Copy and complete, inserting numbers greater than 90.

i $\sin 30° = \sin\square°$

ii $\sin 75° = \sin\square°$

iii $\sin 0° = \sin\square°$

e Use the graph to give an estimate for the solutions to $\sin x = 0.75$.

6 Here is a sketch of the graph of $y = \sin x$ for $0° \leqslant x \leqslant 360°$.

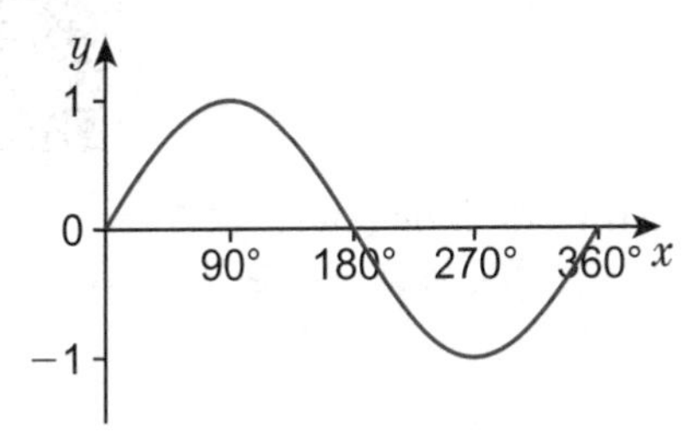

Describe the symmetry of the curve.

7 The graph of $y = \sin x$ repeats every 360° in both directions.

a Sketch the graph of $y = \sin x$ for $0° \leqslant x \leqslant 540°$.

b Use your sketch to find

i sin 390°

ii sin 480°.

c The exact value of sin 45° is $\frac{1}{\sqrt{2}}$.

Write down the exact value of

i sin 405°

ii sin 495°.

d Explain how you worked out your answers to part **c**.

> **Q7a hint** Include x-values 0°, 90°, 180°, 270°, 360°, 450°, 540° and y-values 1, 0.5, 0, −0.5, −1.

8 a Write down four values of x such that $\sin x = -\frac{1}{\sqrt{2}}$.

b Write down four values of x such that $\sin x = -\frac{\sqrt{3}}{2}$.

c Check each of your answers using your calculator.

9 **Exam-style question**

Here is a sketch of $y = \sin x$.

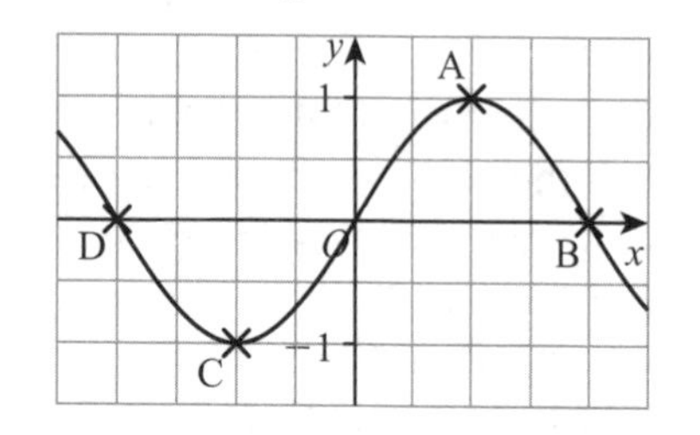

Write down the coordinates of each of the labelled points. **(4 marks)**

10 Solve the equation $10 \sin x = 3$ for all values of x in the interval 0° to 720°.

11 Solve the equation $8 \sin \theta = 6.5$ for all values of θ in the interval 0° to 720°.

13.3 Graph of the cosine function

1 Find the value of $\cos \theta$ in each diagram.

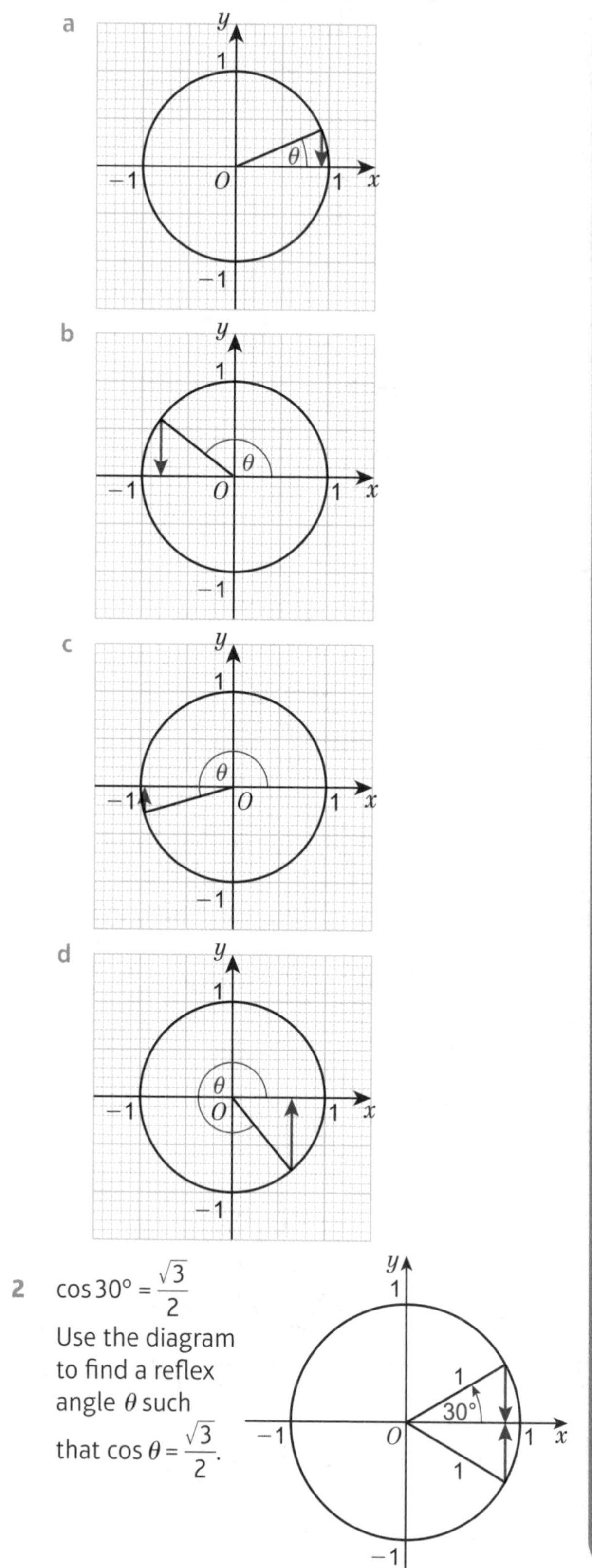

2 $\cos 30° = \frac{\sqrt{3}}{2}$

Use the diagram to find a reflex angle θ such that $\cos \theta = \frac{\sqrt{3}}{2}$.

3 $\cos 60° = 0.5$

a Find a reflex angle θ such that $\cos\theta = 0.5$.

b Find an obtuse angle such that $\cos\theta = -0.5$.

4 As θ increases from 0° to 90°, $\cos\theta$ decreases from 1 to 0.

Copy and complete these statements in the same way.

a As θ increases from 90° to 180°, $\cos\theta$ ....

b As θ increases from 180° to 270°, $\cos\theta$ ....

c As θ increases from 270° to 360°, $\cos\theta$ ....

5 Here is the graph of $y = \cos x$ for $0° \leqslant x \leqslant 360°$.

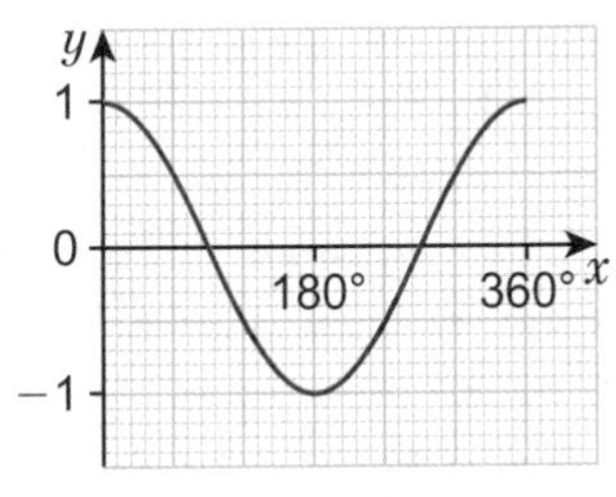

a Use the graph to find

i $\cos 90°$ ii $\cos 240°$.

b Describe the symmetry of the curve.

c Copy and complete

i $\cos 0° = \cos\square°$

ii $\cos 30° = \cos\square°$

iii $\cos 150° = \cos\square°$

6 The graph of $y = \cos x$ repeats every 360° in both directions.

a Sketch the graph of $y = \cos x$ for $0° \leqslant x \leqslant 720°$.

b Use your graph to find

i $\cos 420°$ ii $\cos 600°$.

c The exact value of $\cos 45°$ is $\frac{1}{\sqrt{2}}$.

Write down the exact value of

i $\cos 405°$ ii $\cos 675°$.

Q6a hint Include x-values 0°, 90°, 180°, 270°, 360°, …, 720° and y values 1, 0.5, 0, −0.5, −1.

7 Use your sketch from **Q6** to find four values of x such that

a $\cos x = -\frac{1}{\sqrt{2}}$ b $\cos x = -\frac{\sqrt{3}}{2}$.

Check your answers using a calculator.

8 **Exam-style question**

The diagram shows a sketch of the graph $y = \cos x°$

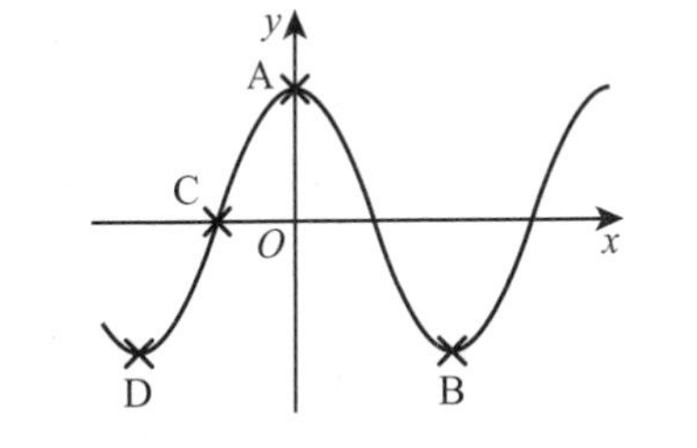

Write down the coordinates of points A, B, C and D. **(4 marks)**

9 $8\cos x = 4.64$

a Use $\cos^{-1}$ on your calculator to find one value of x.

b Use your answer to part **a** and your sketch from **Q6** to solve $8\cos x = 4.64$ for values of x in the interval 0° to 720°.

10 Solve the equation $7\cos\theta = -4.3$ for values of θ in the interval 0° to 720°.

13.4 The tangent function

1 Find the value of $\tan\theta$ in each diagram.

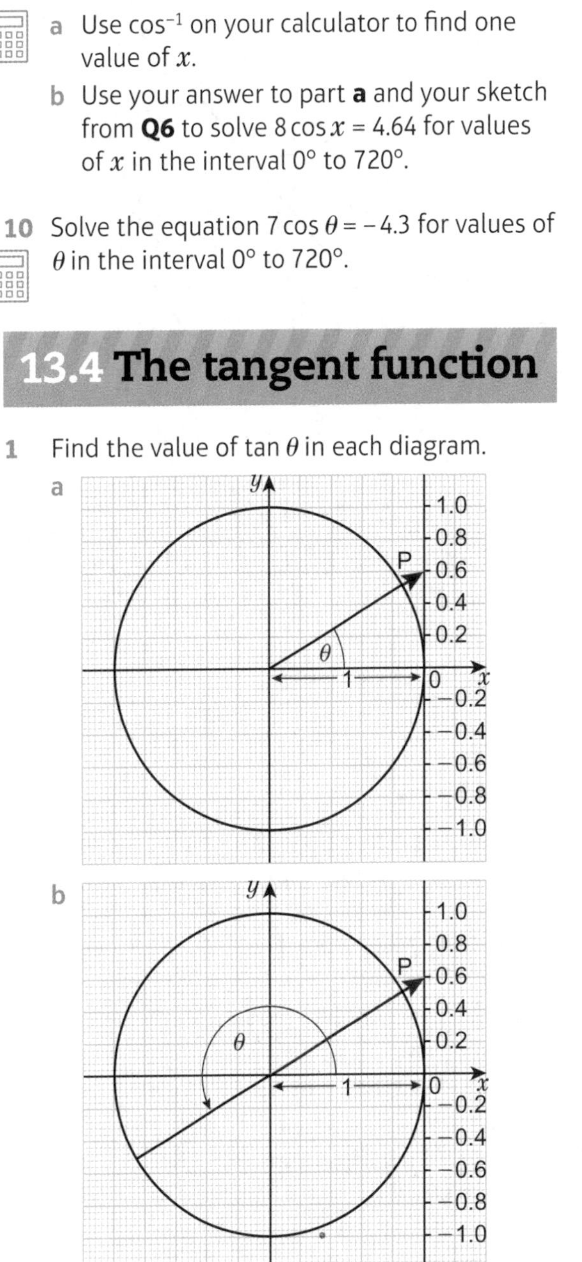

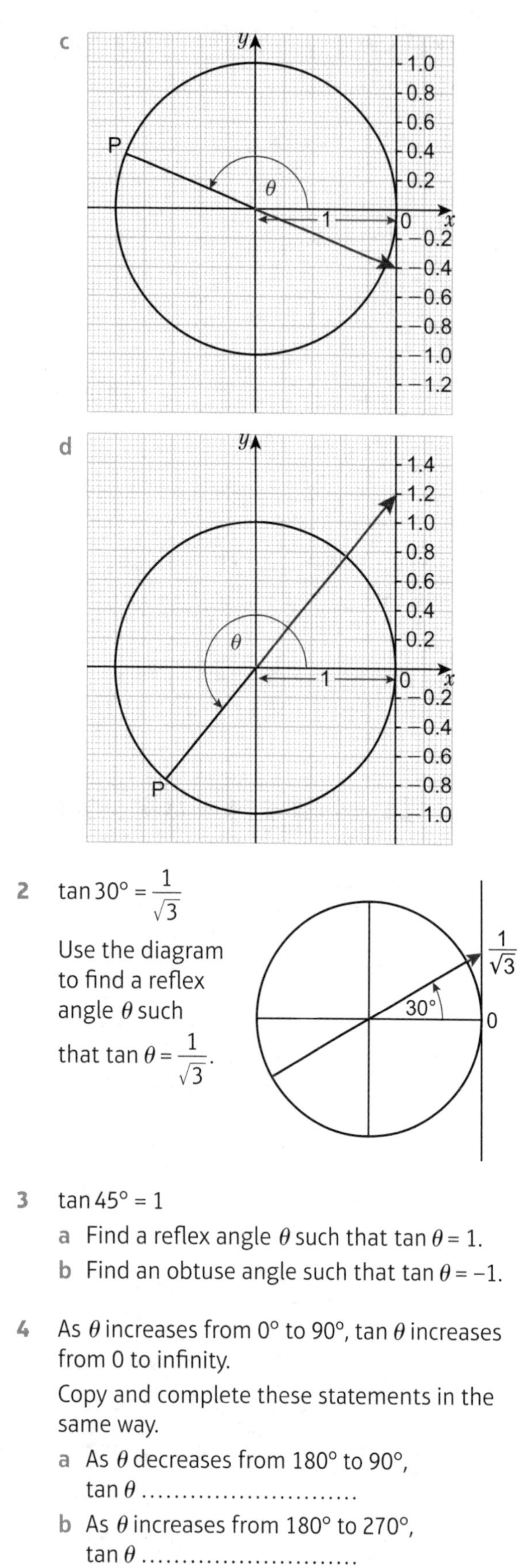

2 $\tan 30° = \frac{1}{\sqrt{3}}$

Use the diagram to find a reflex angle θ such that $\tan \theta = \frac{1}{\sqrt{3}}$.

$\frac{1}{\sqrt{3}}$

30°

0

3 $\tan 45° = 1$

a Find a reflex angle θ such that $\tan \theta = 1$.

b Find an obtuse angle such that $\tan \theta = -1$.

4 As θ increases from 0° to 90°, $\tan \theta$ increases from 0 to infinity.

Copy and complete these statements in the same way.

a As θ decreases from 180° to 90°, $\tan \theta$ ....

b As θ increases from 180° to 270°, $\tan \theta$ ....

c As θ decreases from 360° to 270°, $\tan \theta$ ....

5 Here is the graph of $y = \tan x$ for $0° \leqslant x \leqslant 360°$.

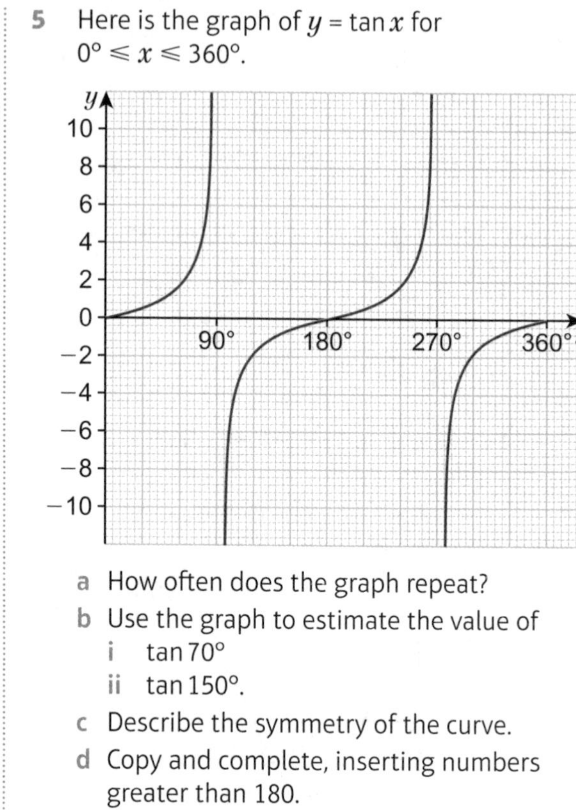

a How often does the graph repeat?

b Use the graph to estimate the value of

i $\tan 70°$

ii $\tan 150°$.

c Describe the symmetry of the curve.

d Copy and complete, inserting numbers greater than 180.

i $\tan 45° = \tan \square°$

ii $\tan 80° = \tan \square°$

iii $\tan 135° = \tan \square°$

6 a Sketch the graph of $y = \tan x$ for $0° \leqslant \theta \leqslant 540°$.

b Use your sketch to find

i $\tan 495°$ ii $\tan 420°$.

c The exact value of $\tan 30°$ is $\frac{1}{\sqrt{3}}$.

Write down the exact value of

i $\tan 390°$ ii $\tan 150°$.

d Explain how you worked out your answers to part **c**.

7 a Write down four values of x such that $\tan x = \frac{1}{\sqrt{3}}$.

b Write down four values of x such that $\tan x = -\frac{1}{\sqrt{3}}$.

c Check your answers using a calculator.

8 $4 \tan x = 9$

a Use $\tan^{-1}$ on your calculator to find one value of x.

b Use your answer to part **a** and your sketch from **Q6** to solve $4 \tan x = 9$ for values of x in the interval 0° to 540°.

9 Solve the equation 8 tan θ = 7 for values of θ in the interval 0° to 720°.

10 **Exam-style question**

a Sketch the graph of $y = \tan x$ in the interval 0° to 720°.

b Given that $\tan 60° = \sqrt{3}$ solve the equation $\sqrt{3}\tan x = 3$ in the interval 0° to 720°. **(4 marks)**

Exam hint
You are expected to use your sketch from part **a** to solve the equation in part **b**.

13.5 Calculating areas and the sine rule

1 **a** Write h, the perpendicular height of the triangle, in terms of a and θ.

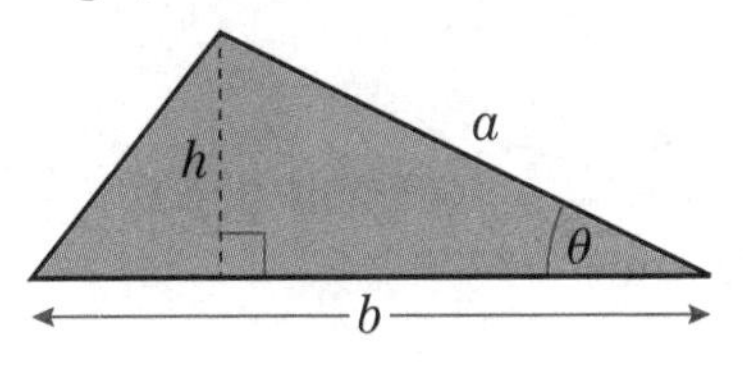

b Write a formula in terms of a and θ to calculate the area of the triangle.

2 Find the area of each triangle.

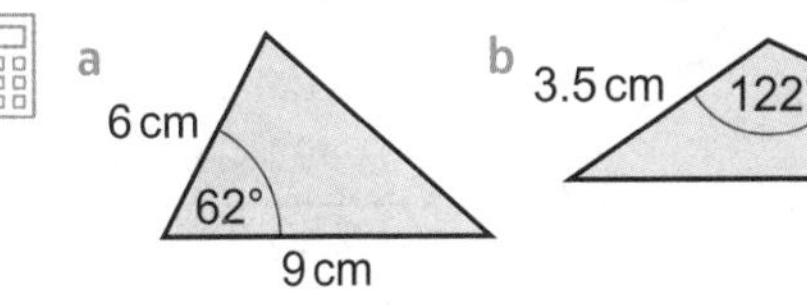

3 The area of triangle ABC is 52.96 cm^2.
Work out the length of AB.

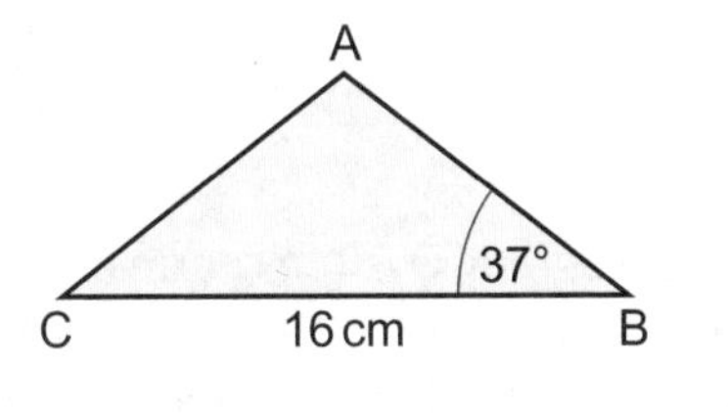

4 **a** Find the area of triangle AOB in this circle.

b Find the area of the sector AOB.

c Find the area of the shaded segment of the circle.

A B 55° 8 cm O

5 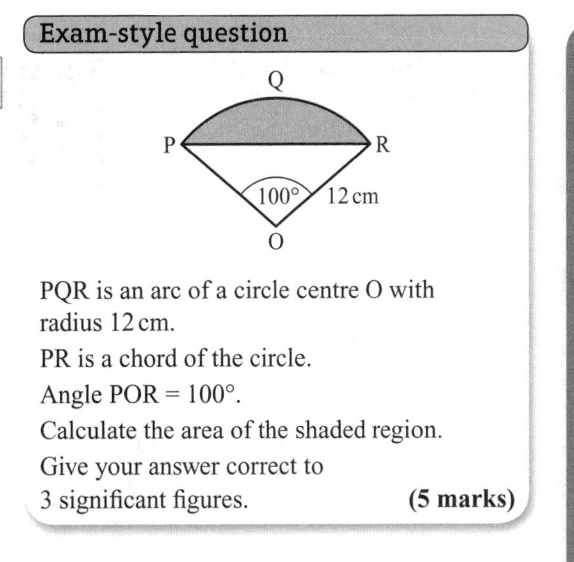

Exam-style question

PQR is an arc of a circle centre O with radius 12 cm.
PR is a chord of the circle.
Angle POR = 100°.
Calculate the area of the shaded region.
Give your answer correct to 3 significant figures. **(5 marks)**

6 **P** **a** Calculate angle AOB.
Give your answer correct to 1 decimal place.

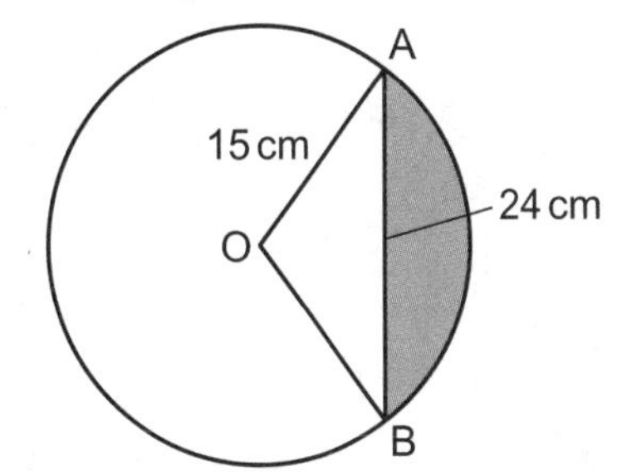

b Work out the area of the shaded segment.
Give your answer correct to 3 significant figures.

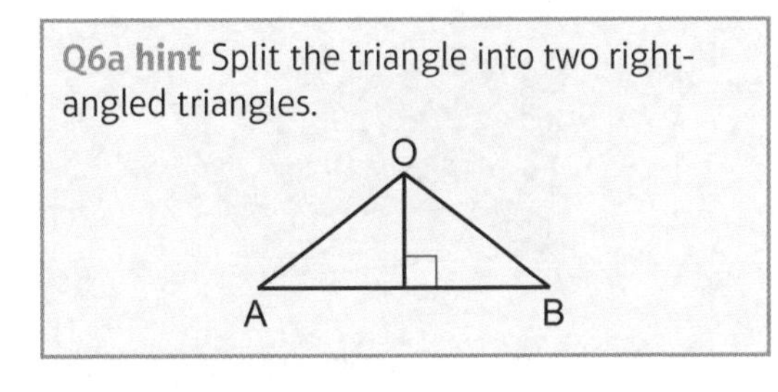

Q6a hint Split the triangle into two right-angled triangles.

7 **P** In the diagram, O is the centre of the circle.
Work out the area of the shaded segment.
Give your answer correct to 3 significant figures.

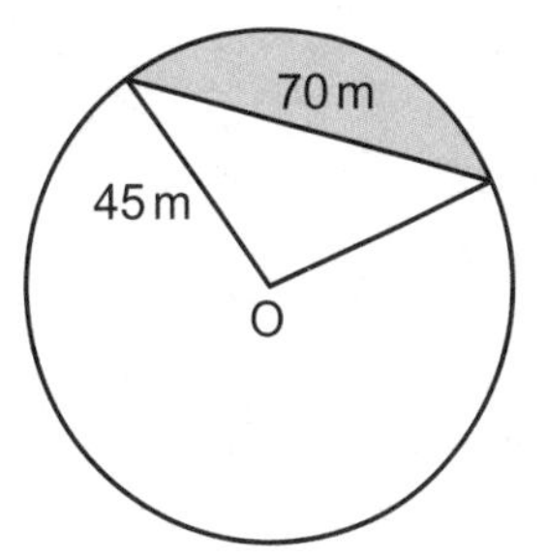

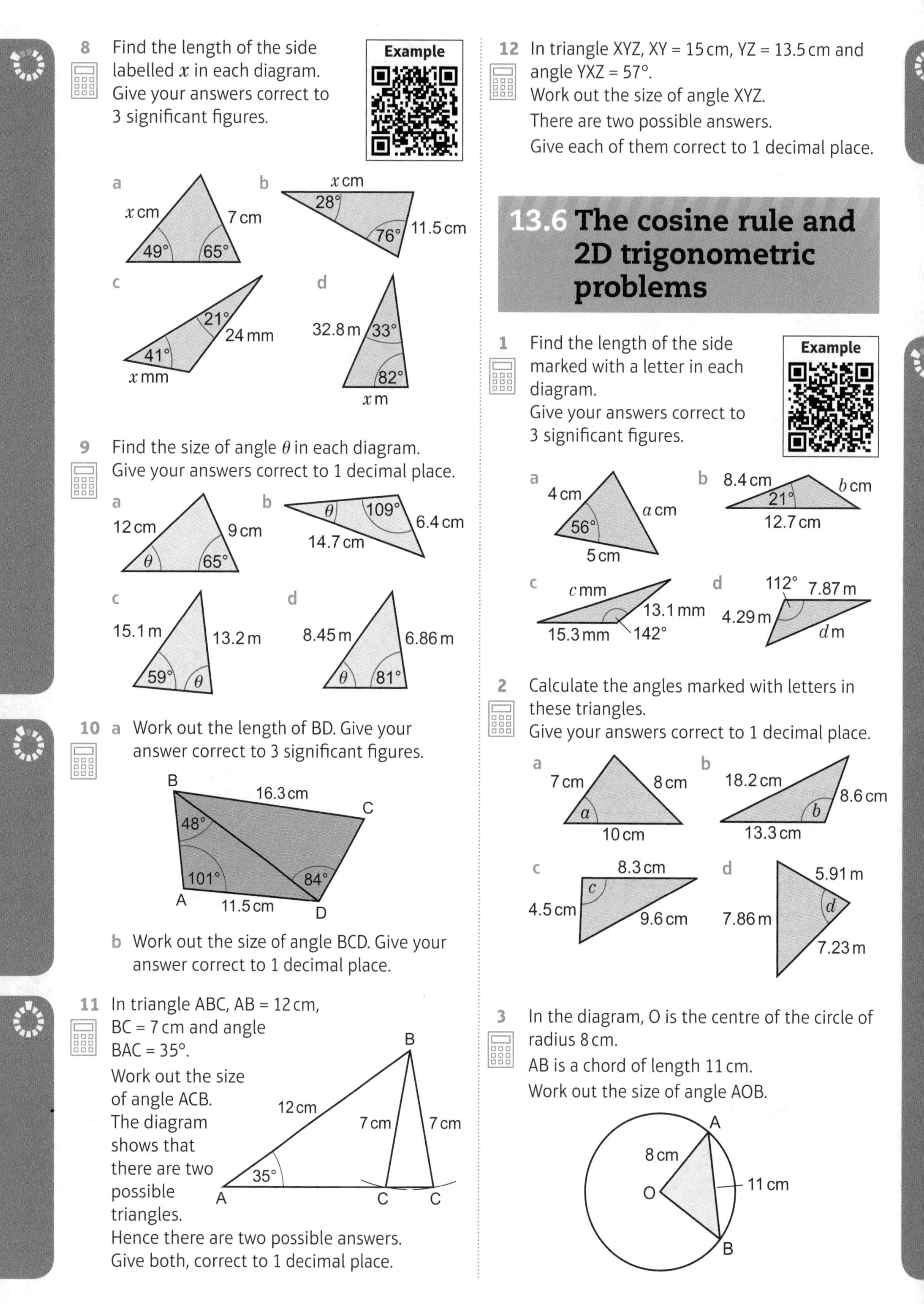

8 Find the length of the side labelled x in each diagram. Give your answers correct to 3 significant figures.

Example

a x cm, 7 cm, 49°, 65°

b x cm, 28°, 76°, 11.5 cm

c 21°, 24 mm, 41°, x mm

d 32.8 m, 33°, 82°, x m

9 Find the size of angle θ in each diagram. Give your answers correct to 1 decimal place.

a 12 cm, 9 cm, θ, 65°

b θ, 109°, 6.4 cm, 14.7 cm

c 15.1 m, 13.2 m, 59°, θ

d 8.45 m, 6.86 m, θ, 81°

10 **a** Work out the length of BD. Give your answer correct to 3 significant figures.

B, 16.3 cm, C, 48°, 101°, 84°, A, 11.5 cm, D

b Work out the size of angle BCD. Give your answer correct to 1 decimal place.

11 In triangle ABC, AB = 12 cm, BC = 7 cm and angle BAC = 35°.
Work out the size of angle ACB.
The diagram shows that there are two possible triangles.
Hence there are two possible answers.
Give both, correct to 1 decimal place.

B, 12 cm, 7 cm, 7 cm, 35°, A, C, C

12 In triangle XYZ, XY = 15 cm, YZ = 13.5 cm and angle YXZ = 57°.
Work out the size of angle XYZ.
There are two possible answers.
Give each of them correct to 1 decimal place.

13.6 The cosine rule and 2D trigonometric problems

1 Find the length of the side marked with a letter in each diagram.
Give your answers correct to 3 significant figures.

Example

a 4 cm, a cm, 56°, 5 cm

b 8.4 cm, 21°, b cm, 12.7 cm

c c mm, 13.1 mm, 15.3 mm, 142°

d 112°, 7.87 m, 4.29 m, d m

2 Calculate the angles marked with letters in these triangles.
Give your answers correct to 1 decimal place.

a 7 cm, 8 cm, a, 10 cm

b 18.2 cm, 8.6 cm, b, 13.3 cm

c 8.3 cm, c, 4.5 cm, 9.6 cm

d 5.91 m, d, 7.86 m, 7.23 m

3 In the diagram, O is the centre of the circle of radius 8 cm.
AB is a chord of length 11 cm.
Work out the size of angle AOB.

A, 8 cm, O, 11 cm, B

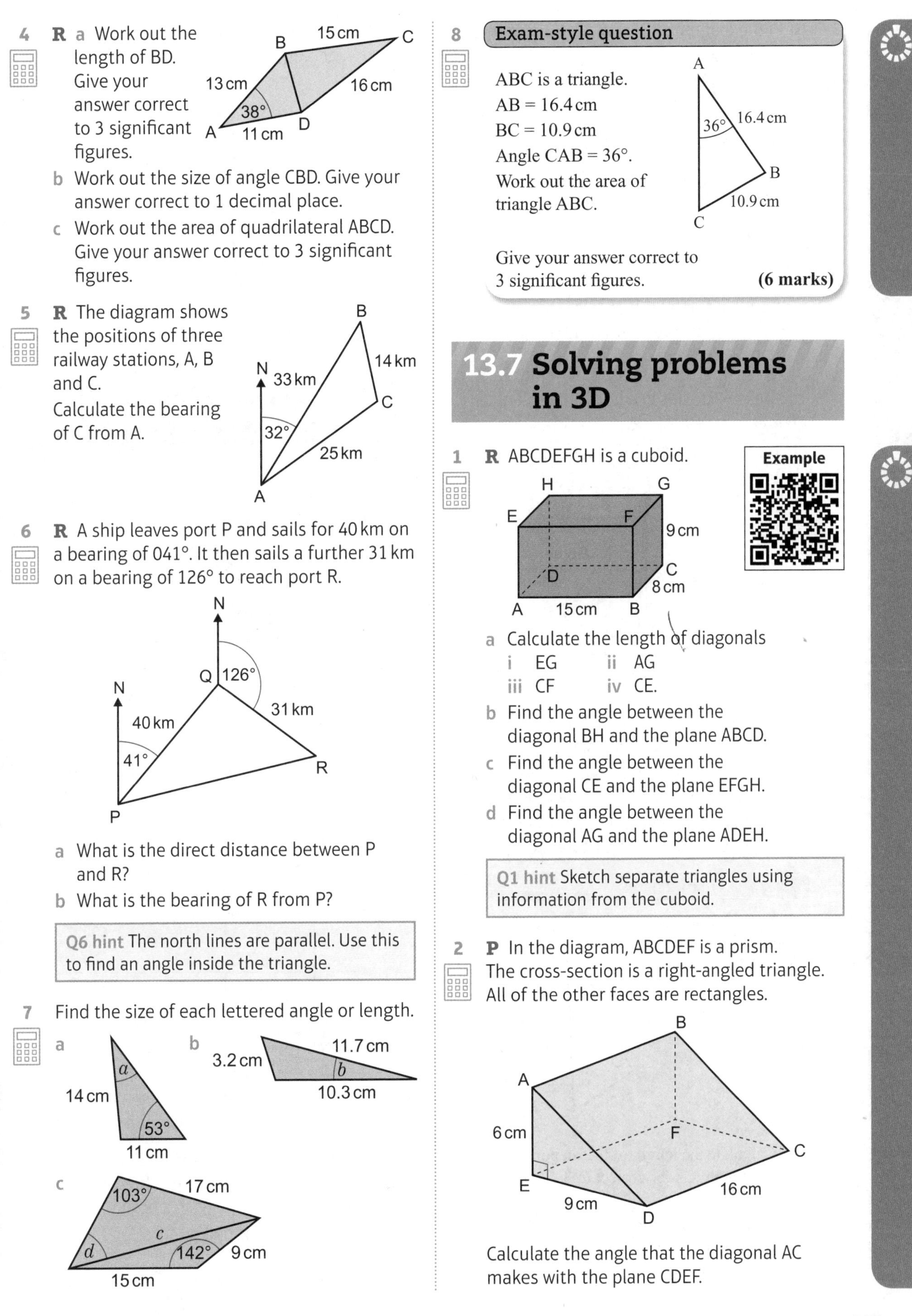

4 **R** a Work out the length of BD. Give your answer correct to 3 significant figures.

b Work out the size of angle CBD. Give your answer correct to 1 decimal place.

c Work out the area of quadrilateral ABCD. Give your answer correct to 3 significant figures.

5 **R** The diagram shows the positions of three railway stations, A, B and C.
Calculate the bearing of C from A.

6 **R** A ship leaves port P and sails for 40 km on a bearing of 041°. It then sails a further 31 km on a bearing of 126° to reach port R.

a What is the direct distance between P and R?

b What is the bearing of R from P?

Q6 hint The north lines are parallel. Use this to find an angle inside the triangle.

7 Find the size of each lettered angle or length.

8 **Exam-style question**

ABC is a triangle.
AB = 16.4 cm
BC = 10.9 cm
Angle CAB = 36°.
Work out the area of triangle ABC.

Give your answer correct to 3 significant figures. **(6 marks)**

13.7 Solving problems in 3D

1 **R** ABCDEFGH is a cuboid.

a Calculate the length of diagonals
i EG ii AG
iii CF iv CE.

b Find the angle between the diagonal BH and the plane ABCD.

c Find the angle between the diagonal CE and the plane EFGH.

d Find the angle between the diagonal AG and the plane ADEH.

Q1 hint Sketch separate triangles using information from the cuboid.

2 **P** In the diagram, ABCDEF is a prism.
The cross-section is a right-angled triangle.
All of the other faces are rectangles.

Calculate the angle that the diagonal AC makes with the plane CDEF.

3 **R** In the diagram, ABCD is a tetrahedron.

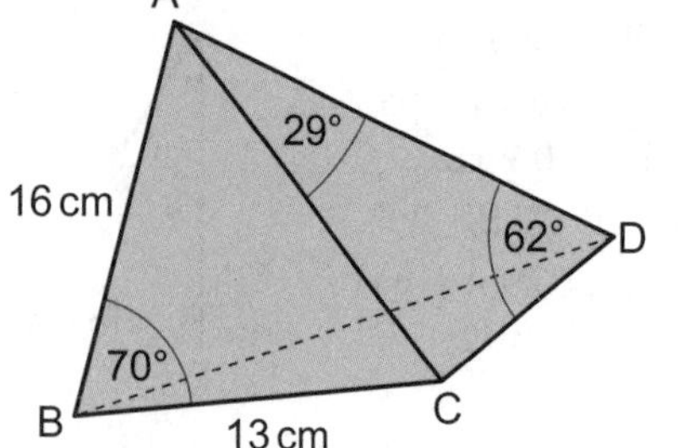

a Work out the length of AC.

b Work out the length of CD.

c Given that BD = 17 cm, calculate angle BCD.

4 **R** ABCDE is a square-based pyramid.
The base BCDE lies in a horizontal plane.
AB = AC = AD = AE = 18 cm
AM is perpendicular to the base.

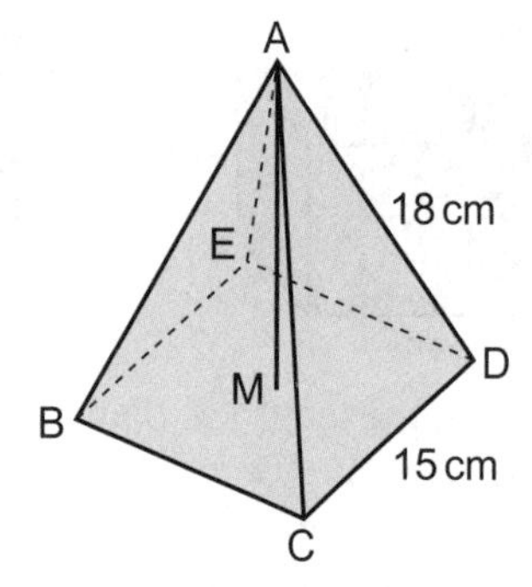

a Calculate the length of
i BD ii BM iii AM.

b Calculate the angle that AD makes with the base, correct to the nearest degree.

c Calculate the angle between AM and the face ABC, correct to the nearest degree.

5 **Exam-style question**

The diagram shows a square-based pyramid ABCDE. Each triangular face is an isosceles triangle.

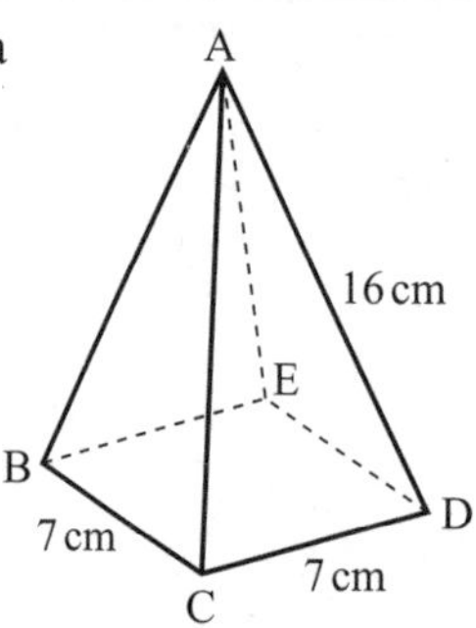

a Calculate the length of the diagonal BD. Give your answer correct to 3 significant figures.

b Calculate the area of triangle ABD. Give your answer correct to 3 significant figures. **(6 marks)**

6 **P** ABCDE is a pyramid with a rectangular base.
AB = AC = AD = AE = 35 cm

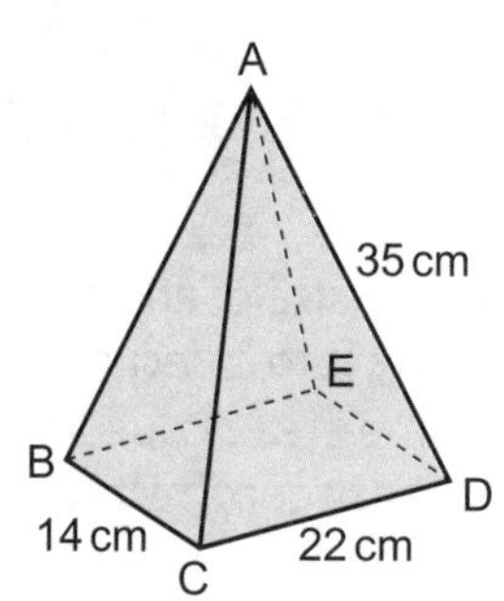

Calculate the size of angle BAD correct to the nearest degree.

13.8 Transforming trigonometric graphs 1

1 Here is the graph of $y = \cos x$ for $-180° \leqslant x \leqslant 180°$.

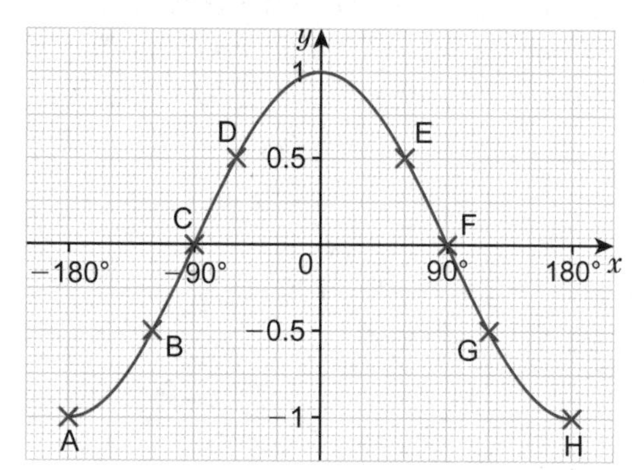

a Copy the table.

i Write in the values of x and $\cos x$ at points A to H on the graph.

ii For each x-value, write the value of $-\cos x$.

	x	$\cos x$	$-\cos x$
A	−180°	−1	1
B	−120°	−0.5	0.5
C			

b Sketch the graph of $y = -\cos x$ for $-180° \leqslant x \leqslant 180°$.

c Describe how the graph of $y = \cos x$ is transformed to give the graph of $y = -\cos x$.

2 a Use your table from **Q1**. Add a column for $\cos(-x)$. Find the cosine values from the graph to fill in the $\cos(-x)$ column.

b Sketch the graph of $y = \cos(-x)$ for $-180° \leqslant x \leqslant 180°$.

c What do you notice?

3 Here is the graph of $y = \tan x$ for $-180° \leqslant x \leqslant 180°$.

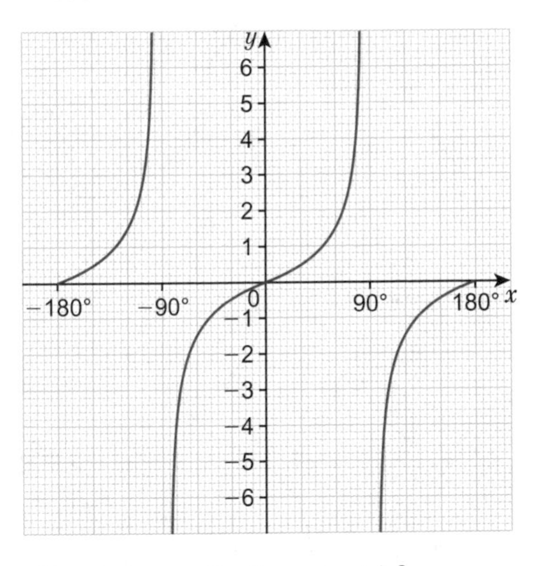

Sketch the graph of $y = -\tan(x)$ for $-180° \leqslant x \leqslant 180°$.

4 a Look at your graph of $y = \cos(-x)$ for $-180° \leqslant x \leqslant 180°$ in **Q2**.
What transformations will turn the graph of $y = \cos(-x)$ into the graph of $y = -\cos(-x)$?
b Sketch the graph of $y = -\cos(-x)$.
c Can the transformation in part **a** be described in any other way?

5 a Sketch the graph of $y = \sin x$ for $-180° \leqslant x \leqslant 180°$.
b Sketch the graph of $y = -\sin(-x)$.

6 Explain why the graph of $y = -\sin(-x)$ is the same as the graph of $y = \sin x$.

7 a Describe the transformation that maps the graph of $y = \tan x$ to the graph of $y = \tan(-x)$.
b Sketch the graph of $y = \tan x$ and the graph of $y = \tan(-x)$ for the interval 0 to 360°.

8 a Describe the transformation that maps the graph of $y = \sin x$ to the graph of $y = -\sin x$.
b Sketch the graphs of $y = \sin x$ and $y = -\sin x$ for the interval −180° to 180°.

9 a Describe the transformation that maps the graph of $y = \tan x$ to the graph of $y = -\tan(-x)$.
b Sketch the graphs of $y = \tan x$ and $y = -\tan(-x)$ for the interval −180° to 180°.

10 **Exam-style question**

Here is a sketch of the graph of $y = -\cos x$.

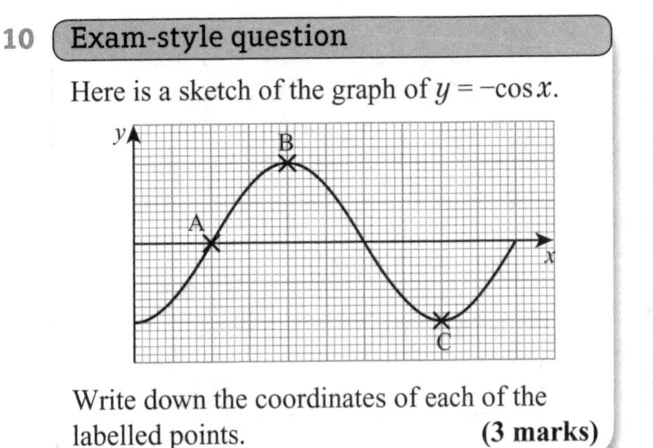

Write down the coordinates of each of the labelled points. **(3 marks)**

13.9 Transforming trigonometric graphs 2

1 a Copy the graph of $y = \cos x$ for $-180° \leqslant x \leqslant 180°$ from **Q1** in lesson **13.8**.
b Add 1 to the y-coordinate at each of the labelled points.
c Draw the cosine graph that passes through the new points. Label it $y = \cos x + 1$.
d Describe the transformation from the graph of $y = \cos x$ to this graph.
e Now subtract 1 from the y-coordinate at each of the labelled points on the original graph.
f Draw the cosine graph that passes through the new points.
g Describe the transformation from the graph of $y = \cos x$ to this graph.
h Write down the equation of the graph.

2 Write down the equation of each graph.

Q2 hint First decide whether it is a sin, cos or tan graph.

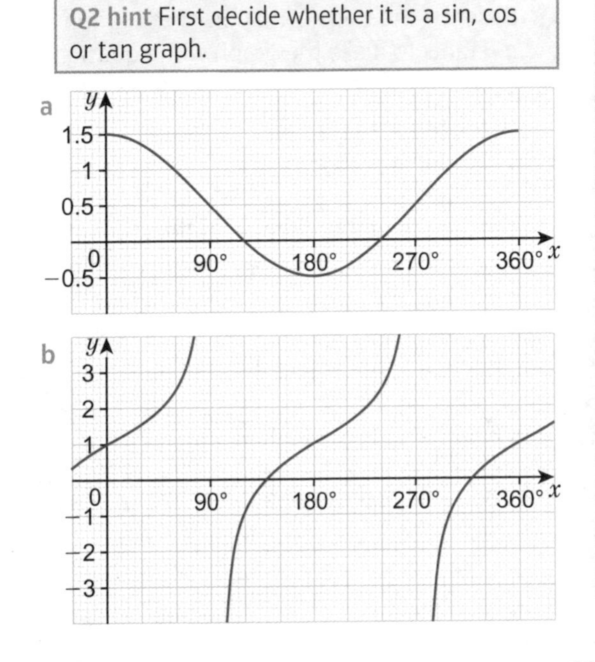

c

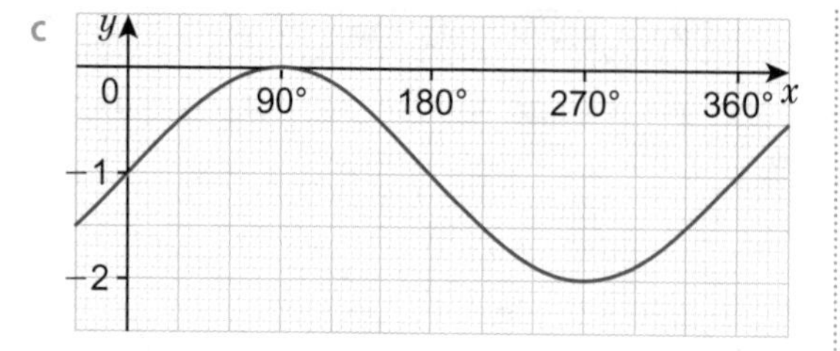

3 Here is the graph of $y = \sin x$ for $0° \leqslant x \leqslant 360°$.

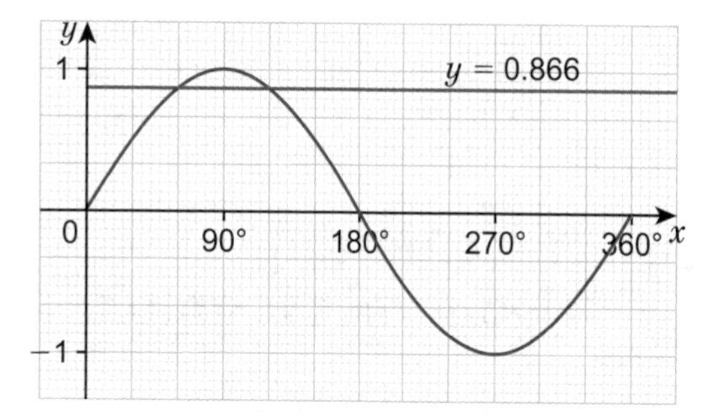

a Copy and complete this table of values for $\sin(x + 60°)$.

x	0	30	60	90
$\sin(x + 60°)$	$\sin 60° = \square$			

b Sketch the graph of $y = \sin(x + 60°)$

c Describe the transformation that takes the graph of $y = \sin x$ to the graph of $y = \sin(x + 60°)$.

4 Describe the transformation of the graph of $y = \sin x$ to make the graph with equation

a $y = \sin(x + 30°)$

b $y = \sin(x + 50°)$

c $y = \sin(x - 60°)$

5 Describe the transformation of the graph of $y = \tan x$ to make the graph with equation

a $y = \tan(x + 60°)$

b $y = \tan(x + 20°)$

c $y = \tan(x - 30°)$.

6 Match each graph with one of these equations.

A $y = \tan(x + 30°)$ **B** $y = \sin(x - 45°)$

C $y = \cos(x - 45°)$

a

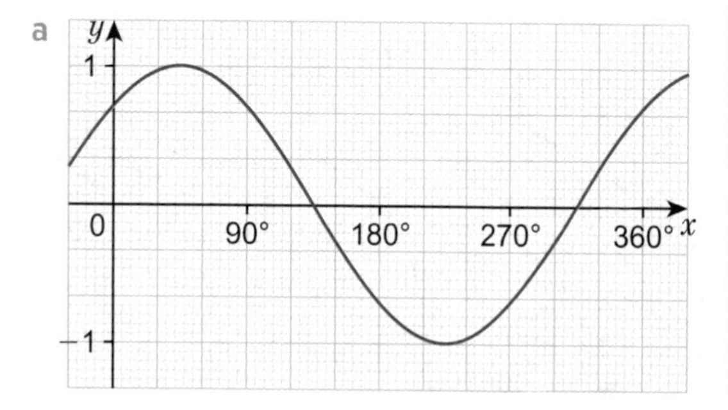

b

c

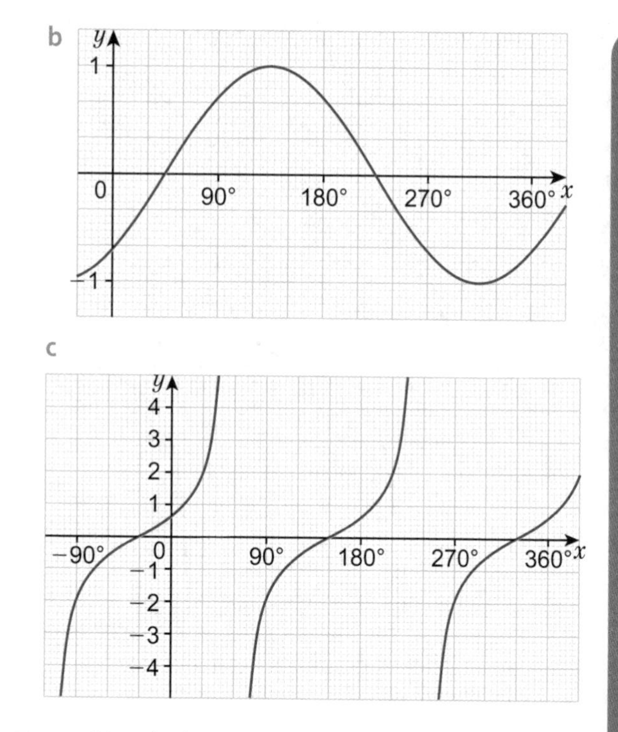

7 a Sketch the graph of $y = \cos x$ for $0° \leqslant x \leqslant 360°$.

b Copy and complete the table of values for $y = 2\cos x$.

x	0	30°	60°	90°	120°
$\cos x$	1	$\frac{\sqrt{3}}{2}$	0.5		
$2\cos x$	2	$\sqrt{3}$			

c On the same axes, sketch the graph of $y = 2\cos x$.

8 Sketch the graphs of these functions for $0° \leqslant x \leqslant 360°$.

a $y = -2\cos x$

b $y = 0.5\tan x$

c $y = 3\sin x$

9 a Copy your sketch graph of $y = \cos x$ for $0° \leqslant x \leqslant 360°$ from **Q7**.

b Copy and complete the table of values for $y = \cos(2x)$.

x	0°	30°	60°	90°	120°
$\cos(2x)$					

c Sketch the graph of $y = \cos(2x)$ on the same axes.

10 Sketch the graphs of these functions for $0° \leqslant x \leqslant 360°$.

a $y = \sin 2x$

b $y = \cos 3x$

c $y = \tan 4x$

11 **Exam-style question**

The diagram shows part of a sketch of the curve $y = \cos x°$

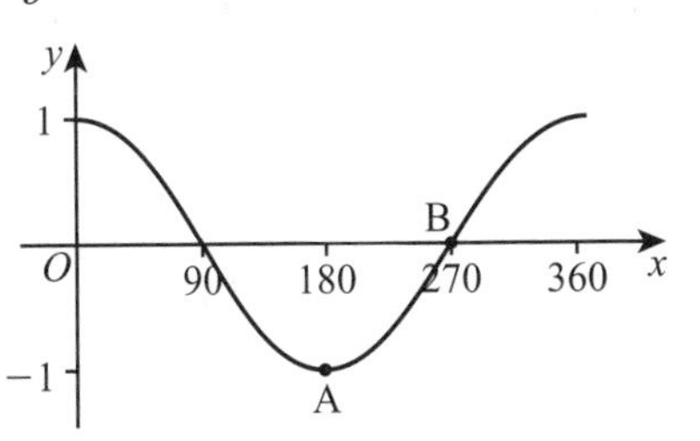

a Write down the coordinates of the points A and B. **(2 marks)**

Here is a sketch of the curve $y = a \sin bx + c$, for $0° \leqslant x \leqslant 360°$.

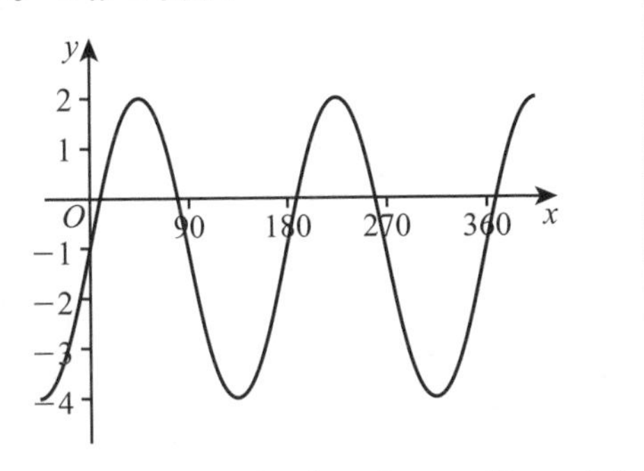

b Find the values of a, b and c. **(3 marks)**

13 Problem-solving

Solve problems using these strategies where appropriate:

- **Use pictures or lists**
- **Use smaller numbers**
- **Use bar models**
- **Use x for the unknown**
- **Use a flow diagram**
- **Use arrow diagrams**
- **Use geometric sketches.**

1 A box of pens contains 12 black, 9 red, 5 green and 20 blue.
A pen is picked at random.
What is the probability of

a picking red or green

b picking blue or green

c picking red or black or green

d not picking green?

2 **R** Triangle XYZ has angles 96°, 39° and 45°.
Can you tell if triangle XYZ is congruent to triangle ABC in the diagram below?
Explain your answer.

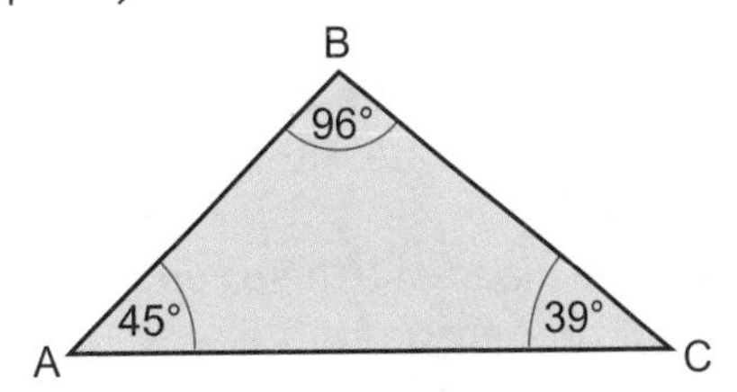

3 Fiona invests £1250 at 4.4% per annum compound interest. Ian invests £900 at 5% per annum compound interest.
After 2 years, what is the difference between the amounts in the two accounts?

4 **R** A town roundabout is being replanted.
The roundabout has a diameter of 8 m and has been divided into 6 equal sectors.

a What is the area of each sector?
Give your answer to 2 d.p.

One sector is split into three sections A, B and C, as shown in the diagram.

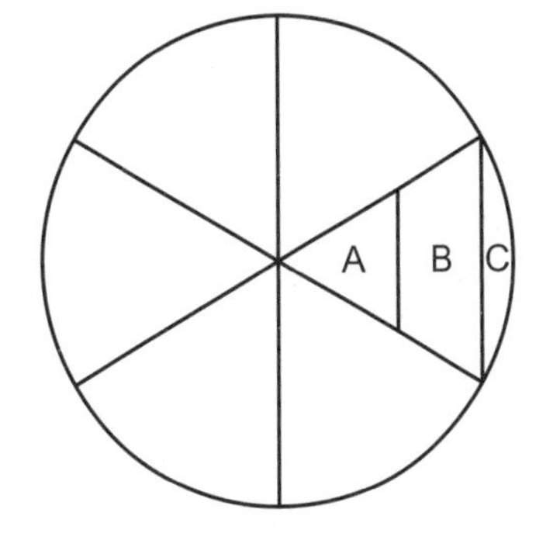

Section A extends from the centre to the midpoint of the radii and will be planted with tall grasses.
Section B will be planted with flowers.
Section C is a segment and will be covered in slate.

b What is the area of section A + section B?
Give your answer to 2 d.p.

c One bag of slate covers 1 m². If only one bag is available, will there be enough slate to cover the whole of section C?

5 Omar needs to make a triangular cover to fit over his deck. He knows two lengths of the triangle and one of the angles.
Use the sine rule to help Omar find the other two angles to 1 d.p.

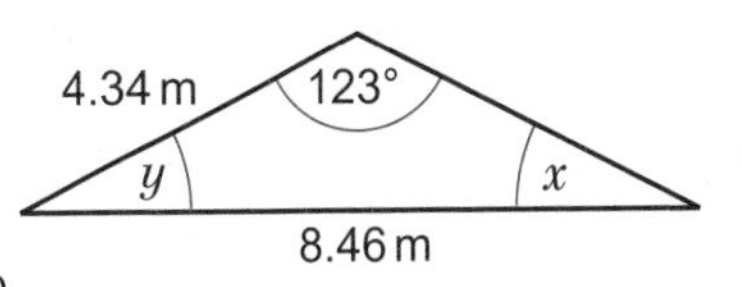

6 **R** In a woodcraft shop a crate is used to store offcut pieces of dowel. The offcuts are all different lengths but cannot stick out above the top of the crate as another crate will be stored on top.

What is the longest piece of dowel that can be placed in the crate without it sticking out over the top?

Look at the diagram to help you.

Work out your answer to 1 d.p.

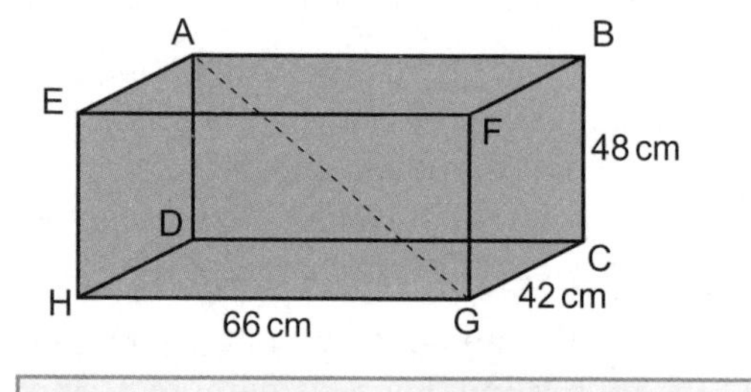

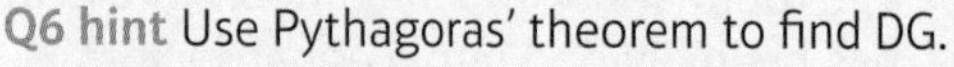
Q6 hint Use Pythagoras' theorem to find DG.

7 **Exam-style question**

Match each of the sketch graphs below to one of these equations.

$y = -x^2$

$y = \sin x$

$y = x^3 + 1$

$y = x^3 + x$

$y = x^2 - 2$

$y = \cos x$

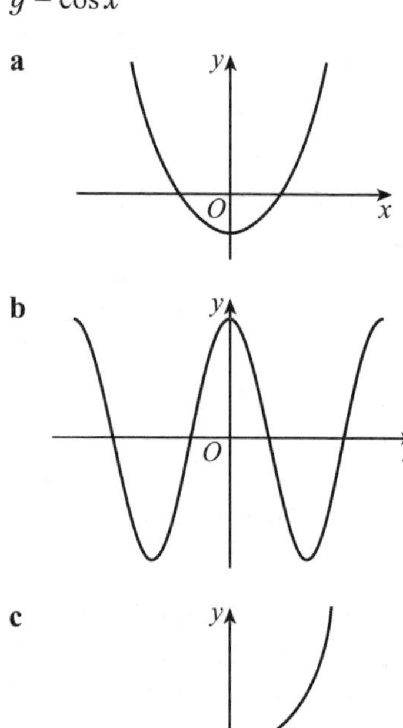

(3 marks)

8 A scientist stands on the roof of a building 100 ft above the ground, to the nearest foot. The scientist fires a zip line down from the roof at an angle of 62°, to the nearest degree.

a How long does the zip line need to be to exactly reach the ground, for the upper and for the lower bound of the angle? Give your answers to 2 d.p.

b What is the difference between the upper bound length and the lower bound length?

9 A cruise ship visits two islands (Q and R) before returning to port (P).

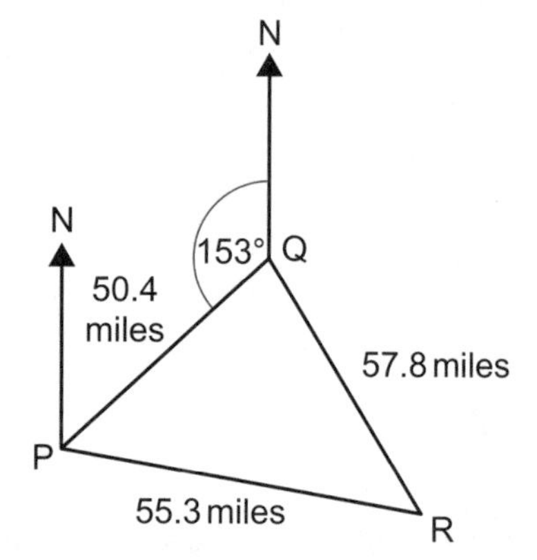

What is the bearing from island Q to island R? Give your answer to the nearest whole degree.

10 A circular stage has a special floor covering. One segment has been damaged and needs to be replaced. To compare quotes for replacing the shaded segment, the stage owners need to find its area.

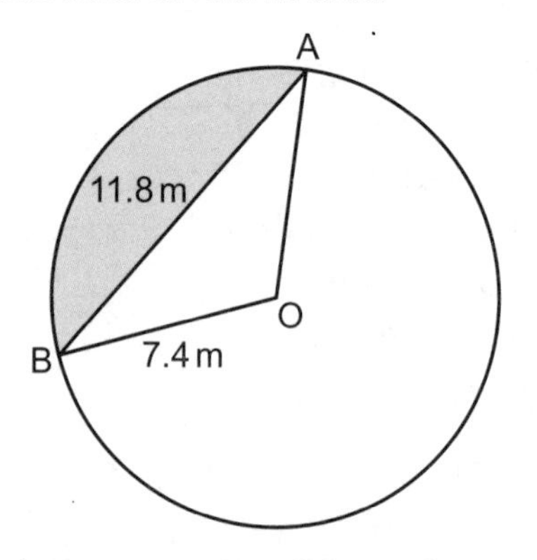

O is the centre point of the circle.

Chord AB = 11.8 m

The radius of the circle = 7.4 m

a Find the angle AOB.

b Find the area of sector AOB.

c Find the area of the damaged segment.

Give all your answers to 1 d.p.

14 FURTHER STATISTICS

14.1 Sampling

1 **R** Muhammad wants to find out about the films people watch.
He wonders whether to ask 50 people coming out of a cinema.
   a Is this sample likely to be representative of the population? Explain.
   b Hamish suggests that he pick 50 people at random from the telephone book.
   Is this sample likely to be representative of the population? Explain.

2 **R** Explain whether each of these samples is biased.
   a A supermarket wants to find out what customers think of their product range. They ask the first 30 customers through the tills on a Monday morning.
   b A sample of voters is asked what they think of the current government. They are chosen randomly from the electoral roll.
   c A science teacher asks the students in his lunchtime science club what they think of science.
   d A company wants to know whether their product is being well advertised. They carry out a street survey in 30 towns across the UK.

3 Elizabeth wants to generate 20 random numbers.
She rolls an ordinary dice twice to give her a 2-digit random number.
   a Use this method to generate 5 random numbers.
   b Will all possible 2-digit numbers be included using this method?

4 A head teacher wants to select 20 students randomly from a list of 100 students.
She uses a calculator to generate random numbers between 1 and 100.
Describe in words how she could use these numbers to select the students.

5 **R** Dr Hyden wants to collect some data on the lengths of appointments at the local health centre.
One day there are 500 patients, of whom 210 are female.
   a Dr Hyden wants a sample of 10% of the patients.
   How many patients should she ask?
   b Describe how Dr Hyden could select a simple random sample.
   c She decides to ask 25 males picked at random and 25 females picked at random.
      i What percentage of the males are selected?
      ii What percentage of the females are selected?

6 The table shows the numbers of students studying different languages.

Language	Number of students
French	120
German	140
Spanish	90
Mandarin	110
Latin	40

   a Habiba chooses a sample of 50 students. Show that this is 10% of the total number of students.
   b Work out 10% of each subject group.
   c Show that taking 10% of each subject group gives a sample of 50 students in total.

7 A library wants to find out what users think of their new lending service.
There are 870 women and 630 men registered with the library.
   a Explain why a stratified sample should be used.
   b The librarian wants to survey 10% of the users.
   How many women and how many men should be asked?
   c The librarian decides to ask 100 users.
   How many of each gender should he ask?

8 **Exam-style question**

A gym has 2000 members.

Each member has one of three types of membership: Bronze, Silver or Gold.

The table shows the number of members in each group.

Type of membership	Number of members
Bronze	450
Silver	980
Gold	570

The manager wants to carry out a survey on a stratified sample of 200 of these members.

Find the number of members from each group that should be in the sample. **(3 marks)**

Q8 hint Show all your calculations clearly. You could add columns to the table.

9 **P** In a colony of birds, 50 are caught and tagged. A month later another 50 birds are caught. Two of them are tagged birds.

a What fraction of the birds in the second sample are tagged birds?

b Assume the fraction of tagged birds in the second sample is the same as the fraction of tagged birds in the colony.
Estimate how many birds are in the colony.

10 **P** A scientists captures and tags 100 penguins from a colony of penguins. There are approximately 1200 penguins in the colony. The scientist returns a month later and captures 60 penguins.
How many would he expect to find tagged?

14.2 Cumulative frequency

1 The frequency table shows the ages of 90 customers in a spa. Copy and complete the cumulative frequency table.

Age, y (years)	Frequency	Cumulative frequency
$20 < y \leqslant 30$	6	6
$30 < y \leqslant 40$	18	6 + 18 = □
$40 < y \leqslant 50$	22	
$50 < y \leqslant 60$	34	
$60 < y \leqslant 70$	10	

2 This frequency table gives the heights of 50 seedlings.

Draw a cumulative frequency table for this data.

Height, h (cm)	Frequency
$0.5 < h \leqslant 0.7$	2
$0.7 < h \leqslant 0.9$	5
$0.9 < h \leqslant 1.1$	18
$1.1 < h \leqslant 1.3$	12
$1.3 < h \leqslant 1.5$	9
$1.5 < h \leqslant 1.7$	4

3 **Exam-style question**

This table gives the weights of 30 new-born babies.

Example

Weight, w (kg)	Frequency
$2.7 < w \leqslant 3.0$	1
$3.0 < w \leqslant 3.3$	0
$3.3 < w \leqslant 3.6$	5
$3.6 < w \leqslant 3.9$	7
$3.9 < w \leqslant 4.2$	12
$4.2 < w \leqslant 4.5$	5

a Draw a cumulative frequency diagram. **(2 marks)**

b Use the diagram to find an estimate for the median weight. **(1 mark)**

c Estimate the range. **(2 marks)**

4 a Draw a cumulative frequency diagram for the seedlings data in **Q2**.

b Find an estimate for the median height of the seedlings.

5 The times taken by 50 students to complete their maths homework are shown in the table.

Time, m (minutes)	Frequency
$10 < m \leqslant 15$	1
$15 < m \leqslant 20$	7
$20 < m \leqslant 25$	11
$25 < m \leqslant 30$	12
$30 < m \leqslant 35$	12
$35 < m \leqslant 40$	7

a Draw a cumulative frequency diagram.

b Estimate the median time taken.

c Estimate the lower quartile of the time taken.

d Estimate the upper quartile.

e Use your answers to parts **c** and **d** to work out an estimate for the interquartile range.

6 **R** The table shows the masses of 100 cakes in grams.

Mass, m (grams)	Frequency
$900 < m \leqslant 950$	2
$950 < m \leqslant 1000$	37
$1000 < m \leqslant 1050$	32
$1050 < m \leqslant 1100$	22
$1100 < m \leqslant 1150$	5
$1150 < m \leqslant 1200$	2

a Draw a cumulative frequency diagram.

b Estimate the median, quartiles and interquartile range.

c Estimate how many cakes weigh less than 1075 g.

d Copy and complete.

90 cakes are estimated to weigh less than ______ grams.

7 **Exam-style question**

The table below shows information about the heights of 60 students.

Height, x (cm)	Number of students
$140 < x \leqslant 150$	4
$150 < x \leqslant 160$	5
$160 < x \leqslant 170$	16
$170 < x \leqslant 180$	27
$180 < x \leqslant 190$	5
$190 < x \leqslant 200$	3

a Draw a cumulative frequency table. **(1 mark)**

b Draw a cumulative frequency graph. **(2 marks)**

c Use your graph to find an estimate of the median height. **(1 mark)**

d Use your graph to find an estimate for the number of students who are less than 185 cm tall. **(2 marks)**

Exam hint

For parts **c** and **d** draw lines on the graph with a ruler to show how you got your answers.

14.3 Box plots

1 Draw a box plot for this data on the ages of customers.

Minimum	LQ	Median	UQ	Maximum
19	25	33	45	61

2 **R** This data shows the lengths of time, in minutes, it took 11 people to walk 1 mile.

9, 11, 12, 12, 15, 15, 16, 16, 17, 18, 18

a Write down the median time taken.

b Find the upper and lower quartiles.

c Draw a box plot for the data.

3 **R** This stem-and-leaf diagram shows the amounts of time 25 people spent in a coffee shop.

1	8 8 9
2	0 3 3 4 5 7 8
3	5 5 7 9 9 9 9 9
4	0 2 2 5 7 8
5	2

Key: 1 | 8 represents 18 minutes

a What was the shortest length of time a customer spent in the coffee shop?

b What is the median length of time?

c Find the lower and upper quartiles of the length of time.

d Work out the interquartile range.

e Draw a box plot.

4 The results in two different maths tests are shown in the comparative box plots.

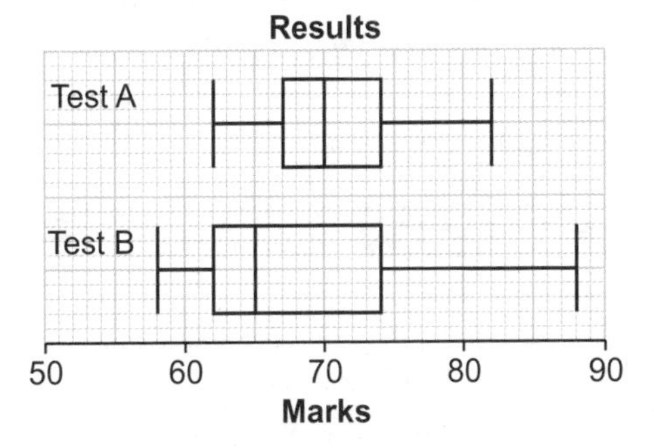

a Which test has the higher median?

b Work out the interquartile range for each test.

c Work out the range for each test.

5 **R** Summary statistics on the ages of people living in two different blocks of flats are given in this table.

	Block A	Block B
Minimum	3	17
LQ	18	35
Median	29	37
UQ	42	42
Maximum	73	91

a Draw comparative box plots for the two blocks of flats.

b Compare the blocks of flats.

Q5b hint Compare the medians, interquartile ranges and ranges.

6 **R** The cumulative frequency graph gives information about how long the batteries last in two different types of mobile phones.

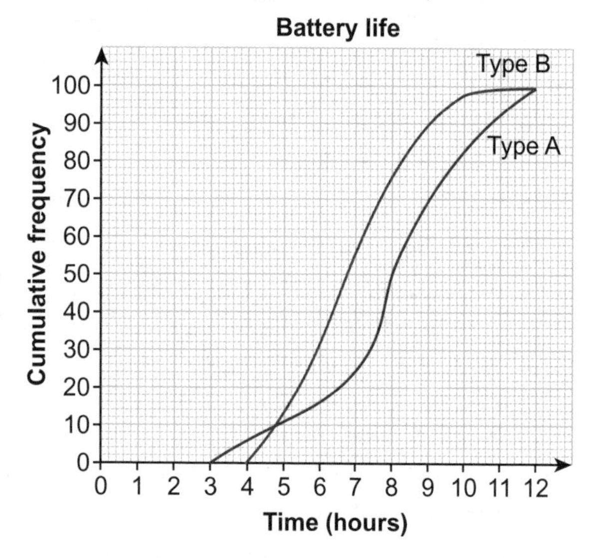

a Use the graph to find the median and quartiles for each type of phone.

b Draw comparative box plots for the two phones.

c Compare the two phones.

7 **Exam-style question**

Harry grows tomatoes.

This year he put his tomato plants into two groups, group A and group B.

Harry gave fertiliser to the tomato plants in group A.

He did not give fertiliser to the tomato plants in group B.

Harry weighed 60 tomatoes from group A.

The cumulative frequency graph shows some information about these weights.

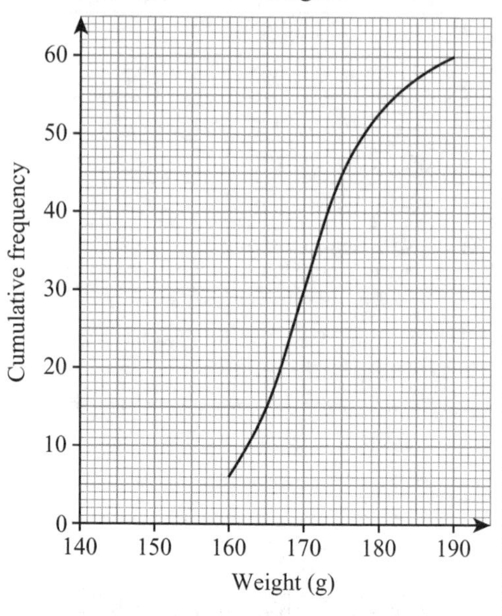

a Use the graph to find an estimate for the median weight. **(1 mark)**

The 60 tomatoes from group A had

- minimum weight of 153 grams
- maximum weight of 186 grams.

b Use this information and the cumulative frequency graph to draw a box plot for the 60 tomatoes from group A.

Use a scale like this:

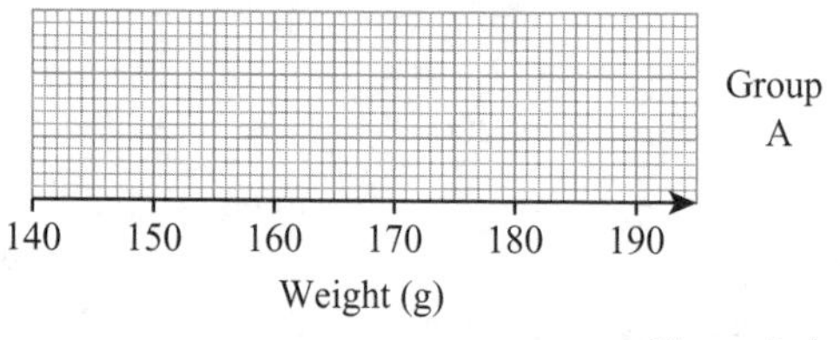

(3 marks)

Harry weighed 60 tomatoes from group B.

He drew this box plot for his results.

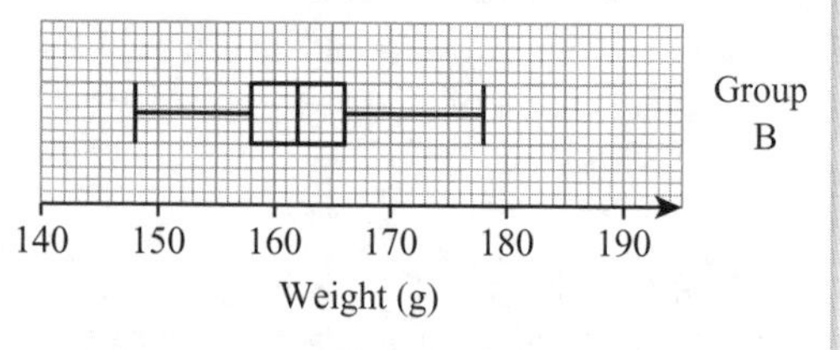

c Compare the distribution of the weights of the tomatoes from group A with the distribution of the weights of the tomatoes from group B. **(2 marks)**

June 2012, Q15, 1MA0/1H

Q7c hint Be sure to put your comparisons in the context of the question.

14.4 Drawing histograms

1 The table shows the ages of 60 patients.

Age, a (years)	Frequency	Class width	Frequency density
$0 < a \leqslant 10$	3	10	$\frac{3}{10} = 0.3$
$10 < a \leqslant 20$	14		
$20 < a \leqslant 40$	17		
$40 < a \leqslant 60$	19		
$60 < a \leqslant 80$	7		

a Work out each class width.

b Work out the frequency density for each class.

2 This table shows the lengths of time that 100 people spent watching TV one evening.

Example

Time, t (hours)	Frequency
$0 \leqslant t \leqslant 0.5$	5
$0.5 < t \leqslant 1$	35
$1 < t \leqslant 2$	56
$2 < t \leqslant 3$	4

Draw a histogram for this data.

3 This table contains data on the weights of 65 women.

Weight, w (kg)	Frequency
$40 < w \leqslant 45$	2
$45 < w \leqslant 55$	17
$55 < w \leqslant 65$	31
$65 < w \leqslant 70$	11
$70 < w \leqslant 90$	4

Draw a histogram for this data.

4 **Exam-style question**

Mrs Morris records the results in the Year 11 maths exam.

Mark, m (%)	Frequency
$0 < m \leqslant 40$	2
$40 < m \leqslant 50$	4
$50 < m \leqslant 60$	12
$60 < m \leqslant 80$	63
$80 < m \leqslant 100$	9

a Work out an estimate for the mean mark of Year 11. **(4 marks)**

b Draw a histogram for the information given in the table. **(3 marks)**

Q4 hint Use the midpoint of each class interval.

14.5 Interpreting histograms

1 **R** The histogram shows the heights of a sample of students.

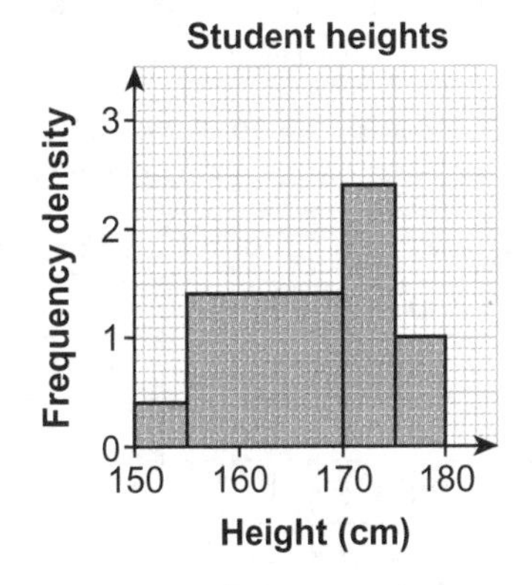

a How many students were between 150 and 155 cm tall?

b How many students were between 170 and 180 cm tall?

c How many students were measured in total?

2 **R** The histogram shows the distances an audience at a theatre had travelled to the production.

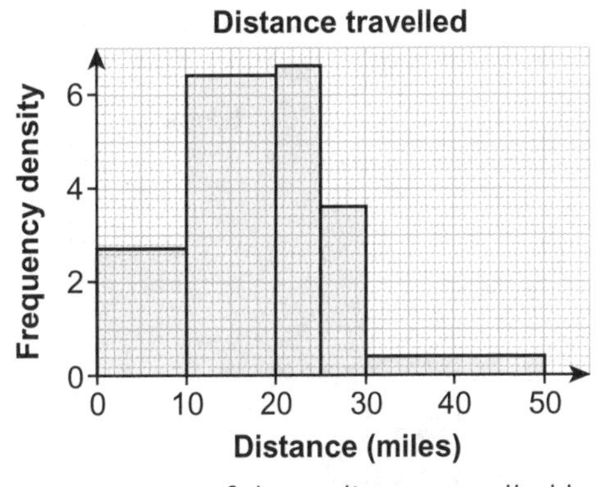

a How many of the audience travelled less than 10 miles?

b Estimate how many of the audience travelled less than 15 miles.

c Estimate how many people travelled between 15 and 40 miles.

3 **Exam-style question**

The incomplete table and histogram give some information about the distances people travel to work.

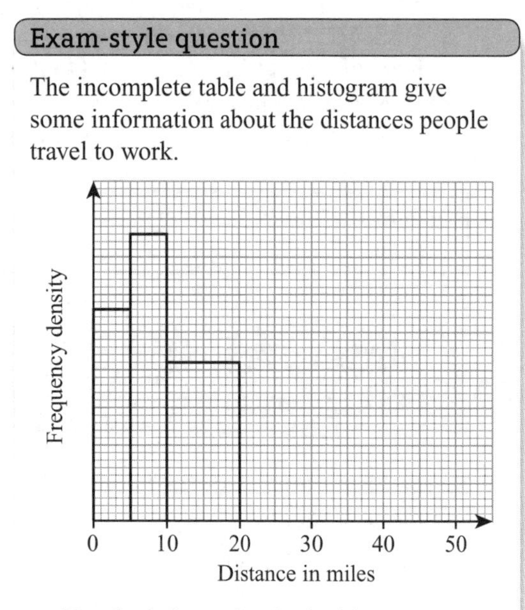

a Use the information in the histogram to complete the frequency table below.

Distance (d) in miles	Frequency
$0 < d \leqslant 5$	140
$5 < d \leqslant 10$	
$10 < d \leqslant 20$	
$20 < d \leqslant 35$	120
$35 < d \leqslant 50$	30

(2 marks)

b Complete the histogram. **(2 marks)**

Exam hint
Draw the bars on the histogram neatly with a ruler. Show your working to calculate the frequencies.

4 **P** The histogram shows the times taken by a number of athletes to run 100 m.

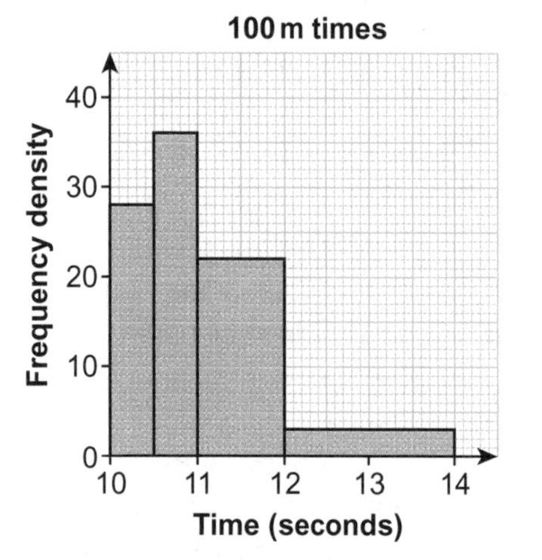

a Draw a grouped frequency table for the data.

b Work out an estimate for the mean time taken.

c How many athletes took longer than 11.5 seconds to run 100 m?

5 Work out an estimate for the median of the data in **Q2**.

6 **R** The histogram shows the weights of all the babies born in one day in a maternity hospital.

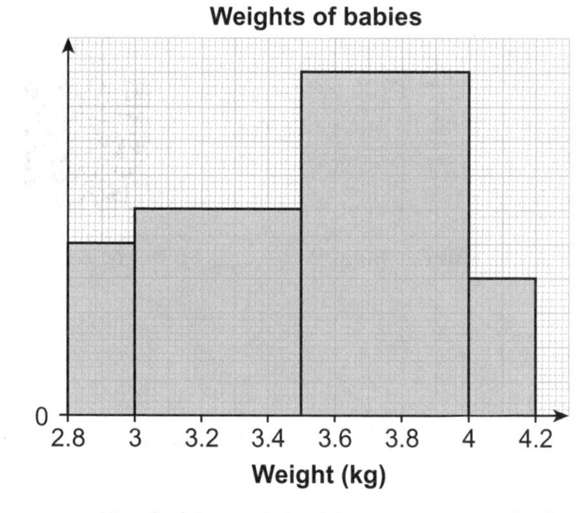

a Five babies weighed between 2.8 and 3 kg. Work out the frequency density for that class.

b How many babies were born in total?

c Work out an estimate for the median weight.

d Draw a frequency diagram for the data in the histogram.

e Work out an estimate for the mean weight from your frequency table.

f How many of the babies were heavier than the mean weight?

Q6b hint Label the frequency density scale.

7 **P** The histogram shows the heights of some sunflowers.

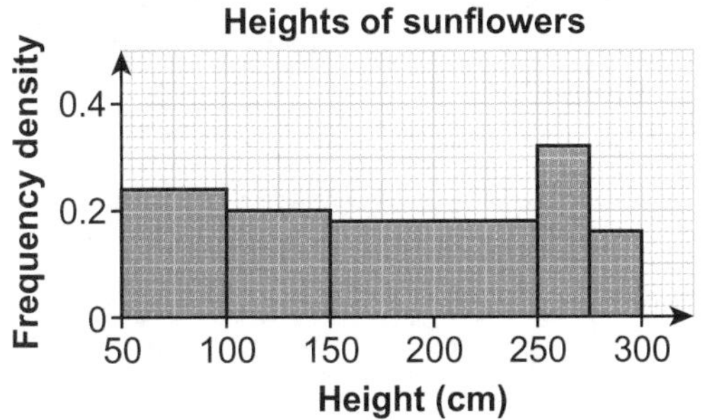

a How many sunflowers are there in total?

b Work out an estimate of the median height.

c Estimate how many sunflowers are taller than 110 cm.

14.6 Comparing and describing populations

1 Every day one week, a passenger group recorded how late two trains were, in minutes.

Train A	3	7	5	6	8
Train B	1	9	2	8	3

Compare the lengths of time the trains were late by.

2 The ages, in years, of the first five clients at two different physiotherapists are recorded.

Back Health	54	49	39	44	63	71	65
Joints Ease	33	39	45	57	62	21	25

a Write down the median age of the clients for each physiotherapist.

b Work out the interquartile range of the age of the clients for each physiotherapist.

c Compare the ages of the clients at the two physiotherapists.

3 The lengths of time, in minutes, students spent playing computer games one day were recorded.

17, 19, 23, 25, 28, 31, 34, 38, 40, 41, 120

a Work out the mean length of time.

b Work out the median length of time.

c Work out the range and interquartile range.

4 **R** Ten male and ten female swimmers compete in a race.

The times, in minutes, to complete the swim are recorded.

Males: 32, 34, 35, 37, 39, 41, 42, 44, 49, 63

Females: 30, 41, 43, 44, 47, 49, 50, 52, 52, 53

a Explain which of the median and interquartile range or mean and range should be used to compare the data.

b Compare the times for males and females.

5 **P** The table shows how far employees at Gatwick Airport travel to work.

Length of journey, x (miles)	Frequency
$0 < x \leqslant 10$	1
$10 < x \leqslant 20$	16
$20 < x \leqslant 30$	25
$30 < x \leqslant 40$	15
$40 < x \leqslant 50$	3

a Draw a cumulative frequency diagram.

b Find the median and interquartile range.

The box plot shows far employees at Heathrow Airport travel to work.

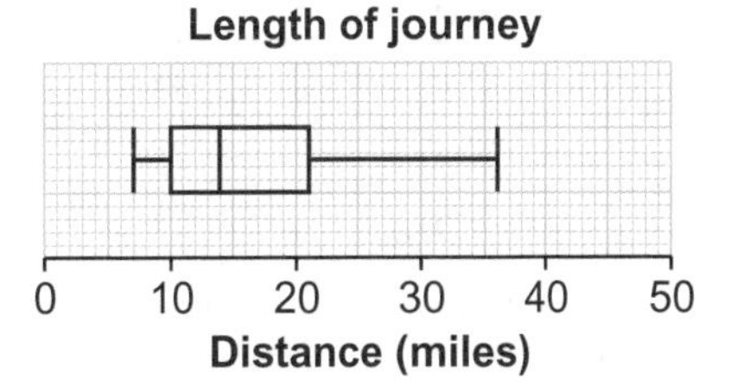

c Compare the lengths of the employees' journeys to work at the two airports.

> **Q5c hint** Compare the medians and interquartile ranges.

6 **P** This back-to-back stem-and-leaf diagram shows the weights of 20 members at two different slimming clubs.

Weight Loss World		Slimmers' Hut
7 5	7	0 1 1 3 5
9 7 6 6 3	8	0 5 5 7 9 9
9 8 6 6 5 5	9	1 1 6 8 9 9 9
9 9 6 6 2 0 0	10	2 4

Key: 5 | 7 represents 75 kg

Key: 7 | 0 represents 70 kg

Compare the weights of the members at the two different clubs.

7 **P** The cumulative frequency graph shows the ages of children at two different nurseries.

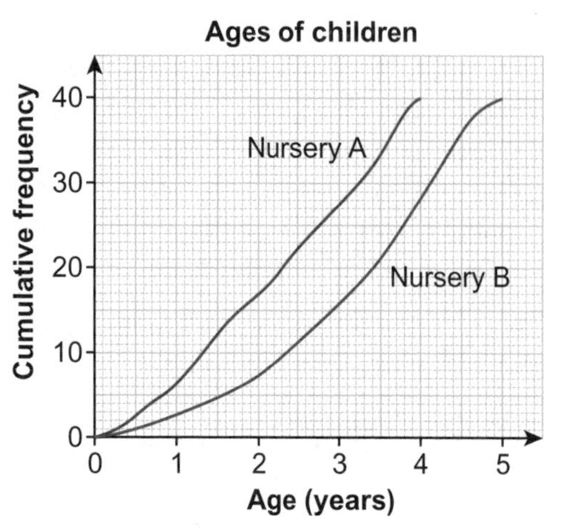

Compare the two sets of data.

8 **Exam-style question**

A survey was carried out on the speeds at which cars travelled past a primary school.

Speed, m (miles per hour)	Cumulative frequency
$0 < m \leqslant 10$	15
$10 < m \leqslant 20$	74
$20 < m \leqslant 30$	45
$30 < m \leqslant 40$	18
$40 < m \leqslant 50$	3

A similar survey outside another school gave a median of 27 and an interquartile range of 15.
Compare the speeds at which cars travelled outside the two schools. **(2 marks)**

Q8 hint Draw a cumulative frequency diagram for the results given in the table.

14 Problem-solving

Solve problems using these strategies where appropriate:

- **Use pictures or lists**
- **Use smaller numbers**
- **Use bar models**
- **Use x for the unknown**
- **Use a flow diagram**
- **Use arrow diagrams**
- **Use geometric sketches.**

1 Gary conducted a survey asking people of different ages if they shopped more often at Shop A or Shop B. This back-to-back stem-and-leaf diagram shows the results.

Shop A		Shop B
9	1	8 8 9 9
8 7 5	2	2 4 4 6 7
7 6 5 3 3 2	3	1 2 2 5
9 6 6 5 1 0 0	4	0
8 8 8 6 3 2	5	1

Key: 9 | 1 represents 19 years

Key: 1 | 8 represents 18 years

Compare the data by finding the median ages and interquartile ranges.

2 Plane A took 11 hours to travel 6022 miles. Plane B travelled at the same average speed on a journey of 3904 miles. Round your answers to the questions to the nearest whole number.

a What was the average speed of the planes?

b About how long did Plane B's journey take?

3 **R** Judy is conducting a survey of her company's employees. There are 250 employees. Judy will survey a random sample of 10 employees. She uses a random number table to work out whom to ask.

9845679256938259793198984293232456...

a Explain one way Judy can generate 10 random numbers.

b Would you consider Judy's survey a good-sized sample? Explain.

4 **R** Ian buys two similar pictures.
The larger picture has a height of 23 cm.
The smaller picture has a length of 7.5 cm and a height of 5.75 cm.
What is the length of the larger picture?

5 The frequency table shows the test scores for Year 9 students in a recent science test.

Test scores, t (%)	Frequency
$0 < t \leqslant 25$	13
$25 < t \leqslant 50$	18
$50 < t \leqslant 75$	34
$75 < t \leqslant 100$	20

To pass the test, students need to score 60% or higher. Draw a cumulative frequency diagram and estimate how many students passed the test.

Q5 hint Use the data to draw a cumulative frequency table.

6 **Exam-style question**

Chilli plants are watered with water or with a water and fertiliser mixture.
The box plot shows data about their heights.

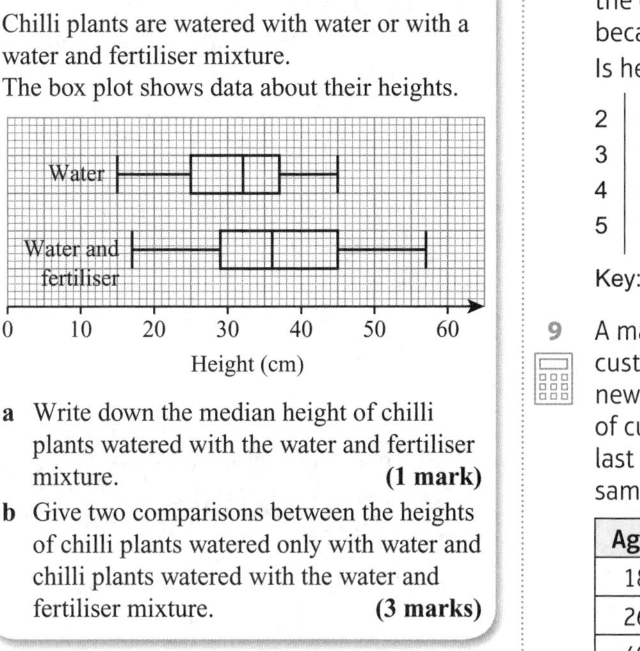

a Write down the median height of chilli plants watered with the water and fertiliser mixture. **(1 mark)**

b Give two comparisons between the heights of chilli plants watered only with water and chilli plants watered with the water and fertiliser mixture. **(3 marks)**

7 This table shows summary statistics from a data set of the heights of sunflowers.

Minimum	LQ	Median	UQ	Maximum
6 ft	7 ft	8 ft	10 ft	12 ft

a Draw a box plot for the data.

b Suppose some more sunflowers were measured and the lower quartile changed to 5 ft.
Describe how your box plot would change.
Describe what would stay the same.

8 **R** Mark says the interquartile range for the data in this stem-and-leaf diagram is 8 because 12 − 4 = 8.
Is he correct? Explain how you know.

2	9 9
3	1 3 3 5 7 9
4	0 2 4 6
5	0 5 6

Key: 2 | 9 represents 29

9 A market researcher wants to find out what customers of different ages think about a new product. The table shows the numbers of customers who bought the product in the last month. She wants to look at a stratified sample of 20% of the customers.

Age (years), a	Number of customers
$18 < a \leqslant 25$	160
$26 < a \leqslant 39$	125
$40 < a \leqslant 60$	85
$61 < a$	70

a How many customers will be surveyed in total?

b How many more customers in the $18 < a \leqslant 25$ age range will be surveyed than in the $40 < a \leqslant 60$ age range?

c Explain why a company might want to use a stratified sample like this.

10 Solve the equation $4 \sin x = 3$ for $0° < x < 720°$.
Give your answers correct to 1 d.p.

15 EQUATIONS AND GRAPHS

15.1 Solving simultaneous equations graphically

1 **R** a Match the equations to the three lines A, B and C shown opposite.

i $x + y = 4$ ii $y = x + 2$ iii $3y + x = 6$

b Hence write down the solutions to these pairs of simultaneous equations.

i $x + y = 4$
$y = x + 2$

ii $x + y = 4$
$3y + x = 6$

iii $3y + x = 6$
$y = x + 2$

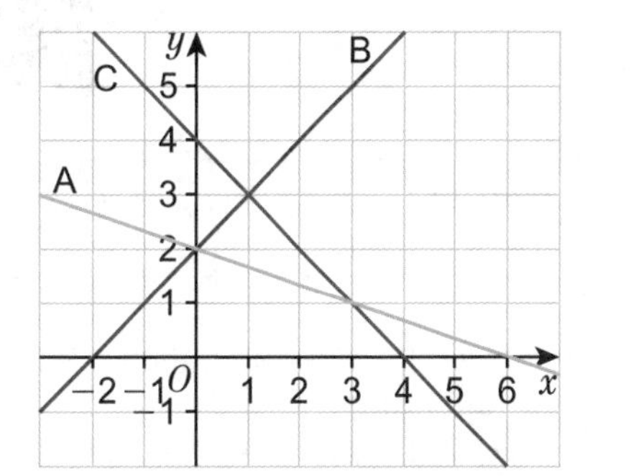

2 Solve the pairs of simultaneous equations by drawing the graphs.

a $y = 4x - 1$
$y = -4x + 1$

b $5y = x - 8$
$x + 5y = 12$

c $0 = 2y - x + 4$
$4y = 13 - x$

d $3x - 2y = -11$
$x + y = -4$

e Show that solving the equations algebraically gives the same solutions to the equations in part **d**.

3 Tim buys 3 rollerball pens and 5 biros for £9.75. In the same shop, Jenny buys 2 rollerball pens and 2 biros for £5.50.
Write a pair of simultaneous equations and solve them graphically to find the cost of

a one rollerball pen

b one biro.

4 **R** Two mobile phone companies offer the following prices.

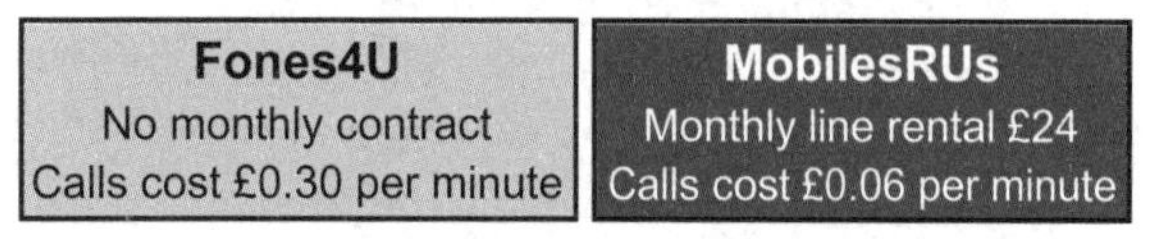

Fones4U	MobilesRUs
No monthly contract	Monthly line rental £24
Calls cost £0.30 per minute	Calls cost £0.06 per minute

a For each company form an equation to calculate the monthly cost, with y = total monthly cost and x = minutes of calls.

b Use a graphical method to work out how many minutes of calls are used if the two companies charge the same.

5 Use a graphical method to find an approximate solution to the pair of simultaneous equations

$y + x^2 = 1$

$y + 2 = x$

Example

Q5 hint Start by rearranging the equations, then plot the graphs.

6 **R** a Solve this pair of simultaneous equations

i graphically

ii algebraically to 2 decimal places.

$4x - 3y = 7$

$2x^2 - y = 3$

b Which method gives the more accurate solution? Explain.

7 a On a suitable grid draw the graph of $x^2 + y^2 = 36$

b On the same grid draw the graph of $y = x + 6$

c What are the coordinates of the points at which the graphs intersect?

8 Use a graphical method to find an estimate for the solution to the simultaneous equations

$x^2 + y^2 = 16$

$x - y = 3$

15.2 Representing inequalities graphically

1 a Write down the inequalities represented by the shaded regions.

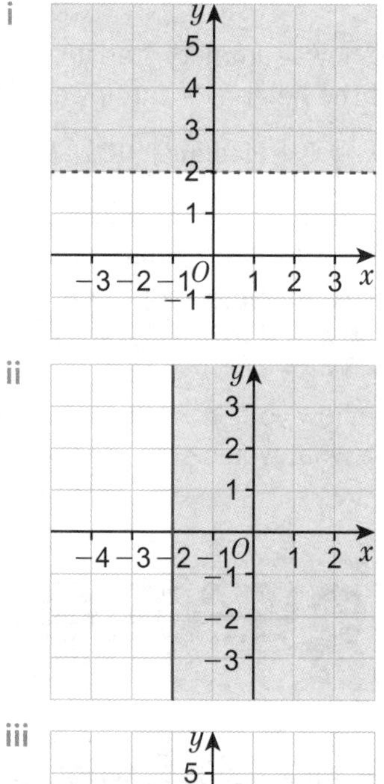

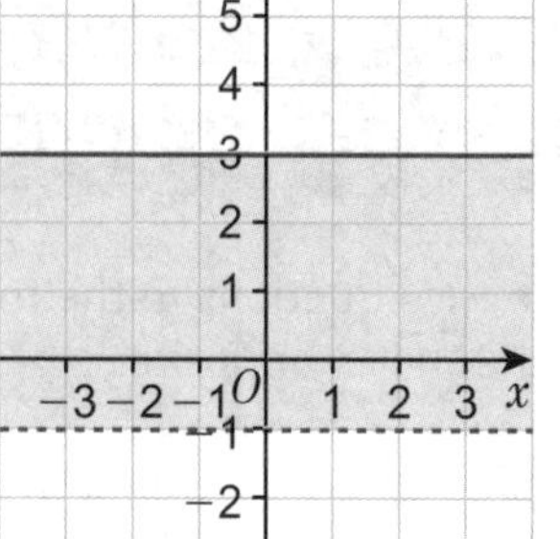

b On a suitable coordinate grid, shade the region of points whose coordinates satisfy

i $x \geqslant -1$ ii $y < 4$
iii $0 > y$ iv $y + 1 < x \leqslant 2$
v $-1 < y$ and $y > 1$ vi $-1.5 \leqslant x \leqslant 3.5$
vii $-4.5 < y \leqslant -2$

2 a Draw a coordinate grid with −4 to 4 on both axes.

b Draw the graph of $y = 3x - 1$

c Does the point (4, 2) satisfy the inequality $y \leqslant 3x - 1$?

d Shade the region of points that satisfy $y \leqslant 3x - 1$

> **Q2c hint** At the point (4, 2), $y = \square$ and $3x - 1 = \square$

3 Draw a coordinate grid with −5 to 5 on both axes.
Shade the regions that satisfy

a $y < x$ b $y \leqslant 2x - 3$
c $y > -\frac{1}{2}x + 2$ d $y \geqslant 4 - 3x$

4 **R** x and y are integers.
On a coordinate grid with −5 to 5 on both axes, mark on all the points with integer coordinates which satisfy all three inequalities:

$y + x \leqslant 3$ $x > -2$ $y \geqslant 1$

Example

5 **Exam-style question**

The lines $y = x + 3$ and $x + y = 7$ are drawn on the grid.

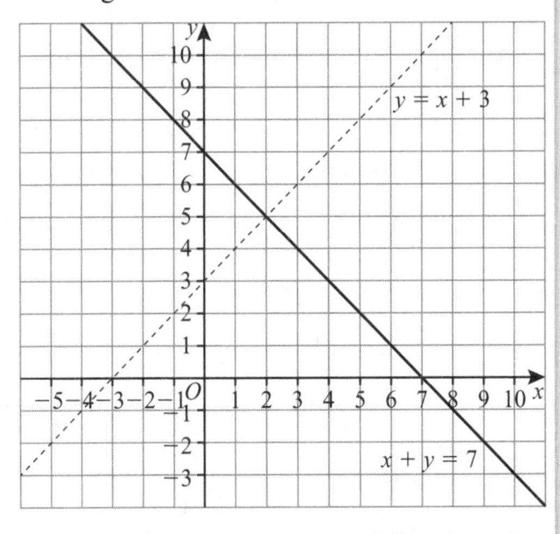

On the grid, mark with a cross (X) each of the points with integer coordinates that are in the region defined by $y < x + 3$, $x + y \leqslant 7$ and $y > 2$ **(3 marks)**

6 **R** The diagrams show a shaded region bounded by three lines.
For each diagram

i write down the equations of the lines

ii write down the three inequalities satisfied by the coordinates of the points in this region.

a

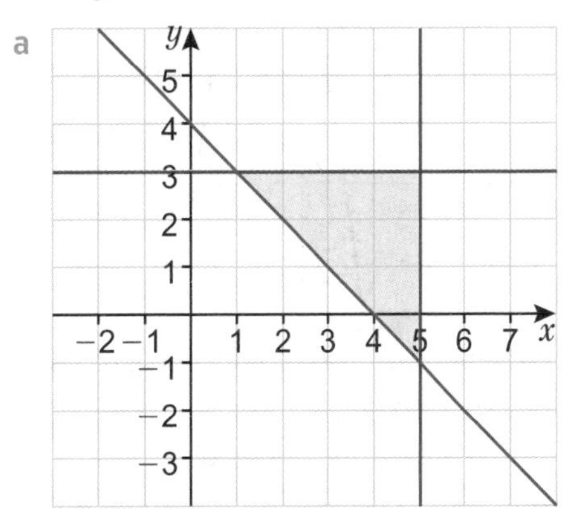

b

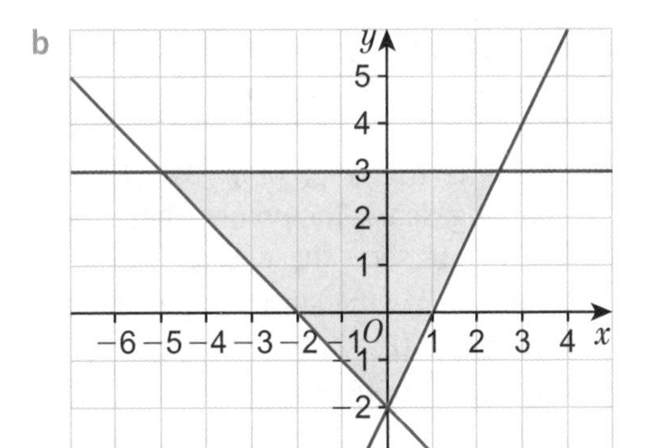

c

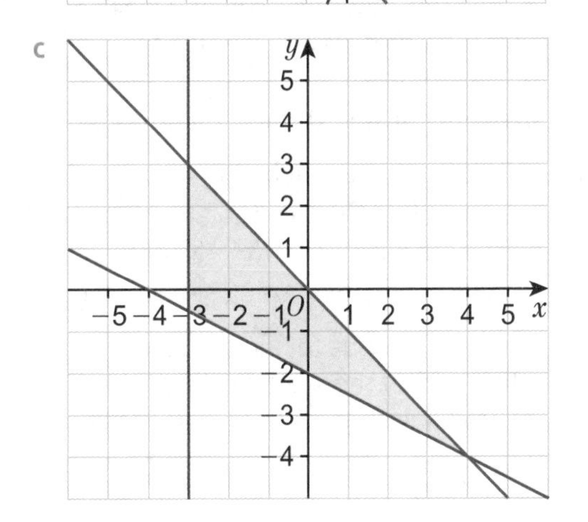

7 **P** How many points with integer coordinates satisfy these inequalities?

$y + 2x > 4$ $y > 2x - 1$ $x > -1$

8 a Draw the graph of $y = 4 - x^2$ for values of x from −4 to +4.

b Draw the line $y = x + 1$ on the same axes.

c Shade the region that satisfies $y < 4 - x^2$ and $y > x + 1$

9 This is the graph of $y = 5 + 3x - 2x^2$

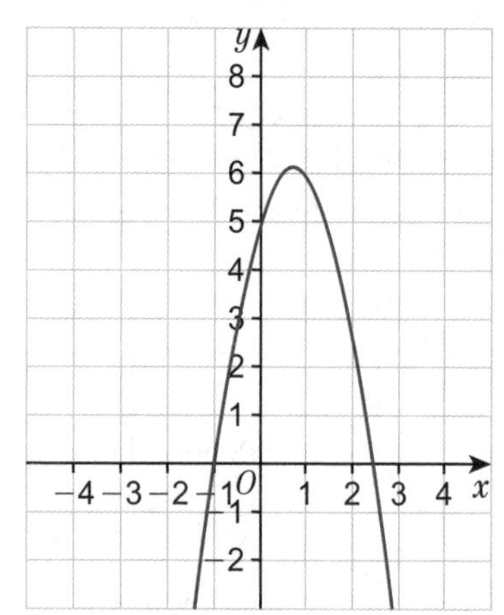

a For what integer values of x is the graph below the x-axis?

b For what integer values of x is $5 + 3x - 2x^2 < 0$?

Q9 hint Write your answer as two inequalities: $x < \square$ and $x > \square$

10 **R** a Sketch the graph of $y = 12 - 2x - 2x^2$, marking clearly the points where the graph intersects the x-axis.

b From the graph identify the values of x for which $0 \geqslant 12 - 2x - 2x^2$

Give your answer using set notation.

11 **P** By sketching the graph of $y = 3x^2 + 6x - 9$, find the values of x which satisfy $0 \geqslant 3x^2 + 6x - 9$

Give your answer using set notation.

12 **R** a Sketch the graph of $y = x^2 + x - 12$

b Hence find the values of x which satisfy the inequality $12 \geqslant x^2 + x$

Give your answer using set notation.

13 **R** Sketch graphs to find the values of x which satisfy these inequalities.

Give your answers using set notation.

a $x^2 + 3x > 4$ b $x + 14 \leqslant 3x^2$ c $x^2 < 25$

15.3 Graphs of quadratic functions

1 Look at the graph of $y = x^2 - 6x + 8$

a Use the graph to find the roots.

b Where does the graph intersect the y-axis?

c Is the turning point a maximum or a minimum?

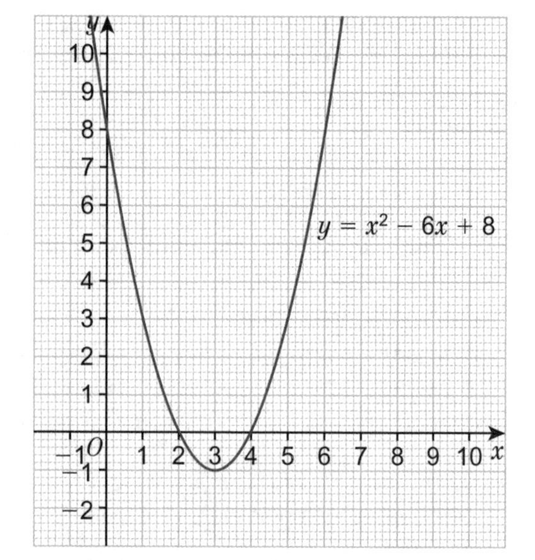

d What are the coordinates of the turning point?

2 a Plot the graph of $y = x^2 + 4x - 21$

b Use your graph to work out the roots of the equation $y = x^2 + 4x - 21$

c Where does the graph intersect the y-axis?

d Does the graph have a maximum or minimum point?

e What are the coordinates of the turning point?

3 a Solve the equations

i $0 = x^2 + 2x - 8$

ii $0 = x^2 - 7x + 6$

b Find the value of y when $x = 0$ for the equations

i $y = x^2 + 2x - 8$

ii $y = x^2 - 7x + 6$

4 **R** Match the graphs to their equations, explaining your reasoning.

a $y = x^2 + x - 20$ b $y = x^2 - x - 20$

c $y = -x^2 + 8x - 15$ d $y = -2x^2 + 13x - 15$

i

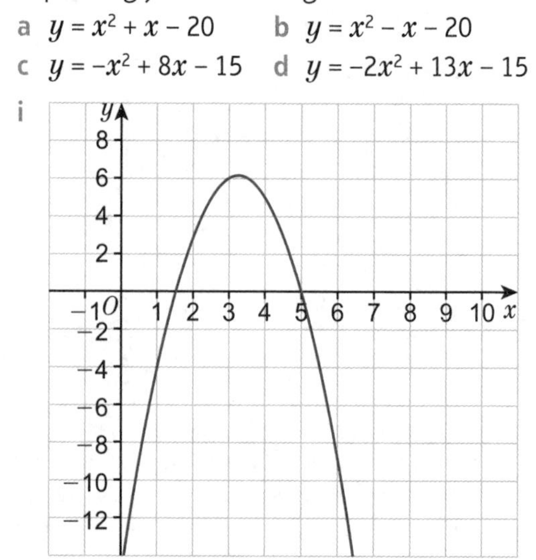

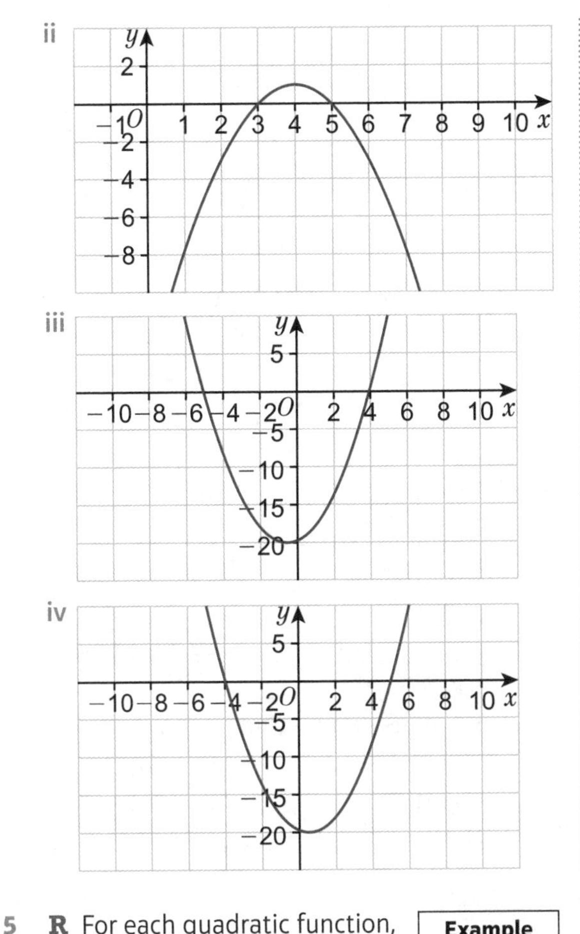

5 **R** For each quadratic function, work out the coordinates of the turning point and state whether it is a maximum or a minimum.

a $y = x^2 - 4x + 7$
b $y = -x^2 - 2x - 5$
c $y = x^2 - 12x + 19$
d $y = 2x^2 + 16x + 40$
e $y = 3x^2 - 18x + 33$
f $y = -4x^2 - 2x + 5$

6 a Factorise the expression $x^2 + 2x - 35$
b Hence write down the coordinates of the roots of $y = x^2 + 2x - 35$
c Where does the graph of $y = x^2 + 2x - 35$ cross the y-axis?
d Write $x^2 + 2x - 35$ in completed square form.
e Hence write down the coordinates of the turning point of $y = x^2 + 2x - 35$
f Is the turning point a maximum or a minimum?
Explain your answer.
g Using your answers to parts **a** to **f**, sketch the graph of $y = x^2 + 2x - 35$

Q6b hint Solve $x^2 + 2x - 35 = 0$

7 Use the method in **Q6** to sketch these graphs.
a $y = x^2 + x - 56$
b $y = -x^2 - 5x + 36$
c $y = -3x^2 - 6x + 9$
d $y = 4x^2 - 9$

8 **Exam-style question**

a Solve the equation $-x^2 + 6x - 5 = 0$ **(2 marks)**

b On a suitable coordinate grid, sketch the graph of $y = -x^2 + 6x - 5$, marking clearly the coordinates of the points of intersection with the axes and the coordinates of the turning point. **(4 marks)**

9 a Write down the coordinates of the turning point of the graph of $y = (x - 2)^2 - 7$
b Substitute $y = 0$ into the equation $y = (x - 2)^2 - 7$ and hence find the coordinates of the roots, giving your answers in surd form.

10 Find the roots of these equations given in completed square form, giving your answers in surd form where appropriate.
a $y = (x - 1)^2 - 4$
b $y = 2(x + 3)^2 - 32$
c $y = 5(x - 2)^2 - 15$

11 By writing the equations in completed square form, calculate the roots of the equations.
Give your answers in surd form.
a $y = x^2 + 6x - 7$
b $y = 2x^2 - 12x + 4$
c $y = -3x^2 - 8x + 1$

12 **R** Give three reasons why the graph shown is *not* $y = -3x^2 + 12x + 36$

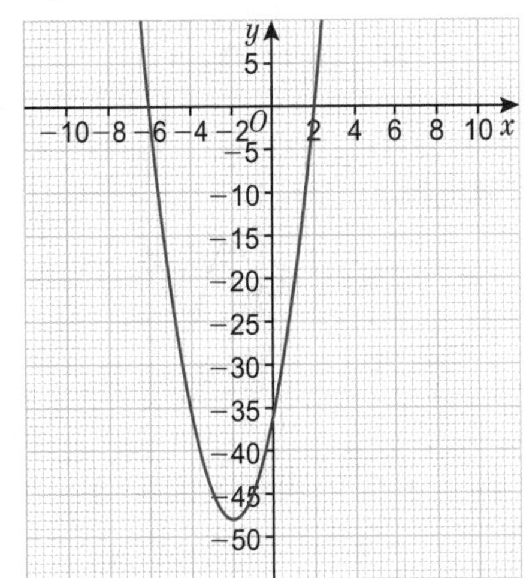

13 **P** Find the equation of this graph.

(4, 6)

(0, −10)

15.4 Solving quadratic equations graphically

1 **R** Match each graph to its equation.
Hence estimate the roots of the equation.

a $y = x^2 - 5x + 2$ b $y = 2x^2 - 2x - 3$
c $y = -x^2 - 3x + 1$ d $y = -3x^2 + 5x + 3$

i

ii

iii

iv

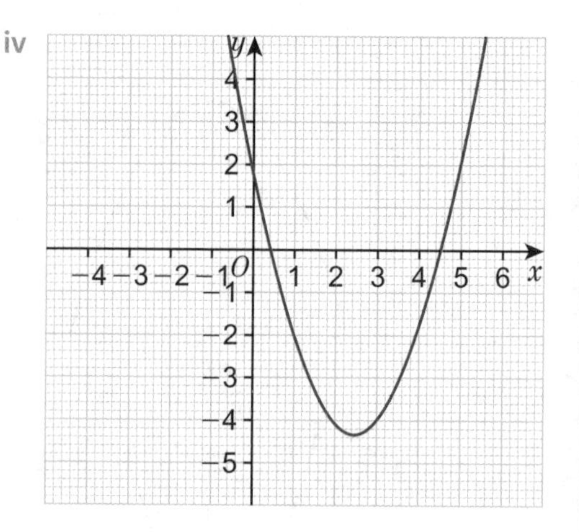

2 a Copy and complete the table of values for $y = 2x^2 + 5x - 4$

x	-2	-1	0	1	2	3
y		-7				

b Plot the graph of $y = 2x^2 + 5x - 4$ on a suitable grid.

c From the graph estimate the roots of the equation $y = 2x^2 + 5x - 4$

3 a Plot the graphs of the following functions. Hence estimate the roots.

i $y = x^2 + 3x - 6$ ii $y = -x^2 + 4x + 3$
iii $y = 3x^2 + 2x - 4$ iv $y = -2x^2 - 5x + 5$

b Use the quadratic formula to find the roots of the equations in part **a** to 3 significant figures.
Check your answers to part **a**.

4 **Exam-style question**

a Complete the table of values for $y = -2x^2 + 9x - 3$

x	−1	0	1	2	3	4	5
y							

(2 marks)

b Plot the graph of $y = -2x^2 + 9x - 3$ **(2 marks)**

c Use your graph to estimate the roots of the equation $y = -2x^2 + 9x - 3$ **(3 marks)**

d Write the expression $-2x^2 + 9x - 3$ in the form $a(x + b)^2 + c$. **(2 marks)**

5 **R** For each function

i find the coordinates of the turning point
ii find the y-intercept
iii sketch the graph.

a $y = x^2 + 3x + 4$
b $y = -x^2 - 5x - 5$
c $y = -2x^2 + 5x - 6$

6 **R** Dan is sketching the graph of
$y = -2(x + 4)^2 - 1$
He is finding it difficult to identify the roots of the equation. Explain why.

7 **R** By completing the square, decide whether these quadratic equations have
- no roots
- two roots
- one repeated root.

a $y = x^2 + 7x + 12$ b $y = x^2 + 5x - 4$
c $y = x^2 - 8x - 13$ d $y = 2x^2 + 8x + 8$
e $y = -x^2 + 3x - 3$ f $y = 4x^2 - 11x + 7$
g $y = -3x^2 - 10x - 9$ h $y = -9x^2 - 6x - 1$

8 **Exam-style question**

a By completing the square, find the roots of the equation $y = x^2 - 9x - 5$, giving your answers in surd form. **(3 marks)**

b Show algebraically that $y = x^2 + 3x + 8$ has no real roots. **(3 marks)**

9 Write an equation for each graph.

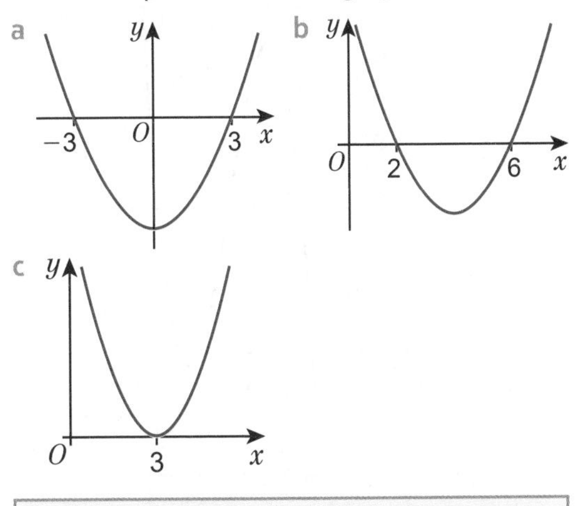

Q9 hint Write $y = (x - a)(x - b)$ and expand.

10 Use the iterative formula and the starting point given to find one root for each quadratic equation.
Give your answers correct to 5 decimal places.

a $y = x^2 - 7x - 5$ $x = \sqrt{5 + 7x}$ $x_0 = 7$
b $y = x^2 - 2x - 6$ $x = \sqrt{2x + 6}$ $x_0 = 3.5$
c $y = x^2 - x - 4$ $x = \frac{4}{x} + 1$ $x_0 = 2.5$

11 a Solve the quadratic equation $x^2 + 2x - 3 = 0$
b Sketch the graph of $y = x^2 + 2x - 3$
c Write the set of values of x that satisfy $x^2 + 2x - 3 > 0$ (where the curve is above the x-axis).
d Write the set of values of x that satisfy $x^2 + 2x - 3 < 0$ (where the curve is below the x-axis).

12 Find the set of values that satisfy each inequality.
a $x^2 + 3x - 4 > 0$
b $x^2 + 2x - 15 < 0$
c $x^2 + 6x + 8 < 0$
d $x^2 + 12x + 35 > 0$
e $x^2 - 5x + 6 > 0$
f $x^2 - 7x + 6 < 0$

15.5 Graphs of cubic functions

1 Expand the expression $(x + 1)(x^2 + 2x + 3)$

2 Copy and complete to expand the expression
$(x + 1)(x + 3)(x - 2) = (x^2 + \square x + \square)(x - 2)$
$= x^3 + \square x^2 - \square x - \square$

3 Expand the expressions
a $(x^2 + 2x + 1)(x + 1)$ b $(x + 1)(x + 2)(x + 3)$
c $(x + 2)(x + 4)(x + 2)$ d $(x - 2)(x + 5)(x - 4)$
e $(x + 4)(x - 1)(x - 5)$ f $x(x - 6)(x - 3)$
g $(x + 2)^2(x - 2)$ h $(x - 3)^3$

4 Here is the graph of $y = x^3 + 3x^2 - 13x - 15$

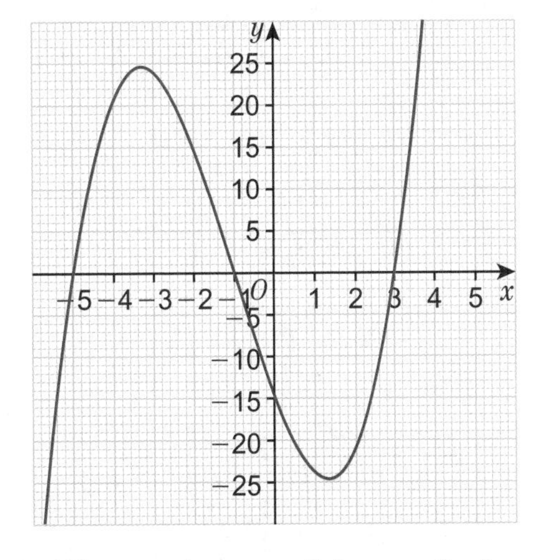

a What are the roots of the equation?
b Where does the graph cross the y-axis?

5 a What are the roots of the equation $y = (x + 2)(x + 4)(x - 6)$?

b Where does the graph of $y = (x + 2)(x + 4)(x - 6)$ cross the y-axis?

c Sketch the graph of $y = (x + 2)(x + 4)(x - 6)$

Example

6 **R** Match each graph to its equation.

a $y = (x - 1)(x + 2)(x - 3)$

b $y = (x + 2)(x - 3)^2$

c $y = (x - 3)(x + 3)(x - 4)$

d $y = -x(x - 2)^2$

e $y = (x + 1)(2 - x)(x + 5)$

f $y = -x^2(x - 2)$

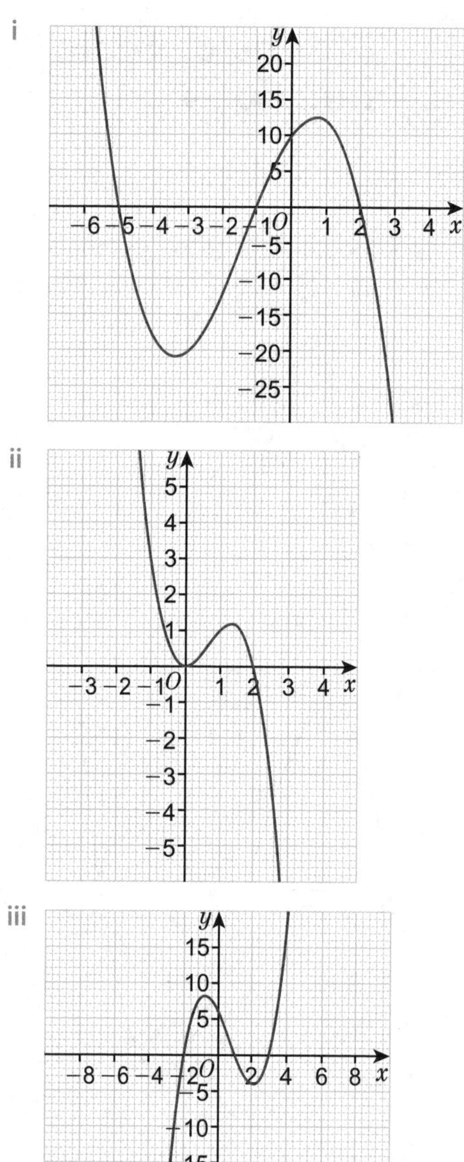

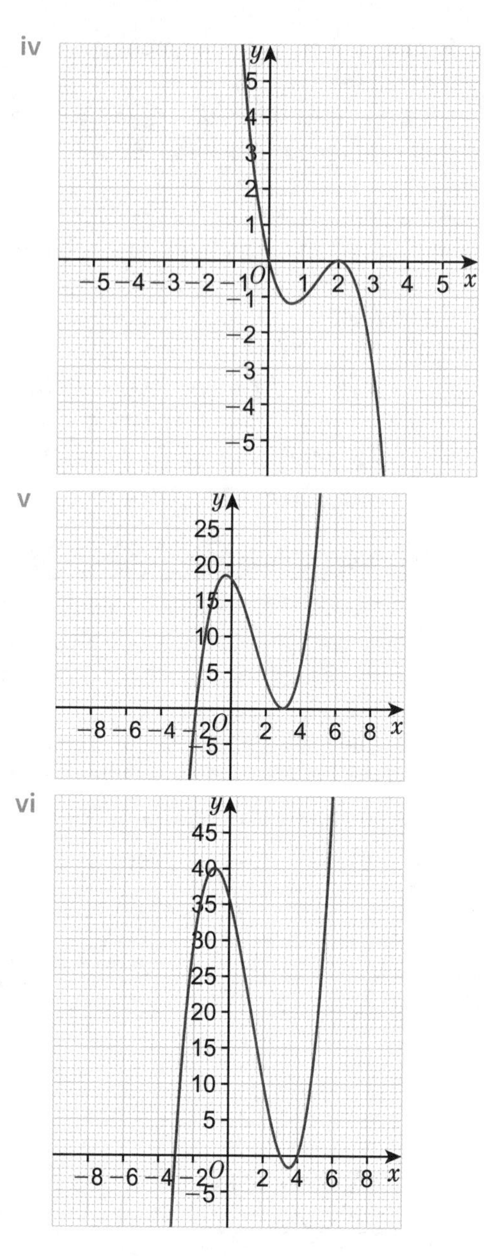

7 How many solutions does each of these cubic equations have?

a $y = (x - 2)(x + 7)(x - 8)$

b $y = (x - 1)^3$

c $y = -x(x - 2)(x + 5)$

d $y = x(x - 4)^2$

e $y = (x^2 - 3x + 4)(x + 5)$

f $y = (x + 6)(x - 2)(3 - x)$

8 Sketch the graphs, marking clearly the points of intersection with the x- and y-axes.

a $y = (x - 4)(x - 2)(x + 1)$

b $y = x(x - 1)^2$

c $y = (2 - x)(x + 2)(x - 5)$

d $y = (x - 3)^2(x + 4)$

e $y = (x - 4)^3$

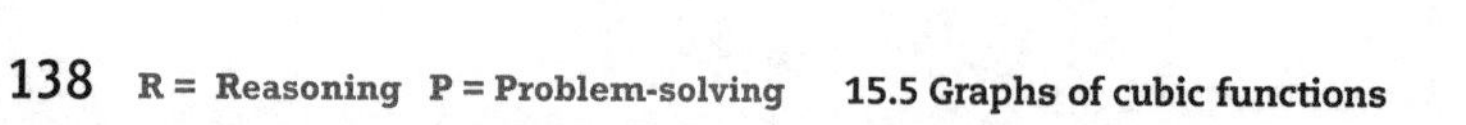

9 **Exam-style question**

Sketch the graph of $y = -x(x + 4)^2$ on a suitable coordinate grid marking clearly the points of intersection with the axes. **(3 marks)**

10 **P** The graph has equation $y = -x^3 + ax^2 + bx + c$

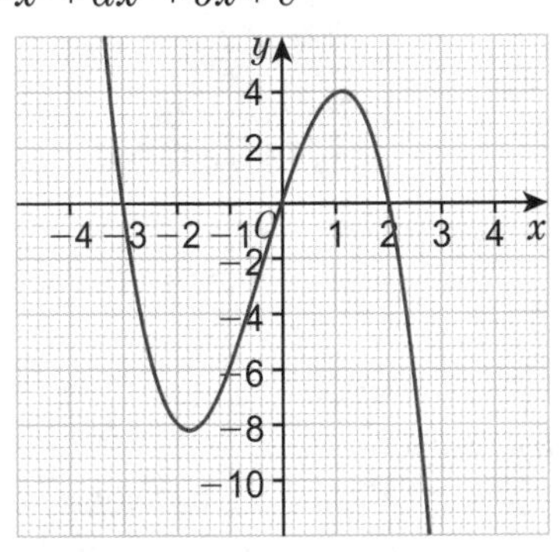

Work out the values of a, b and c.

11 **P** A graph has equation $y = x^3 + ax^2 + bx + c$
It crosses the x-axis at $x = -4$, $x = 1$ and $x = 3$.
Without drawing the graph, work out the values of a, b and c.

12 Use an iterative formula to find the one root of $x^3 - 3x^2 - 2 = 0$ to 4 d.p.
The first steps have been done for you:

$x^3 = 3x^2 + 2$

$x = \sqrt[3]{3x^2 + 2}$

$x_{n+1} = \sqrt[3]{3x_n^2 + 2}$

$x_0 = \square$

$x_1 = \sqrt[3]{3x_0^2 + 2} = \square$

$x_2 = \sqrt[3]{3x_1^2 + 2} = \square$

13 Use an iterative formula to find the negative root of the equation $x^3 - 2x^2 - 7x + 5 = 0$ to 5 d.p.

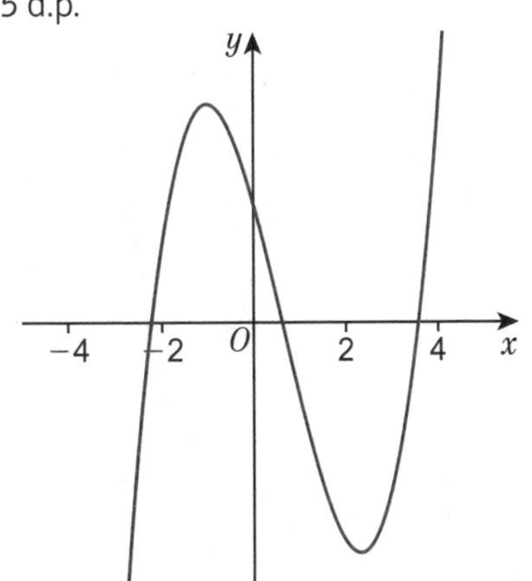

15 Problem-solving

Solve problems using these strategies where appropriate:

- **Use pictures**
- **Use smaller numbers**
- **Use bar models**
- **Use x for the unknown**
- **Use a flow diagram**
- **Use arrow diagrams**
- **Use geometric sketches**
- **Use graphs.**

1 A group of Explorer Scouts are on an orienteering expedition.
They start at point A and are told to travel one mile on a bearing of 050° to point B where they find their next instruction.
Point C is 3 miles from point B on a bearing of 135°.

a Using 3 cm for every mile, draw a diagram to show where the group needs to go to visit both points B and C.

b What is the bearing of C from A?

2 **R** Look at the graph and the shaded region bounded by the three lines.
Write down the three inequalities satisfied by the coordinates of the points in this region.

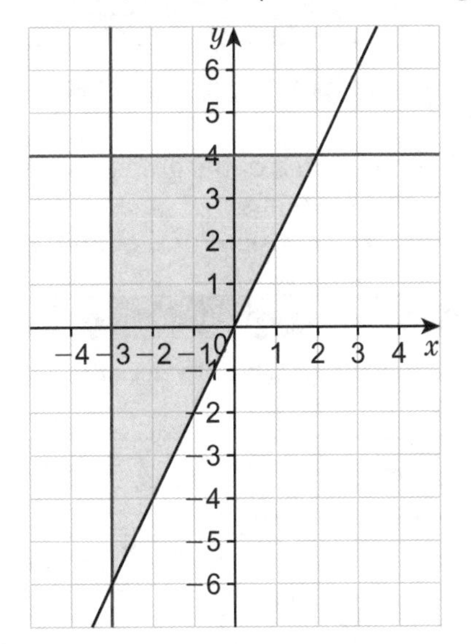

3 **R** This frequency table shows different lengths of material in a craft shop.

Length, l (cm)	Frequency
$10 < l \leqslant 15$	9
$15 < l \leqslant 20$	6
$20 < l \leqslant 25$	12
$25 < l \leqslant 30$	16

Which of the following statements are true about this data?

A The higher class boundary of each class is 30.

B The range is 16 – 9 = 7

C The lower quartile is in the $15 < l \leqslant 20$ class.

D The median is in the $15 < l \leqslant 20$ class.

E The upper quartile is in the $25 < l \leqslant 30$ class.

4 Show the area formed by these inequalities on a grid.

$x < 4$ $\quad y > -2 \quad$ $y < 2x + 1$

5 **R** Kevin has just fitted a new granite area around his sink.

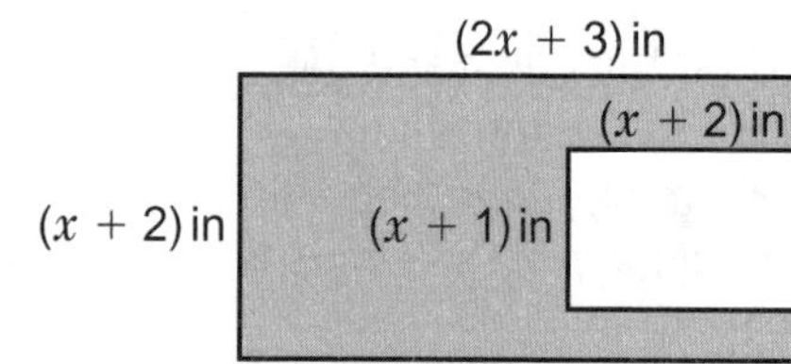

a Find an expression in x for the area of the granite.

b The actual area of the granite is 484 inches square.
Find the dimensions of the sink.

6 **Exam-style question**

a Complete the table of values for $y = x^2 - 2x$.

x	−2	−1	0	1	2	3	4
y							

(2 marks)

b Draw the graph of $y = x^2 - 2x$ for values of x from −2 to 4. **(2 marks)**

c Solve $x^2 - 2x - 2 = 1$. **(2 marks)**

June 2013, Q15, 1MA0/2H

7 **R** Joshua is working on a woodwork project.

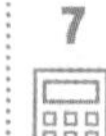

Sean has measured the height and the opposite angle of the piece of wood Joshua needs to cut and has rounded them to the nearest cm and the nearest degree.

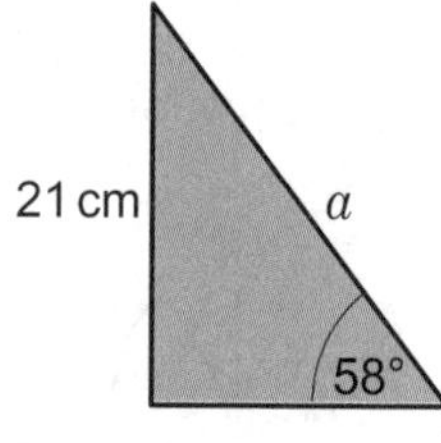

Joshua wants more precision and so cuts the wood to the upper bounds so that he can sand it down if he needs to.

a What is the upper bound of length a, to the nearest mm?

b What is the difference between the upper bound and the lower bound of the length?

8 Work out the minimum value of $n^2 - 6n + 14$.

9 **R** Naomi is making a maths display showing different graphs.
The labels for the graphs have got mixed up.
Match the labels to graphs X, Y and Z.

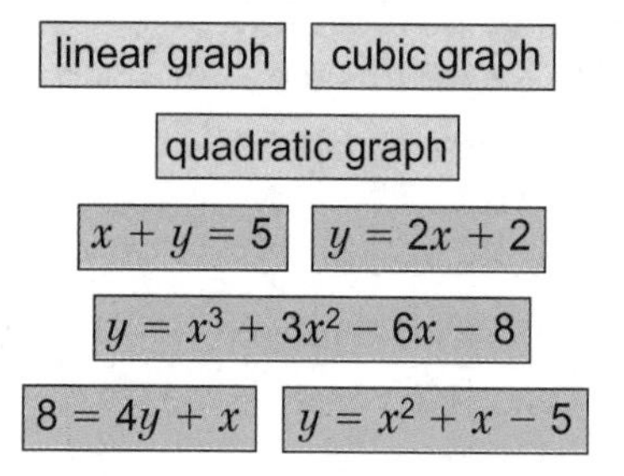

Graph X

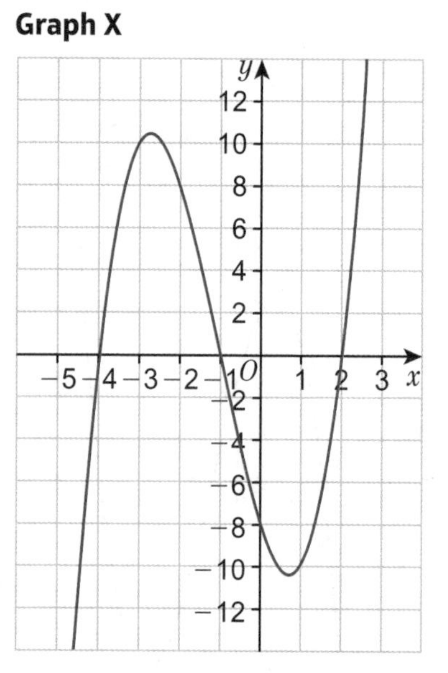

Graph Y

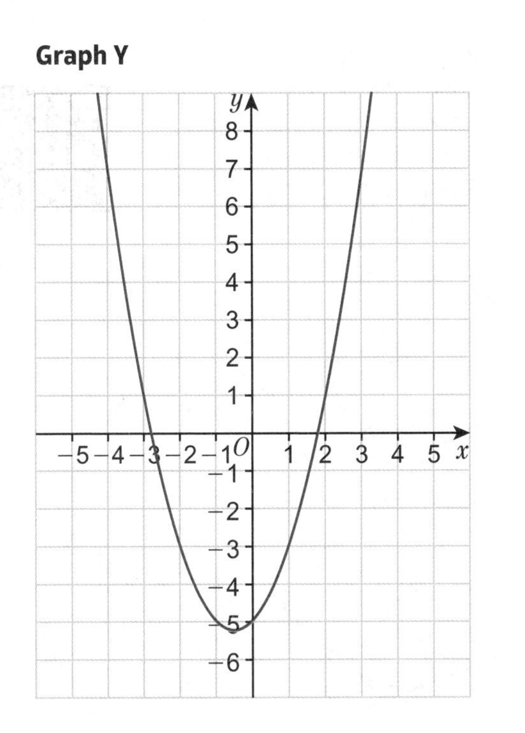

Graph Z

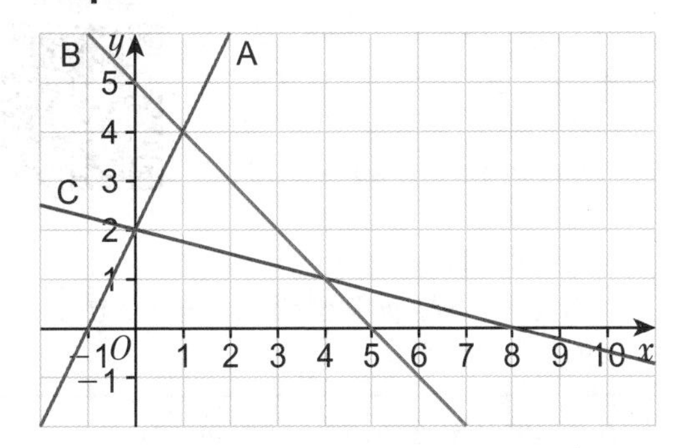

10 Solve these equations simultaneously.

$y - 3 = 4x$

$x^2 + 2y = -6$

16 CIRCLE THEOREMS

16.1 Radii and chords

1 **R** Each diagram shows a circle with centre O. Work out the size of each angle marked with a letter.

a

b

c

2 **R** Kate says that Q cannot be the centre of the circle. Explain why she is wrong.

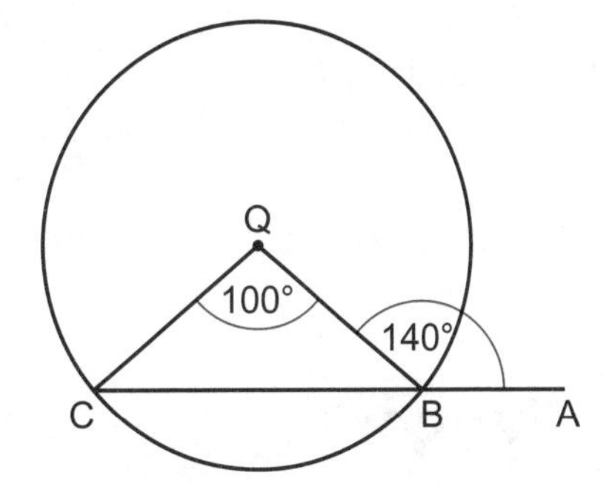

3 **R** O is centre of the circle. AB and CD are straight lines through the centre. Prove that triangles ACO and BDO are congruent.

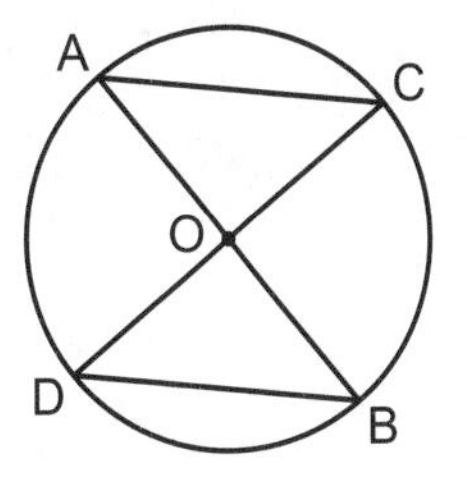

4 **R** O is the centre of a circle.
The length of chord AB is 10 cm.
OM is perpendicular to AB.
OM is 12 cm.

Example

a Work out the length of AM. State any circle theorems that you use.

b What is the length of the radius of the circle?

Q4b hint Use Pythagoras' theorem.

5 **R** O is the centre of a circle.
OA = 20 cm and AB = 24 cm.
M is the midpoint of AB.

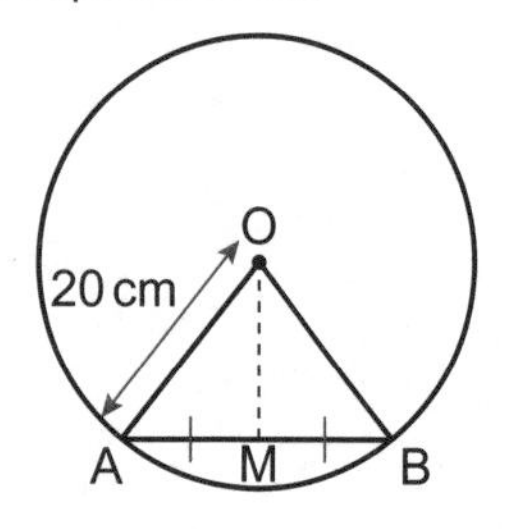

Work out the length of OM.

6 **R** O is the centre of a circle. The radius of the circle is 15 cm. The distance from O to the midpoint of chord AB is 12 cm.
Work out the length of chord AB.

7 **R** O is the centre of a circle. M is the midpoint of chord AB.
Angle OAB = 45°.

a What is angle AMO?

b Work out angle AOM.

c Work out angle AOB.

d Which of the triangles AMO, BOM and ABO are similar?

16.2 Tangents

1 **P** AB is a tangent to the circle with centre O.
AOB = 40°. Work out the size of angle ABO.

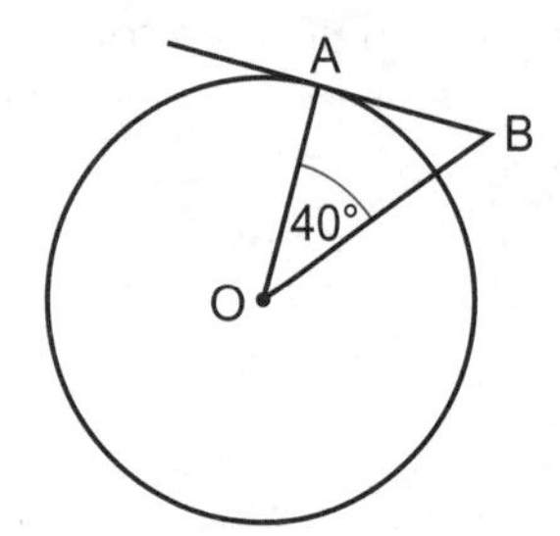

2 **R** The diagrams all show circles, centre O.
Work out the size of each angle marked with a letter.
Give reasons for your answers.

Example

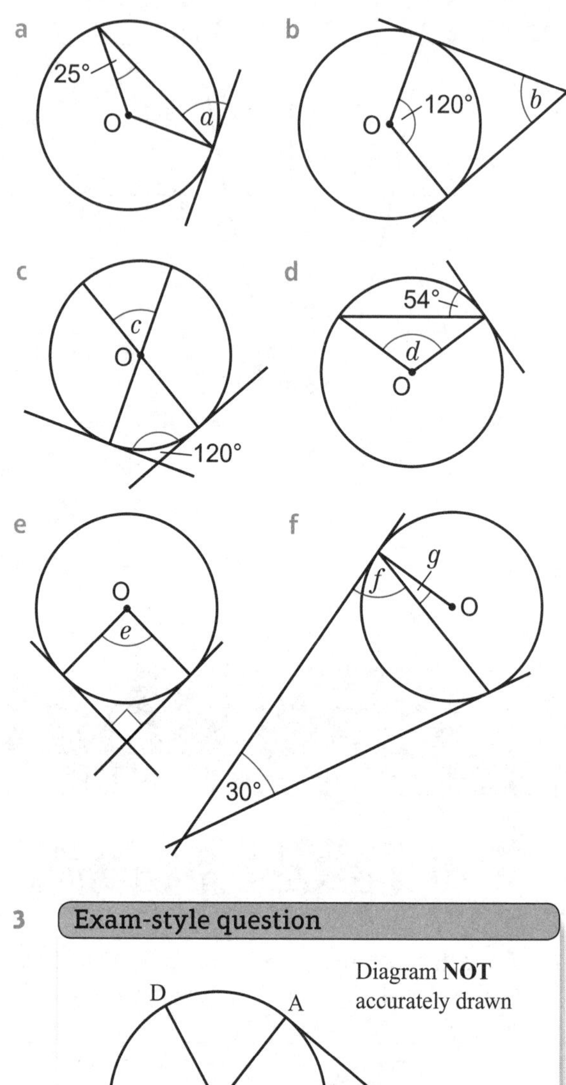

3 **Exam-style question**

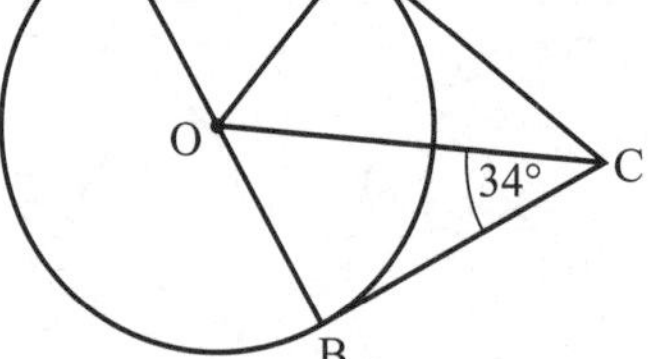

A, B and D are points on the circumference of a circle, centre O.
BOD is a diameter of the circle.
BC and AC are tangents to the circle.
Angle OCB = 34°.
Work out the size of angle DOA. **(3 marks)**

March 2013, Q19, 1MA0/1H

4 **P** A and B are points on the circumference of a circle, centre O.
TA and TB are tangents to the circle.

Show that triangles OAT and OBT are congruent.

5 **R** OA is the radius of a circle with a diameter of 20 cm. AT is a tangent to the circle joining point A to point T. AT = 24 cm.
Calculate the distance from T to the centre of the circle.
State any circle theorems that you use.

> **Q5 hint** Draw a diagram and mark the values on it.

16.3 Angles in circles 1

1 **R** The diagrams show circles, centre O.

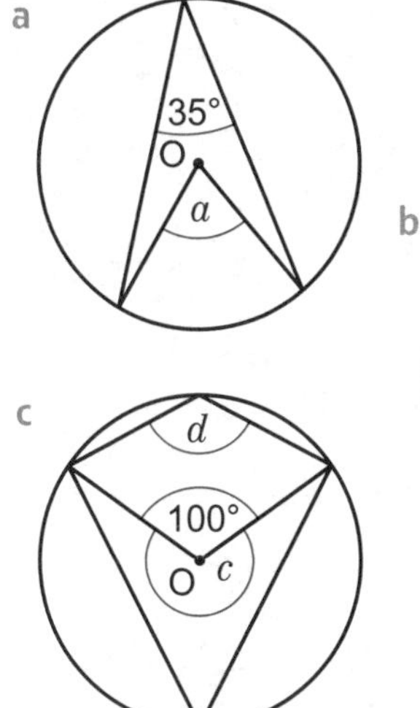

Work out the size of each angle marked with a letter.
Give reasons for your answers.

2 **R** AB is the diameter of a circle.
C is a point on the circumference of the circle.
Use the circle theorem from **Q1** to prove that ABC is a right-angled triangle.

3 **R** The diagrams show circles, centre O.
Work out the size of each angle marked with a letter. Give reasons for your answers.

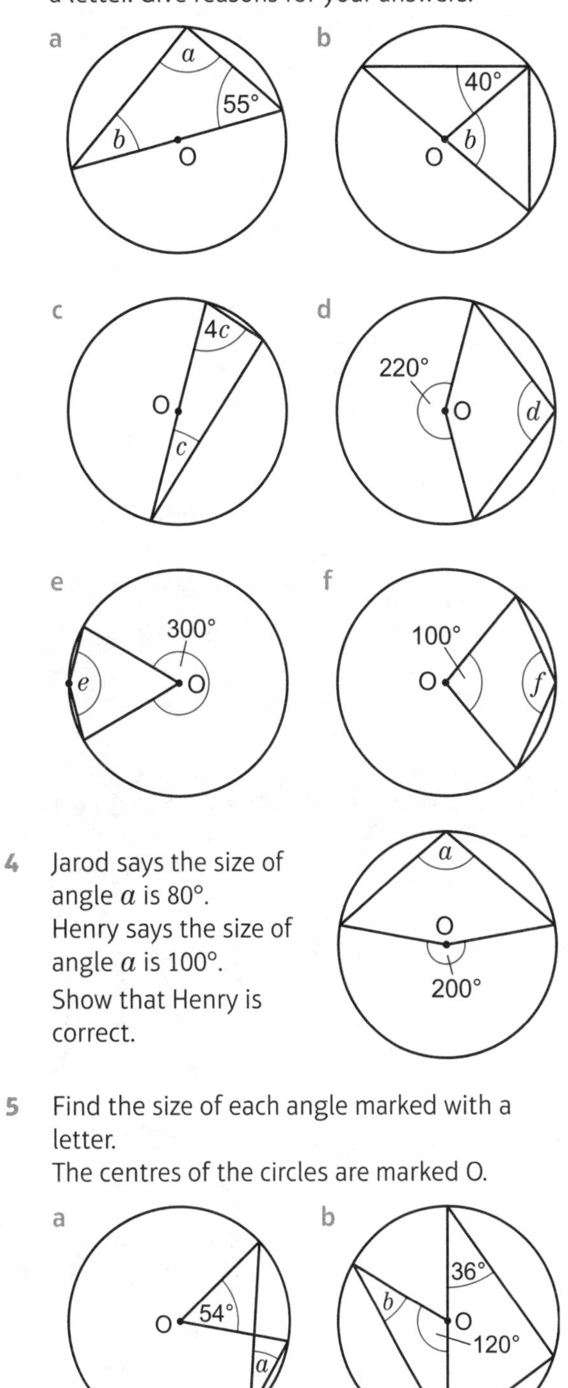

4 Jarod says the size of angle a is 80°.
Henry says the size of angle a is 100°.
Show that Henry is correct.

5 Find the size of each angle marked with a letter.
The centres of the circles are marked O.

6 Shumi says the size of angle a is 10°.
Charlie says the size of angle a is 40°.
Show that Charlie is correct.

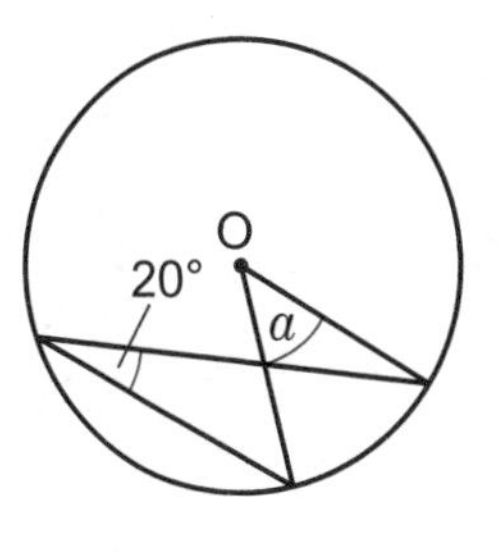

7 **Exam-style question**

Diagram **NOT** accurately drawn

A, B, C and D are points on the circumference of a circle, centre O.

Angle AOC = y.

Find the size of angle ABC in terms of y.

Give a reason for each stage of your working. **(4 marks)**

November 2013, Q22, 1MA0/1H

Exam hint

Each reason given must be a statement of a mathematical rule and not just the calculations you have done.

16.4 Angles in circles 2

1 O is the centre of a circle.

a Work out the size of each angle marked with a letter.

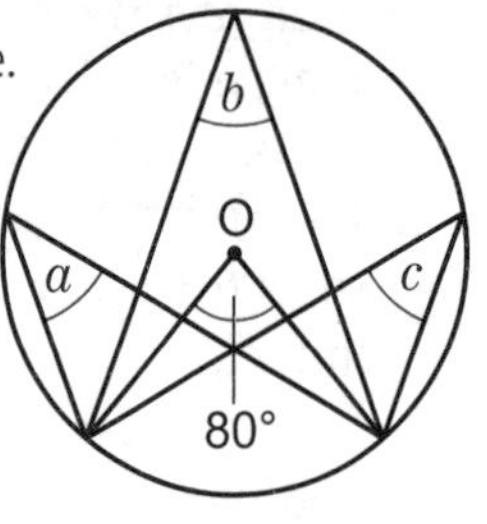

b Work out the size of angles a, b and c in terms of x.

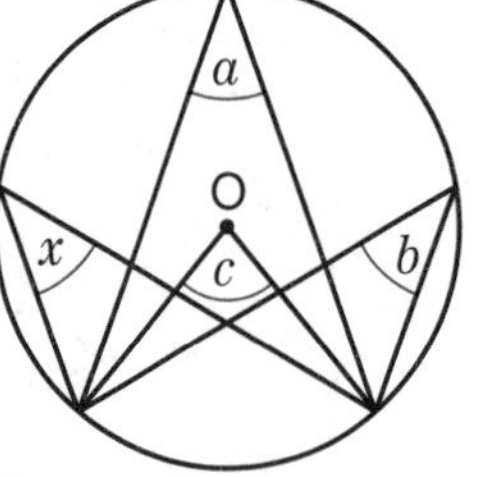

2 **P** a Look at the diagram.

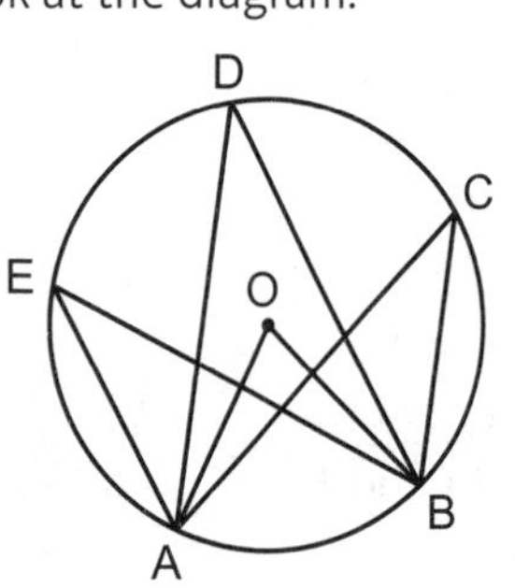

Angle AOB = x. Write down the size of angles

i AEB ii ADB iii ACB

in terms of x.

b What can you say about angles in the same segment?

3 **R** In each diagram, O is the centre of the circle.

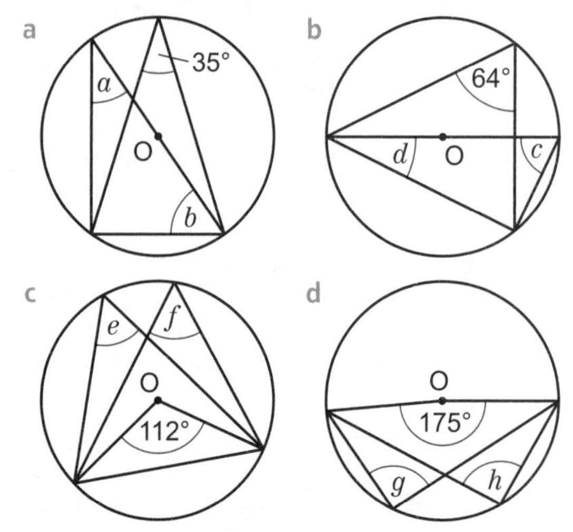

Work out the size of each angle marked with a letter.

Give reasons for each step in your working.

4 **R** In each diagram, O is the centre of the circle.

What is the size of angles a, b and c in each diagram?

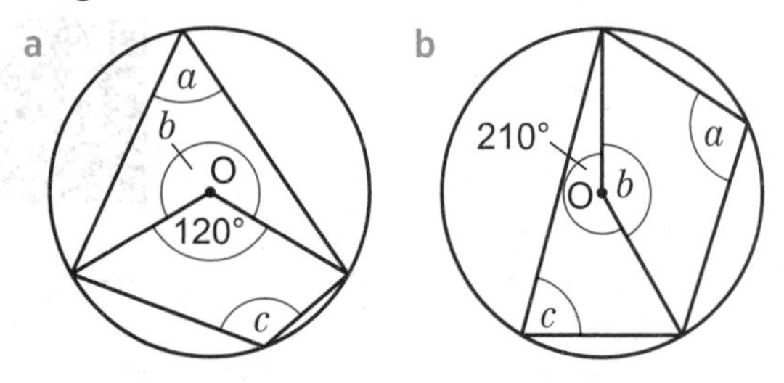

5 **R** ABCD is a cyclic quadrilateral.

O is the centre of the circle.

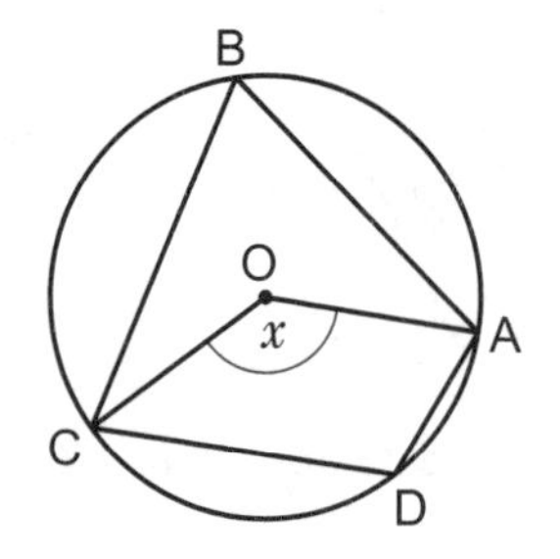

a In terms of x

i what is the reflex angle AOC

ii what is the obtuse angle ADC

iii what is the acute angle ABC?

b Find the sum of the acute angle ABC and the obtuse angle ADC.

6 **R** In each diagram, O is the centre of the circle.

a

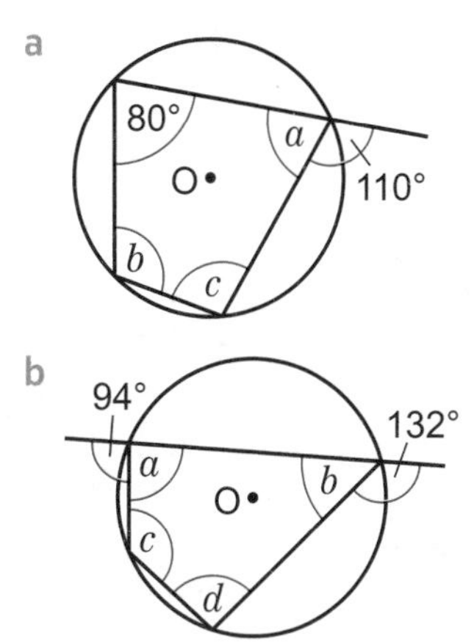

b

94°
132°
a
O
b
c
d

Work out the size of each angle marked with a letter.
Give reasons for each step in your working.

7 Prove that the exterior angle of a cyclic quadrilateral is equal to the opposite interior angle.

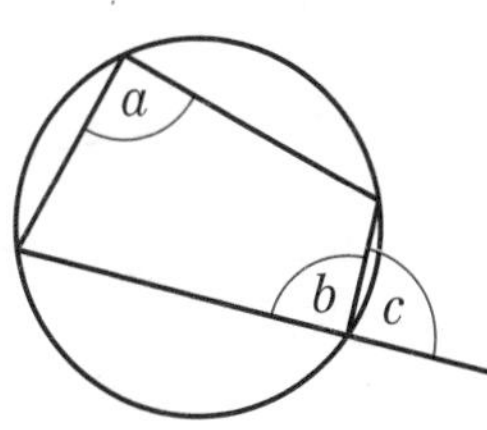

a Write down the size of angle c in terms of b.

b Write down the size of angle a in terms of b.

c Are a and c equal?

8 **R** Work out the size of each angle marked with a letter.
Give reasons for each step in your working.

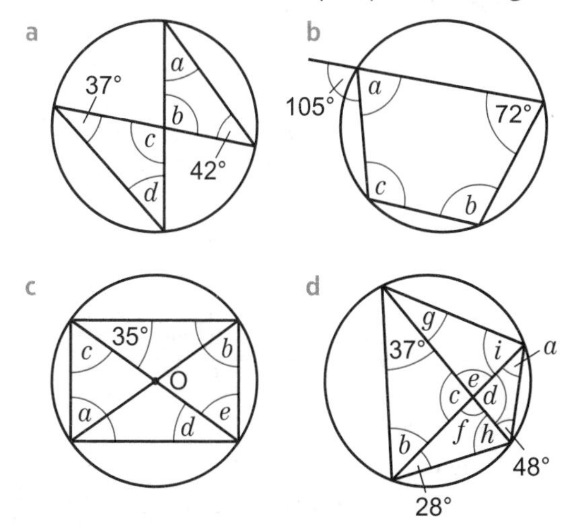

9 **R** O is the centre of a circle. AT is a tangent to the circle.

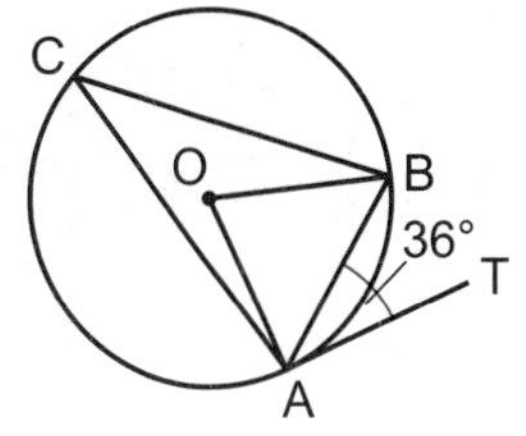

a Copy the diagram.

b Copy the working and complete the reasons.

Angle OAT = 90° because the angle between the radius and the …… is 90°

Angle OAB = 90° – 36° = 54°

OA = OB because ……

Angle OAB = angle OBA because the …… angles of an isosceles triangle are ……

Angle AOB = 180° – (2 × 54°) = 72° because angles in a triangle …… to ……

Angle ACB = 72° ÷ 2 = 36° because the angle at the …… is half the angle at the ……

10 **R** Repeat **Q9** but this time with angle BAT = 50°.
What is the size of angle ACB?

11 **R** Prove that angle ACB = angle BAT.

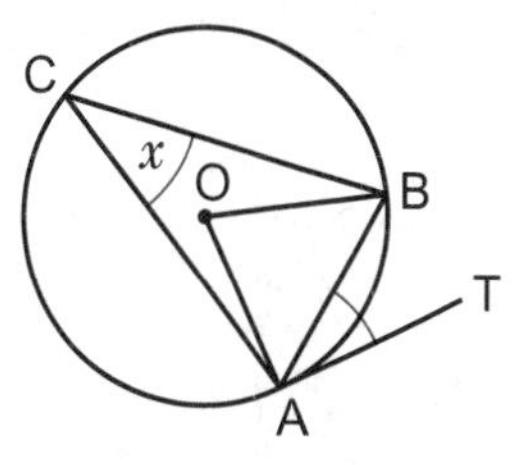

12 **Exam-style question**

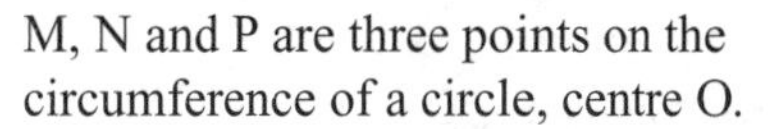

M, N and P are three points on the circumference of a circle, centre O.

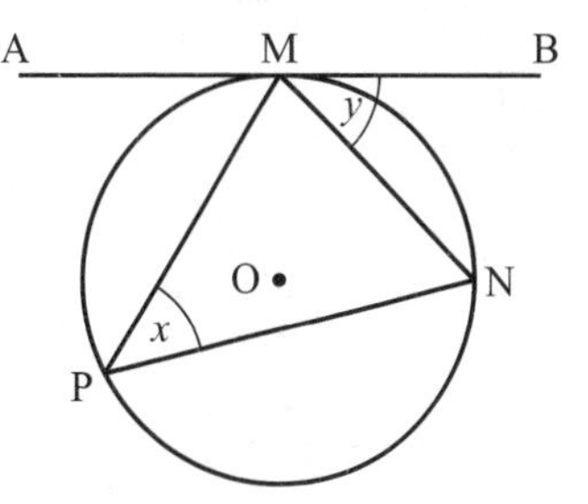

The straight line AMB is the tangent to the circle at M.
Angle MPN = x and angle BMN = y.
Prove that $x = y$ **(5 marks)**

Exam hint
Draw on the diagram the angle MON.

16.5 Applying circle theorems

1 **R** In each diagram, O is the centre of the circle.

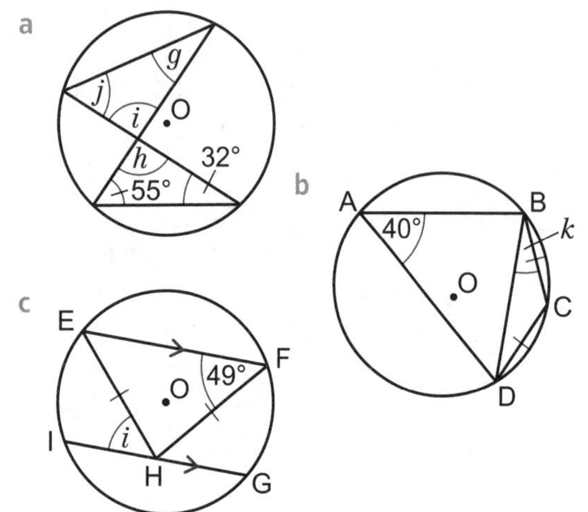

Work out the size of each angle marked with a letter.
Give reasons for each step in your working.

2 **R** In each diagram, AT is a tangent to the circle.

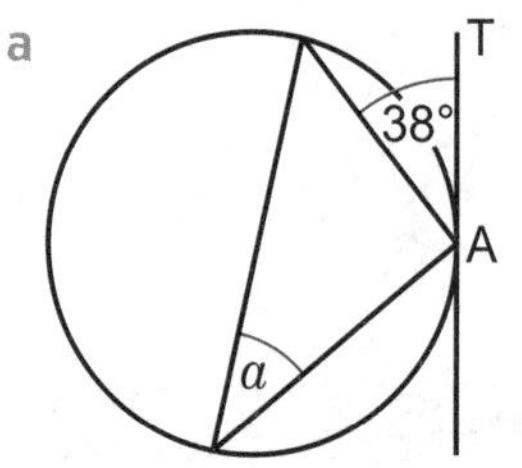

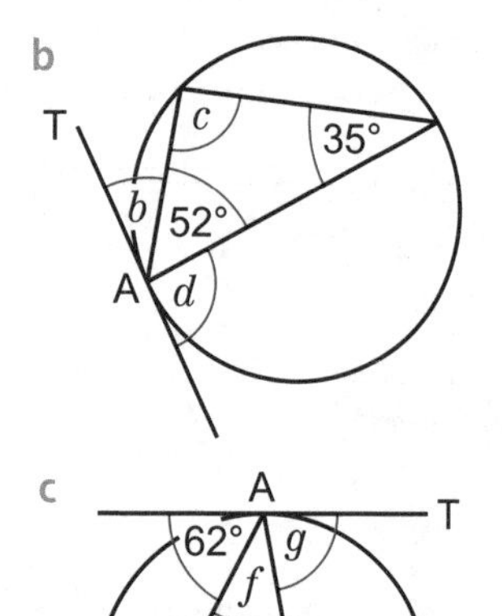

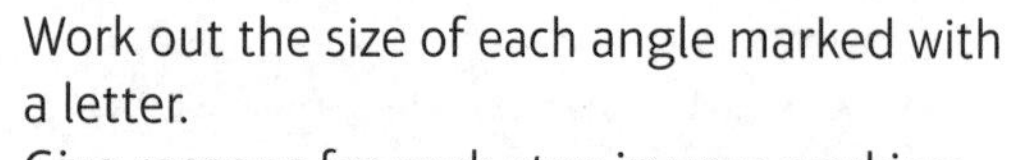

Work out the size of each angle marked with a letter.
Give reasons for each step in your working.

3 **R** Work out the size of each angle marked with a letter.
Give reasons for each step in your working.

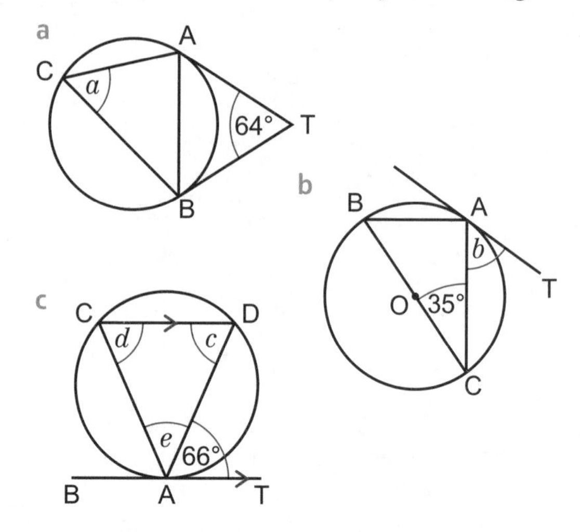

4 **R** O is the centre of the circle.
DAT and BT are tangents to the circle.
Angle CAD = 62° and angle ATB = 20°.

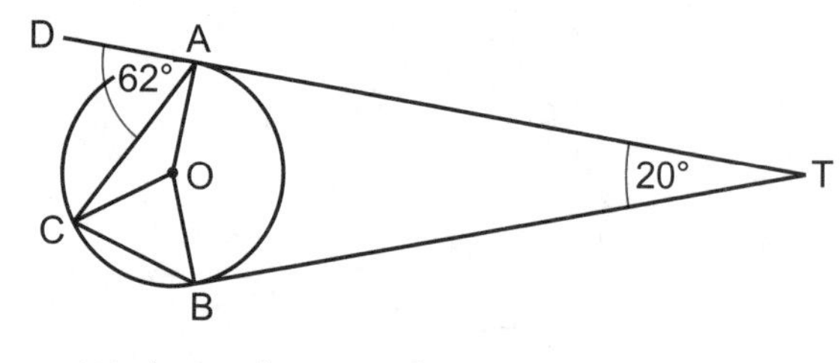

Work out the size of

a angle CAO
b angle AOB
c angle AOC
d angle COB
e angle CBO.

Give reasons for each step in your working.

5 **Exam-style question**

B, C and D are points on the circumference of a circle, centre O.
ABE and ADF are tangents to the circle.

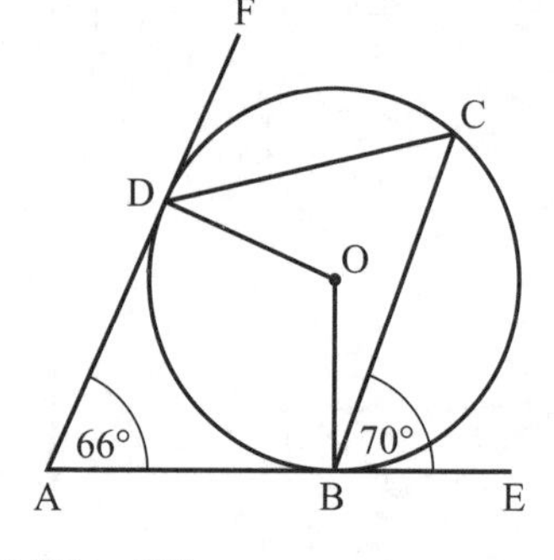

Angle DAB = 66°
Angle CBE = 70°
Work out the size of angle ODC. **(3 marks)**

6 **P** Find the equation of the tangent to the circle $x^2 + y^2 = 100$ at the point B (6, −8).

7 **P** Find the equation of the tangent to the circle $x^2 + y^2 = 25$ at the point C (−3, 4).

8 **P** Find the equation of the tangent to the circle $x^2 + y^2 = 225$ at the point D (−9, 12).

9 **P** Find the equation of the tangent to the circle $x^2 + y^2 = 676$ at the point E (−10, −24).

16 Problem-solving

Solve problems using these strategies where appropriate:

- **Use pictures or lists**
- **Use smaller numbers**
- **Use bar models**
- **Use x for the unknown**
- **Use a flow diagram**
- **Use arrow diagrams**
- **Use geometric sketches**
- **Use graphs**
- **Use logical reasoning.**

1 **R** 3 pairs of football boots cost £54.
8 football shirts cost £152.
There are no special offers or discounts if you buy more shirts or pairs of boots.

a A football coach has £200. How much change will she get if she buys 11 pairs of football boots?

b Another coach has £250. Does he have enough money to buy 7 shirts and 7 pairs of boots?

2 A park fountain is in the centre (F) of a circular pool with a radius of 2.5 m.
A row of lights form a chord across the pool (GH). The shortest distance from the lights to the fountain is 1.8 m.

a Draw a diagram to show this information.

b What is the length of the row of lights? Give your answer to 2 d.p.

3 **R** O is the centre of a circle. CB and CD are tangents to the circle. Find angle x. Explain your reasoning.

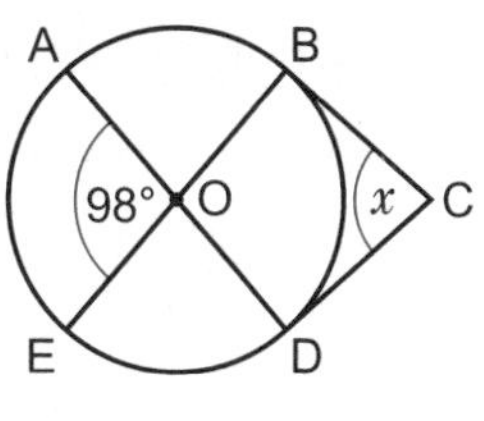

4 **Exam-style question**

A and B are points on the circumference of a circle, centre O.
BC is a tangent to the circle.
AOC is a straight line.
Angle ABO = 35°

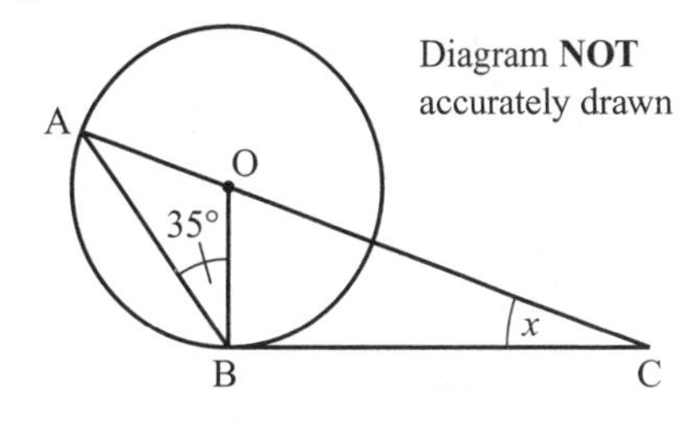

Work out the size of the angle marked x.
Give reasons for your answer. **(5 marks)**

5 **R** The diagram shows a circle, centre O.
CB and CD are tangents to the circle.
Find angle g. Explain your reasoning.

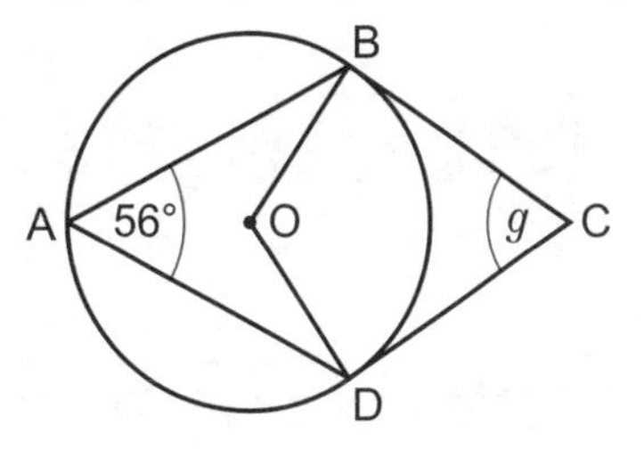

6 **R** Pria says that just from knowing the angle GHO, she can work out all the angles inside triangles GFO and HGO.

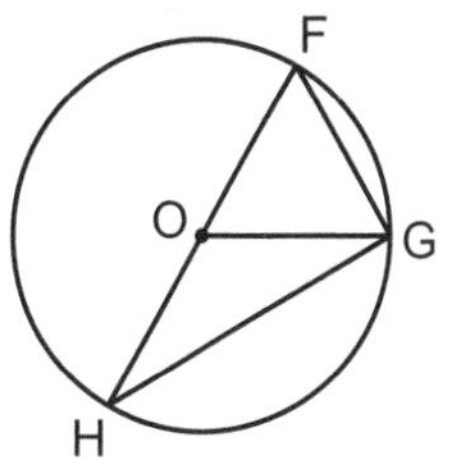

a Prove that Pria is correct.

b If angle GHO = 41°, work out the angles in triangles GFO and HGO.

7 **R** Work out the sizes of angles a, b and c.
Give reasons for your answers.

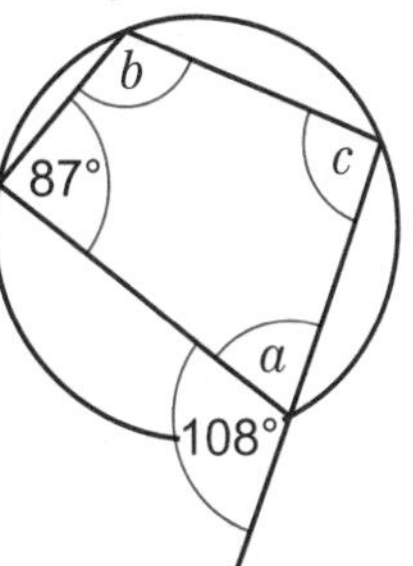

8 **R** 3 avocados and 2 red peppers were bought for £2.95.
2 avocados and 4 red peppers were bought for £3.70.
How much is
a 1 avocado b 1 red pepper?

9 This table shows the lengths of 40 caterpillars.

Length, l (mm)	Frequency
$8 < l \leqslant 10$	7
$10 < l \leqslant 12$	10
$12 < l \leqslant 14$	14
$14 < l \leqslant 16$	9

a Find the frequency density for each class. Explain how you found the answers.
b Draw a histogram for this data. Why do you need to work out the frequency densities to do this?

10 **R** O is the centre of the circle. A, C and E are all points on the circumference.
BCD is a tangent touching the circle at point C.
DEF is a tangent touching the circle at point E.

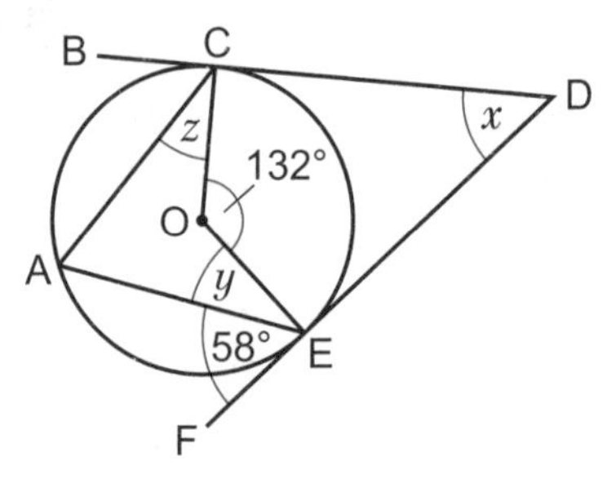

Work out the sizes of angles x, y and z.
Give reasons for any statements you make.

17 MORE ALGEBRA

17.1 Rearranging formulae

1 Make c the subject of the formula
$E = mc^2$

2 Make a the subject of the formula
$P = \sqrt{a + b}$

3 Make x the subject of each formula.

Example

a $A = 4\sqrt{\frac{x}{y}}$ b $K = t\sqrt{\frac{1}{x}}$

c $T = \sqrt{\frac{4x}{2a}}$ d $R = 2(x + y)^2$

4 In each formula change the subject to the letter given in brackets.

a $V = \frac{\pi}{6}d^3$ (d) b $V = \frac{3x^3}{2}$ (x)

c $p = \sqrt[3]{6x}$ (x) d $t = \sqrt[3]{\frac{1}{p}}$ (p)

5 Make x the subject of the formula $A = 2x + px$

Example

6 Make q the subject of the formula
$T = pq + pr - qs$

7 **R** $5ab - 7 = c + 2ab$
a Make a the subject.
b Make b the subject.

8 **R** $L = 4x + xy - 5$
Jordan rearranges the formula to make x the subject.
His answer is $x = \frac{L - 4x + 5}{y}$
a Explain why this cannot be the correct answer.
b What mistake has Jordan made?
c Work out the correct answer.

9 Make x the subject of the formula
$m = \frac{2x - 1}{x}$

Q9 hint First multiply both sides by x.

10 **Exam-style question**

Make y the subject of the formula

$c = \frac{y}{5 - y}$ **(4 marks)**

17.2 Algebraic fractions

1 Write as a single fraction in its simplest form. The first one has been started for you.

a $\frac{a}{5} \times \frac{a}{2} = \frac{a \times a}{5 \times 2} =$

b $\frac{3c}{5} \times \frac{2d}{9}$ c $\frac{3}{2k} \times \frac{8}{11k}$

2 Write as a single fraction in its simplest form. The first one has been started for you.

a $\frac{3x^3}{y} \times \frac{5y^2}{9x} = \frac{{}^1\cancel{3}x^3 \times 5y^{\cancel{2}}}{\cancel{y} \times {}^3\cancel{9}x} =$

b $\frac{10x^2}{3y} \times \frac{9y^4}{20x^5}$ c $\frac{12x^6}{35y^7} \times \frac{21y^2}{20x^3}$

3 Write as a single fraction in its simplest form.

a $\frac{2}{a} \div \frac{3}{a}$ b $xy^2 \div \frac{1}{xy}$

c $\frac{5x^2}{4y^6} \div \frac{15x^4}{8y^3}$ d $\frac{x}{4} \div \frac{x+5}{12}$

Q3b hint Write xy^2 as $\frac{xy^2}{1}$

4 Write as a single fraction in its simplest form.

Example

a $\frac{4y}{9} + \frac{y}{3}$ b $\frac{2y}{3} + \frac{y}{5}$

c $\frac{5y}{3} - \frac{y}{7}$

5 Write down the LCM of

a $2x$ and $3x$ b $2x$ and $6x$

c $5x$ and $6x$ d $7x$ and $3x$

6 a Write $\frac{1}{3x}$ and $\frac{1}{2x}$ as equivalent fractions with denominator the LCM of $3x$ and $2x$.

b Simplify $\frac{1}{3x} + \frac{1}{2x}$

7 Write as a single fraction in its simplest form.

a $\frac{1}{7x} + \frac{1}{2x}$ b $\frac{1}{3x} - \frac{1}{5x}$ c $\frac{7}{10x} - \frac{1}{3x}$

8 a Copy and complete.

$\frac{x+5}{3} = \frac{\square(x+5)}{2 \times 3} = \frac{\square x + \square}{6}$

b Copy and complete.

$\frac{x-7}{2} = \frac{\square(x-7)}{3 \times 2} = \frac{\square x - \square}{6}$

c Use your answers to parts **a** and **b** to work out $\frac{x+5}{3} + \frac{x-7}{2}$

9 Write as a single fraction in its simplest form.

a $\frac{x+1}{3} + \frac{x+4}{5}$ b $\frac{x-2}{3} - \frac{x+6}{7}$

c $\frac{x-6}{5} + \frac{3x+2}{2}$

10 **Exam-style question**

Write as a single fraction in its simplest form

$\frac{4x+5}{7} + \frac{2x-1}{2}$ **(3 marks)**

11 Make a the subject of the formula $\frac{1}{a} - \frac{1}{b} = 1$.

The working has been started for you.

$\frac{1}{a} - \frac{1}{b} = 1$

$\frac{1}{a} = 1 + \frac{1}{b}$

$\frac{1}{a} = \frac{\square}{\square} + \frac{1}{b} =$

12 Scientists use the lens formula to solve problems involving light.

The lens formula is $\frac{1}{f} = \frac{1}{u} + \frac{1}{v}$

where f = focal length, u = object distance and v = image distance.

Make f the subject of the formula.

17.3 Simplifying algebraic fractions

1 Simplify

a $\frac{a}{ab}$ b $\frac{x+2}{5(x+2)}$

c $\frac{x+5}{(x+5)^2}$ d $\frac{(x+4)(x-3)}{(x-3)(x-8)}$

e $\frac{x(x-5)}{(x-5)(x+2)}$ f $\frac{x(x+6)^2}{x^2(x+6)}$

Q1e hint You can only cancel whole brackets.

2 a Factorise $x^2 + 4x$

b Use your answer to part **a** to simplify $\frac{x^2+4x}{x+4}$

3 Simplify fully

a $\frac{x^2-5x}{x}$ b $\frac{6x^2+9x}{2x+3}$ c $\frac{4x+8}{3x^2+6x}$

4 **R** Simplify $\frac{x^2 - 3x}{x^2 - 3}$

Ayesha says, '$(x - 3)$ is a factor of the numerator and the denominator.'

a Is Ayesha correct? Explain.

b Can the fraction be simplified? Explain your answer.

5 Simplify fully

a $\frac{4(x + 1)}{x^2 + 7x + 6}$　　b $\frac{x^2 + 2x - 8}{3(x - 2)}$

6 Simplify fully

Example

a $\frac{x^2 + 7x + 12}{x^2 + 3x - 4}$

b $\frac{x^2 - 7x + 10}{x^2 + 4x - 12}$

c $\frac{x^2 - 16}{(x + 4)^2}$

7 **Exam-style question**

Simplify fully $\frac{x^2 + 12x + 36}{x^2 - 36}$　　**(3 marks)**

8 Simplify fully

a $\frac{2x^2 + x - 15}{3x^2 + 10x + 3}$　　b $\frac{6x^2 + x - 15}{3x^2 + 17x + 20}$

c $\frac{16x^2 - 1}{16x^2 + 8x + 1}$

9 **Exam-style question**

Simplify fully $\frac{3x^2 + 11x + 10}{x^2 - 4x - 12}$　　**(3 marks)**

10 a Copy and complete.

$4 - x = -(\square - \square)$

b Simplify

i $\frac{4 - x}{x - 4}$

ii $\frac{16 - x^2}{x^2 - x - 12}$

11 Simplify fully

a $\frac{25 - x^2}{x^2 - 5x}$　　b $\frac{3x^2 - 27}{x^2 - 6x + 9}$

c $\frac{8x^2 + 20x}{8x^2 + 18x - 5}$

12 Show that

$$\frac{(x^2 - 3x - 10)(x^2 - 2x - 3)(12x^2 + 20x)}{(3x^2 - 4x - 15)(9x - 45)(x^2 + 3x + 2)} = \frac{4x}{9}$$

17.4 More algebraic fractions

1 Write as a single fraction in its simplest form.

a $(x - 2)^2 \times \frac{x + 5}{x - 2}$　　b $\frac{x + 3}{x - 4} \times \frac{x - 4}{x + 6}$

c $\frac{x + 2}{5} \times \frac{3}{4x + 8}$　　d $\frac{4}{x - 3} \div \frac{10}{3x - 9}$

e $\frac{4x + 12}{x - 2} \div \frac{x + 3}{x + 5}$　　f $\frac{(x + 6)^2}{x + 5} \div \frac{x + 6}{x - 1}$

2 a Factorise $x^2 - 16$

b Factorise $x^2 + 7x + 12$

c Write $\frac{x^2 - 16}{6} \times \frac{3}{x^2 + 7x + 12}$ as a single fraction in its simplest form.

3 Write as a single fraction in its simplest form.

a $\frac{x^2 + 3x - 4}{x^2 - 5x + 6} \times \frac{x^2 - 4}{x^2 + 5x - 6}$

b $\frac{15x + 20}{x^2 + 7x + 10} \div \frac{3x^2 - 2x - 8}{x^2 - x - 6}$

4 Write down the LCM of

a x and $x + 1$　　b $x + 1$ and $x + 4$

c $x + 5$ and $x + 7$　　d $x + 2$ and $x - 2$

e $2x + 1$ and $2x - 2$

5 Simplify fully

Example

a $\frac{1}{x + 1} + \frac{1}{x + 4}$

b $\frac{2}{x + 2} + \frac{5}{x - 2}$

c $\frac{6}{x + 3} - \frac{2}{x - 1}$　　d $\frac{1}{2x + 2} - \frac{1}{2x + 5}$

6 **Exam-style question**

Write as a single fraction in its simplest form

$\frac{1}{x - 3} - \frac{5}{x + 4}$　　**(3 marks)**

7 a Factorise

i $3x + 6$　　ii $5x + 10$

b Write down the LCM of $3x + 6$ and $5x + 10$.

c Write $\frac{1}{3x + 6} + \frac{1}{5x + 10}$ as a single fraction in its simplest form.

Q7b hint Look at the factorised form of each expression:

$a(x + y)$　　$b(x + y)$　　LCM = $ab(x + y)$

8 a Factorise $x^2 - 25$

b Write $\frac{1}{x-5} + \frac{1}{x^2-25}$ as a single fraction in its simplest form.

9 Write as a single fraction in its simplest form.

a $\frac{1}{2x^2+5x+3} - \frac{1}{2x+3}$

b $\frac{1}{x^2+5x+4} + \frac{1}{3x+12}$

c $\frac{1}{x^2+7x+10} + \frac{2}{x^2-2x-8}$

d $\frac{2}{6-x} - \frac{5}{36-x^2}$

10 Write $\frac{1}{3x} + \frac{1}{3(x+1)} + \frac{1}{9}$ as a single fraction in its simplest form.

11 Show that

$$\frac{1}{x^2+9x+20} + \frac{1}{4x+16} = \frac{x+9}{A(x+5)(x+4)}$$

and find the value of A.

17.5 Surds

1 Simplify

a i $\sqrt{12}$

ii $\sqrt{75}$

b Use your answers to part **a** to simplify $4\sqrt{12} + 3\sqrt{75}$

2 Simplify

a $2\sqrt{20} + 4\sqrt{80}$

b $3\sqrt{150} - 2\sqrt{54}$

c $3\sqrt{32} + 6\sqrt{8} - \sqrt{200}$

3 Factorise these expressions.
The first one has been started for you.

a $\sqrt{45} + 9 = 3\sqrt{\square} + 9 = 3(\square + \square)$

b $4 + \sqrt{12}$ c $15 - \sqrt{50}$ d $\sqrt{63} - \sqrt{54}$

4 Expand and simplify

a $\sqrt{3}(1+\sqrt{3})$ b $(\sqrt{2}+3)(2+\sqrt{2})$

c $(4-\sqrt{3})(2+\sqrt{3})$ d $(2-\sqrt{3})^2$

e $(4+\sqrt{6})^2$ f $(5-\sqrt{7})^2$

5 **Exam-style question**

Expand $(3-\sqrt{2})^2$.
Write your answer in the form $a + b\sqrt{c}$, where a, b and c are integers. **(2 marks)**

6 **R** a Work out the area of each shape.
Write your answers in the form $a + b\sqrt{3}$

i Rectangle with sides $7+\sqrt{3}$ and $4-\sqrt{3}$

ii Square with side $1+\sqrt{12}$

b Would the perimeter of each shape be rational or irrational? Explain.

7 Rationalise the denominators.
The first one has been started for you.

a $\frac{4+\sqrt{3}}{\sqrt{3}} \times \frac{\sqrt{3}}{\sqrt{3}} = \frac{4\times\sqrt{3}+\sqrt{3}\times\sqrt{3}}{\sqrt{3}\times\sqrt{3}} =$

b $\frac{5-\sqrt{2}}{\sqrt{2}}$ c $\frac{2+\sqrt{6}}{\sqrt{6}}$ d $\frac{6-\sqrt{5}}{\sqrt{5}}$

8 **Exam-style question**

Write $\frac{9+\sqrt{48}}{\sqrt{3}}$ in the form $a + b\sqrt{3}$, where a and b are integers. **(3 marks)**

9 **R** a Expand and simplify $(4+\sqrt{7})(4-\sqrt{7})$

b Is your answer rational or irrational?

c How can you tell if your answer will be rational or irrational?

d Which of these will have rational answers when expanded?

i $(5+\sqrt{3})(3-\sqrt{3})$

ii $(5+\sqrt{3})(5+\sqrt{3})$

iii $(5+\sqrt{3})(5-\sqrt{3})$

Check by expanding the brackets.

e Rationalise the denominator of $\frac{1}{5+\sqrt{3}}$

Q9e hint Multiply the numerator and denominator by $(5-\sqrt{3})$.

10 Rationalise the denominators.
Give your answers in the form $a \pm \sqrt{b}$ or $a \pm b\sqrt{c}$ where a, b and c are rational.

a $\frac{1}{1+\sqrt{5}}$ b $\frac{1}{3-\sqrt{2}}$ c $\frac{2}{2+\sqrt{3}}$

d $\frac{5}{1+\sqrt{7}}$ e $\frac{\sqrt{3}}{4-\sqrt{3}}$ f $\frac{2+\sqrt{5}}{7-\sqrt{5}}$

11 a Solve $x^2 + 4x - 3 = 0$ by using the quadratic formula.

b Solve the equation $x^2 + 6x + 3 = 0$ by completing the square.

c Solve the equation $x^2 - 10x + 5 = 0$.

Write all your answers in surd form.

17.6 Solving algebraic fraction equations

1 Solve these equations.
Give your answer as a simplified fraction.

a $\frac{4}{x} + \frac{3}{x} = 14$

b $\frac{5}{x+1} - \frac{1}{x+1} = 3$

c $10 = \frac{4}{x-2} - \frac{6}{x-2}$

2 Solve these quadratic equations.

a $\frac{3}{x} = \frac{x-1}{2}$ b $\frac{3}{x} = \frac{2x+5}{4}$

c $\frac{3x+7}{4} = -\frac{1}{x}$ d $\frac{2x+1}{3} = \frac{5}{x}$

Q2a hint First multiply both sides by the LCM ($2x$) and simplify. Then multiply out the bracket and solve by factorising.

3 Solve this quadratic equation.

$\frac{5}{x+3} + \frac{2}{2x-1} = 1$

Example

4 a Show that the equation $\frac{2}{x+4} + \frac{x}{2x-1} = 1$ can be rearranged to give $x^2 - x - 2 = 0$

b Solve $x^2 - x - 2 = 0$

5 Solve these quadratic equations.

a $\frac{3}{x+2} + \frac{4}{x-3} = 2$ b $\frac{2}{x+1} + \frac{4}{3x-1} = 3$

c $\frac{3x}{4x-2} + \frac{6}{2x+6} = 1$ d $\frac{6}{x+2} - \frac{1}{x-2} = 3$

e $\frac{4}{x-3} - \frac{6}{x+1} = 1$

6 Solve these quadratic equations.
Give your answers correct to 2 decimal places.

a $\frac{3}{x-1} + \frac{1}{x+3} = 1$ b $\frac{3}{x} - \frac{1}{x+2} = 4$

c $\frac{1}{x+4} + \frac{1}{x-2} = 3$ d $\frac{2}{x+4} - \frac{1}{x+2} = 5$

7 **Exam-style question**

Find the exact solutions of

$x + \frac{3}{x} = 10$ **(3 marks)**

17.7 Functions

1 a $L = 2x$ and $x = 4y$. Write L in terms of y.

b $T = \frac{x}{2}$ and $x = \frac{1}{5}z$. Write T in terms of z.

c $A = x^2$ and $x = y - 2$. Write A in terms of y.

2 $f(x) = \frac{12}{x}$. Work out

a $f(2)$ b $f(-3)$ c $f(\frac{1}{2})$ d $f(-36)$

3 **R** $g(x) = 3x^2$. Robert says that $g(4) = 144$.

a Explain what Robert did wrong.

b Work out $g(4)$.

4 $h(x) = 4x^2$. Work out

a $h(4)$ b $h(-2)$ c $h(\frac{1}{2})$ d $h(-3)$

5 $f(x) = x^2 + 3x$ and $g(x) = x^3$. Work out

a $f(1) + g(1)$ b $f(3) - g(2)$ c $f(2) \times g(2)$

d $\frac{g(10)}{f(2)}$ e $2f(16)$ f $2f(4) - g(-2)$

6 $g(x) = 3x + 7$. Work out the value of a when

a $g(a) = 22$ b $g(a) = 9$ c $g(a) = 0$

7 $f(x) = 2x^2 + 1$. Work out the values of a when

a $f(a) = 33$ b $f(a) = 1$

c $f(a) = 11$ d $f(a) = 25$

8 $f(x) = x(x - 2)$ and $g(x) = (x + 1)(x + 3)$.
Work out the values of a when

a $f(a) = 0$ b $g(a) = 0$

c $f(a) = 8$ d $g(a) = 8$

9 $f(x) = 3x + 2$. Write out in full

a $f(x) + 3$ b $f(x) - 8$ c $3f(x)$

d $5f(x)$ e $f(3x)$ f $f(6x)$

10 $h(x) = 2x^2 - 8$. Write out in full

a $h(x) - 2$ b $4h(x)$

c $h(3x)$ d $h(-x)$

11 **R** $f(x) = x^2 + 3$ and $g(x) = 3x - 7$. Work out

a $gf(2)$ b $gf(5)$ c $fg(3)$ d $fg(6)$

Q11a hint First work out $f(2)$ and then substitute your answer into $g(x)$.

12 **R** $f(x) = 2x + 4$, $g(x) = 5 - x$ and $h(x) = x^2 - 3$.
Work out

a $gf(x)$ b $fg(x)$ c $hf(x)$

d $fh(x)$ e $hg(x)$ f $gh(x)$

13 Find the inverse of each function.

Example

a $x \rightarrow 2x - 5$

b $x \rightarrow \frac{x}{4} + 3$

c $x \rightarrow 4(x + 5)$

d $x \rightarrow 5(x - 1) + 3$

14 **R** $f(x) = 3(x + 2)$ and $g(x) = 3(x - 2)$

a Find $f^{-1}(x)$. b Find $g^{-1}(x)$.

c Work out $f^{-1}(x) + g^{-1}(x)$.

d If $f^{-1}(a) + g^{-1}(a) = 1$, work out the value of a.

17.8 Proof

1 Show that

Example

a $(x + 2)^2 - 4x \equiv x^2 + 4$

b $x^2 - 7x + 25 \equiv (x - 5)^2 + 3x$

c $(x + 4)^2 - 4 \equiv (x + 6)(x + 2)$

d $16 - (x - 1)^2 \equiv (5 - x)(3 + x)$

2 **R** a Show that $(x - 2)(x + 2) \equiv x^2 - 4$

b Use your rule to work out

i 48×52 ii 298×302

3 **R** The diagram shows a garden in the shape of a rectangle of length $x + 7$ and width $x + 3$.

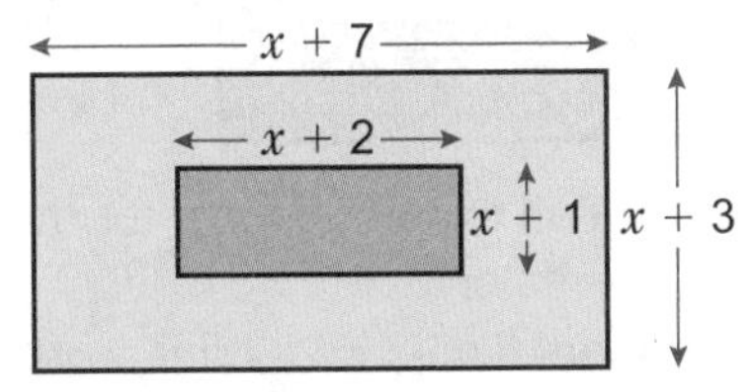

a Write an expression for the area of the garden.

There is a patio in the centre of the garden. The patio is a rectangle of length $x + 2$ and width $x + 1$.

b Write an expression for the area of the patio.

c Show that the area of the garden, excluding the patio, is $7x + 19$.

4 **Exam-style question**

The diagram shows a large rectangle of length $(5x - 4)$ cm and width x cm.
A smaller rectangle of length $2x$ cm and width 3 cm is cut out and removed.
The area of the shape that is left is 40 cm^2.

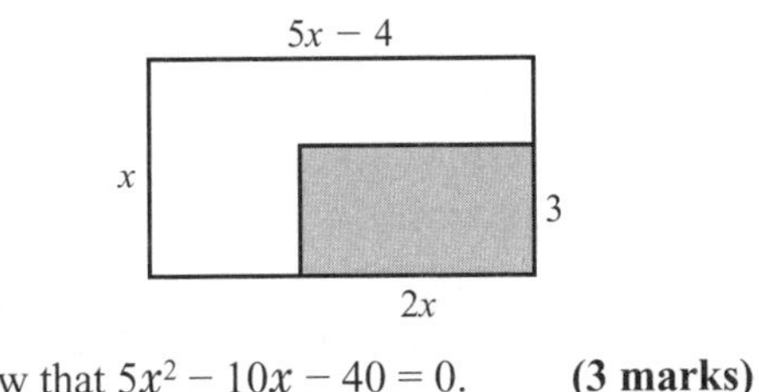

Show that $5x^2 - 10x - 40 = 0$. **(3 marks)**

5 Give a counter example to prove that these statements are *not* true.

a No prime numbers are even.

b The square of a number is always greater than the number itself.

c The product of two numbers is always greater than their sum.

d The sum of two square numbers is always even.

6 **R** Prove that the sum of any two odd numbers is always even.

Q6 hint Let $2n$ be any even number. Let $2n + 1$ be any odd number.

7 **R** a The nth odd number is $2n - 1$. Explain why the next odd number is $2n + 1$.

b Prove that the product of two consecutive odd numbers is 1 less than a multiple of 4.

8 **R** Prove that the product of any odd number and any even number is even.

9 **R** Given that $2(x - a) = x + 8$, where a is an integer, show that x must be an even number.

10 **R** a Work out

i $\frac{1}{3} - \frac{1}{5}$ ii $\frac{1}{5} - \frac{1}{7}$ iii $\frac{1}{2} - \frac{1}{4}$

b Use your answers to part **a** to write down the answer to $\frac{1}{9} - \frac{1}{11}$.

c Explain how you can quickly calculate $\frac{1}{100} - \frac{1}{102}$.

d i Simplify $\frac{1}{x} - \frac{1}{x + 2}$

ii Explain how this proves your answer from part **c**.

11 **R** Show that

$$\frac{1}{x^2 + 4x} - \frac{1}{x^2 - 2x} = \frac{A}{x(x + 4)(x - 2)}$$

and find the value of A.

12 **R** Prove that $n^2 - n$ is even for all values of n.

13 a Write an expression for the product of three consecutive integers, n, $n + 1$ and $n + 2$.

b Hence show that $n^3 + 3n^2 + 2n$ is even.

14 **Exam-style question**

Prove algebraically that the difference between the squares of any two consecutive odd integers is equal to 4 times the integer between them. **(4 marks)**

17 Problem-solving

Solve problems using these strategies where appropriate:

- **Use pictures or lists**
- **Use smaller numbers**
- **Use bar models**
- **Use x for the unknown**
- **Use a flow diagram**
- **Use arrow diagrams**
- **Use geometric sketches**
- **Use graphs**
- **Use logical reasoning.**

1 **R** Anya gets a cab from the station. The cost of the cab ride was £12.40. The cab driver charged £4 for the first mile and then 30p for every tenth of a mile after that.

a What is the formula the cab driver used to find the total (T) cost of the cab ride? Use x for one tenth of a mile.

b How far was the cab ride?

2 Rhian buys a house for £135 000. The value of her house increases by 8% in the first year and 6% in the second year, and then decreases by 3% in the third year.
At the end of the third year, is Rhian's house worth more or less than she originally paid for it? How much more or less?

3 **R** Clare and Emma are asked to rearrange this formula to make d the subject:

$b = \frac{2 + 9d}{d}$

Clare's answer is $d = \frac{2 + 9d}{b}$

Emma says that Clare's answer cannot be correct.

a Why does Emma say this?

b Emma correctly makes d the subject of the formula. What is Emma's answer?

4 Find the sizes of angles x, y and z.

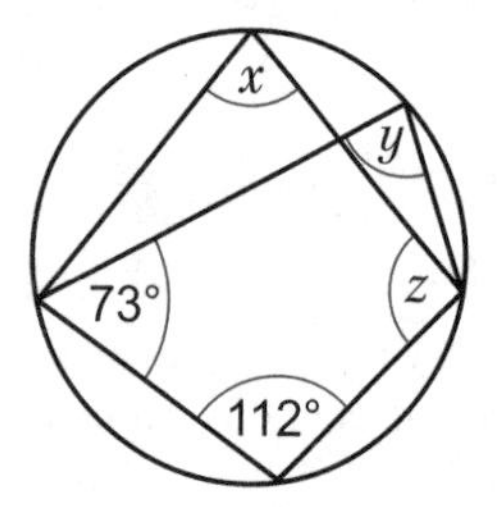

5 **R** Two photographs are mathematically similar.
The area of the larger photograph is 260 cm^2.
The area of the smaller photograph is 65 cm^2.
If the length of the larger photograph is 20 cm, what is the length of the smaller photograph?

6 **Exam-style question**

Simplify fully $\frac{x^2 + 4x + 3}{x^2 + 5x + 6}$ **(3 marks)**

7 **R** Terri is asked to work out value of the expression $\sqrt{20} + \sqrt{180} - \sqrt{48}$, giving her answer in surd form.

a Terri gets the answer $\sqrt{152}$. Explain why she is not correct.

b Work out the correct answer.

8 A rectangular hole is cut out of a rectangular piece of card.
The dimensions are shown on the diagram.

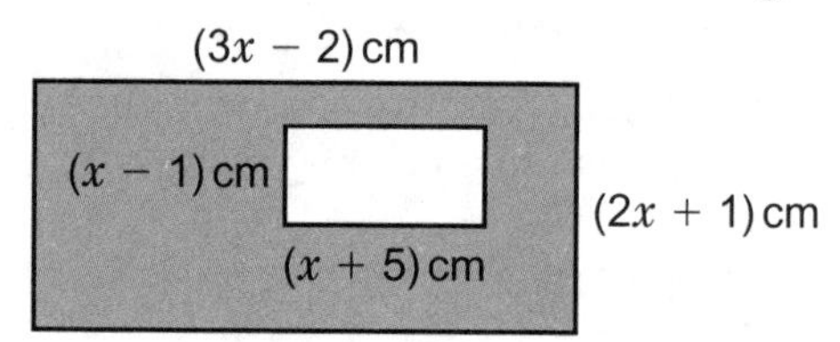

a Work out an expression in x for the shaded area.

b If the shaded area is 283 cm^2, work out the value of x.

9 **R** Two integers have a difference of 2. Prove that the difference between their squares is divisible by 4.

10 Rationalise the denominator and simplify

$\frac{12}{3 - \sqrt{5}}$

Write your answer in the form $a + b\sqrt{c}$ where a, b and c are integers.

18 VECTORS AND GEOMETRIC PROOF

18.1 Vectors and vector notation

1 On squared paper, draw and label these vectors.

a $\mathbf{a} = \begin{pmatrix}3\\2\end{pmatrix}$ b $\mathbf{b} = \begin{pmatrix}-1\\4\end{pmatrix}$ c $\mathbf{c} = \begin{pmatrix}-2\\-5\end{pmatrix}$

d $\overrightarrow{AB} = \begin{pmatrix}-3\\1\end{pmatrix}$ e $\overrightarrow{CD} = \begin{pmatrix}4\\0\end{pmatrix}$

2 The point A is (3, 4), the point B is (2, 7) and the point C is (–2, 5).
Write as column vectors

a $\overrightarrow{AB}$ b $\overrightarrow{BC}$ c $\overrightarrow{AC}$

Example

3 Which of these vectors are equal?

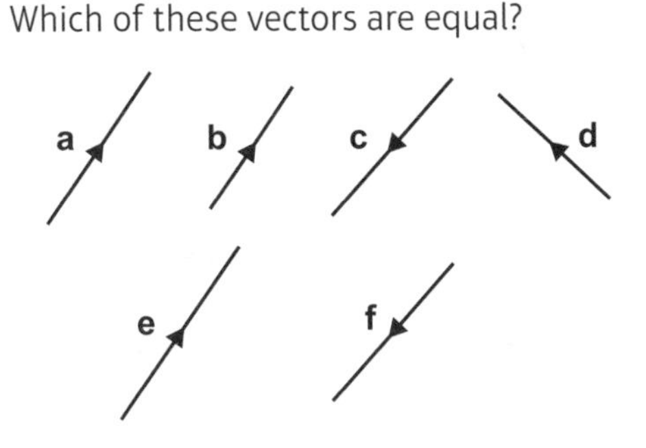

4 Find the magnitude of the vector $\overrightarrow{AB} = \begin{pmatrix}-2\\7\end{pmatrix}$.
Give your answer to 3 significant figures.

5 Work out the magnitude of these vectors.
Where necessary, leave your answer as a surd.

a $\mathbf{a} = \begin{pmatrix}5\\12\end{pmatrix}$ b $\mathbf{b} = \begin{pmatrix}-3\\4\end{pmatrix}$ c $\mathbf{c} = \begin{pmatrix}-1\\-1\end{pmatrix}$

d $\overrightarrow{AB} = \begin{pmatrix}2\\9\end{pmatrix}$ e $\overrightarrow{CD} = \begin{pmatrix}9\\-12\end{pmatrix}$

6 **R** In triangle ABC, $\overrightarrow{AB} = \begin{pmatrix}16\\-63\end{pmatrix}$ and $\overrightarrow{AC} = \begin{pmatrix}-33\\-56\end{pmatrix}$.

a Work out the length of the side AB of the triangle.

b Show that triangle ABC is isosceles.

Q6 hint Draw a sketch of the triangle.

7 **Exam-style question**

A is the point (–2, 5) and B is the point (7, 0).

a Write $\overrightarrow{AB}$ as a column vector. **(1 mark)**

b Find the length of vector $\overrightarrow{AB}$. **(2 marks)**

8 **R** $\overrightarrow{AB} = \begin{pmatrix}-2\\1\end{pmatrix}$. B is the point (3, –4).
Work out the coordinates of A.

18.2 Vector arithmetic

1 **R** The points A, B, C and D are the vertices of a quadrilateral where A has coordinates (3, 2).
$\overrightarrow{AB} = \begin{pmatrix}2\\3\end{pmatrix}$, $\overrightarrow{BC} = \begin{pmatrix}3\\1\end{pmatrix}$ and $\overrightarrow{CD} = \begin{pmatrix}-2\\-3\end{pmatrix}$.

a Draw quadrilateral ABCD on squared paper.

b Write $\overrightarrow{AD}$ as a column vector.

c What type of quadrilateral is ABCD?

d What do you notice about $\overrightarrow{BC}$ and $\overrightarrow{AD}$?

2 **R** The points A, B, C and D are the vertices of a rectangle.
A has coordinates (2, –1), $\overrightarrow{AB} = \begin{pmatrix}4\\0\end{pmatrix}$ and $\overrightarrow{AD} = \begin{pmatrix}0\\-3\end{pmatrix}$.

a Draw rectangle ABCD on squared paper.

b Write as a column vector

i $\overrightarrow{CB}$

ii $\overrightarrow{BC}$

What do you notice?

c What do you notice about

i $\overrightarrow{AB}$ and $\overrightarrow{DC}$

ii $\overrightarrow{AD}$ and $\overrightarrow{BC}$?

3 In quadrilateral ABCD, $\overrightarrow{AB} = \begin{pmatrix}2\\3\end{pmatrix}$, $\overrightarrow{BC} = \begin{pmatrix}1\\-3\end{pmatrix}$, $\overrightarrow{CD} = \begin{pmatrix}-2\\-3\end{pmatrix}$ and $\overrightarrow{DA} = \begin{pmatrix}-1\\3\end{pmatrix}$.
What type of quadrilateral is ABCD?

4 Exam-style question

P is the point (5, 6). $\overrightarrow{PQ} = \begin{pmatrix}3\\1\end{pmatrix}$

a Find the coordinates of Q. **(1 mark)**

R is the point (7, 4).

b Express $\overrightarrow{PR}$ as a column vector. **(3 marks)**

$\overrightarrow{RT} = \begin{pmatrix}3\\-5\end{pmatrix}$

c Calculate the length of PT.
Give your answer to 3 significant figures. **(3 marks)**

Q4 hint Sketch a diagram.

5 The diagram shows the vectors **a** and **b**.

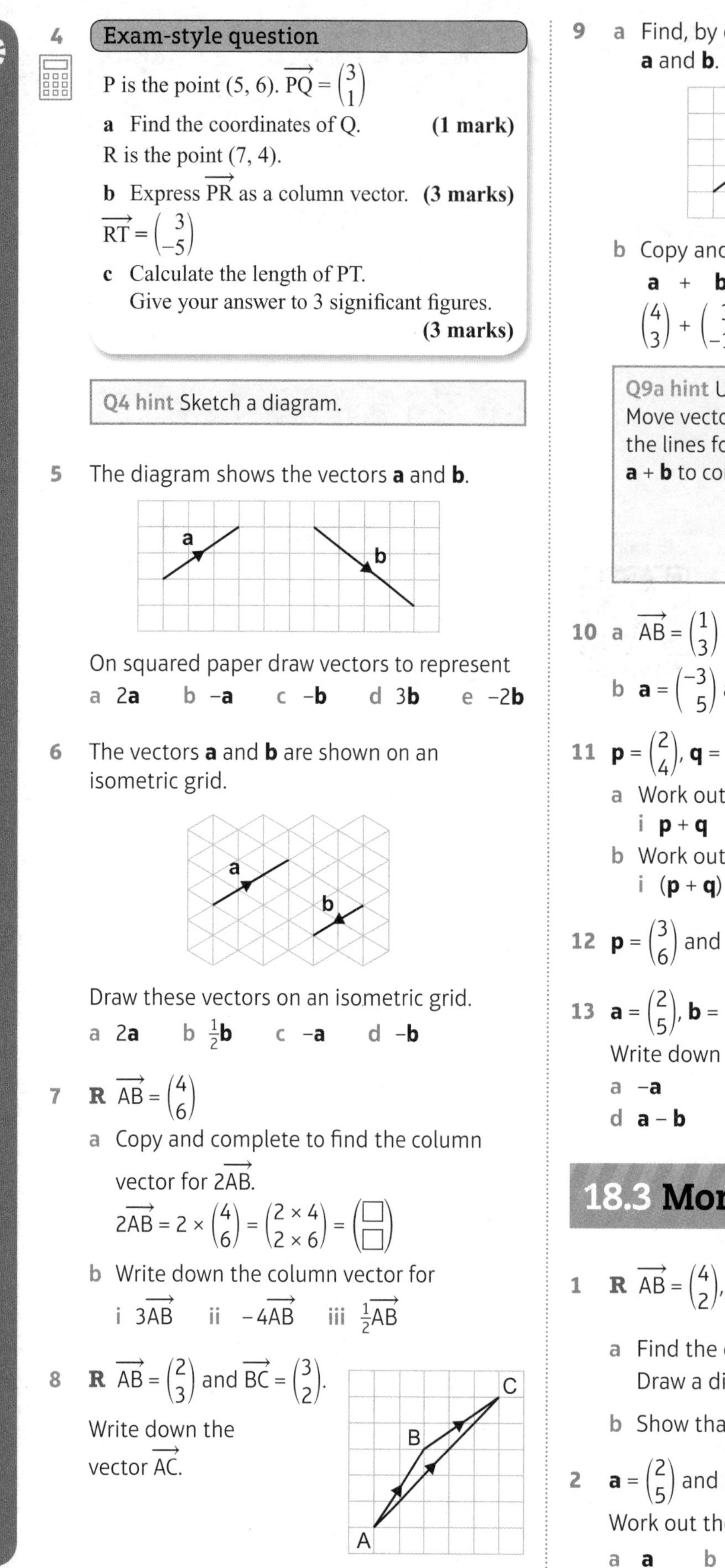

On squared paper draw vectors to represent

a 2**a** b −**a** c −**b** d 3**b** e −2**b**

6 The vectors **a** and **b** are shown on an isometric grid.

Draw these vectors on an isometric grid.

a 2**a** b $\frac{1}{2}$**b** c −**a** d −**b**

7 **R** $\overrightarrow{AB} = \begin{pmatrix}4\\6\end{pmatrix}$

a Copy and complete to find the column vector for $2\overrightarrow{AB}$.

$2\overrightarrow{AB} = 2 \times \begin{pmatrix}4\\6\end{pmatrix} = \begin{pmatrix}2 \times 4\\2 \times 6\end{pmatrix} = \begin{pmatrix}\square\\\square\end{pmatrix}$

b Write down the column vector for

i $3\overrightarrow{AB}$ ii $-4\overrightarrow{AB}$ iii $\frac{1}{2}\overrightarrow{AB}$

8 **R** $\overrightarrow{AB} = \begin{pmatrix}2\\3\end{pmatrix}$ and $\overrightarrow{BC} = \begin{pmatrix}3\\2\end{pmatrix}$.

Write down the vector $\overrightarrow{AC}$.

9 a Find, by drawing, the sum of the vectors **a** and **b**.

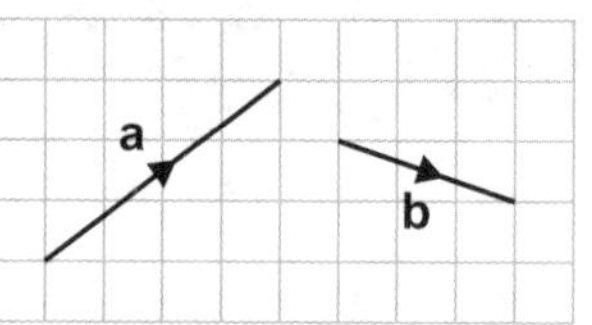

b Copy and complete this vector addition.

a + **b** = **a** + **b**

$\begin{pmatrix}4\\3\end{pmatrix} + \begin{pmatrix}3\\-1\end{pmatrix} = \begin{pmatrix}\square\\\square\end{pmatrix}$

Q9a hint Use the triangle law of addition. Move vector **b** to the end of vector **a** so that the lines follow on. Draw and label the vector **a** + **b** to complete the triangle.

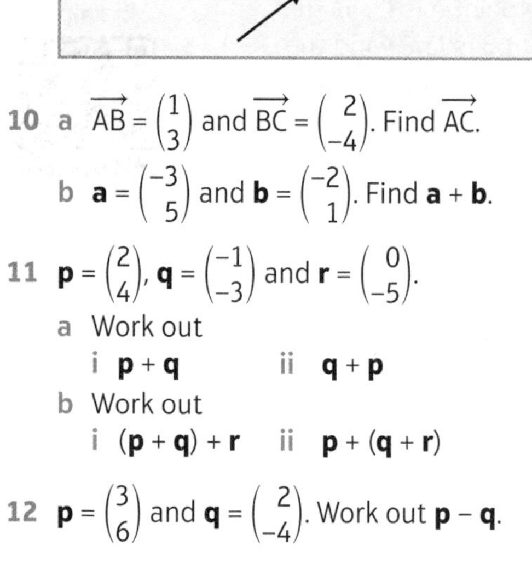

10 a $\overrightarrow{AB} = \begin{pmatrix}1\\3\end{pmatrix}$ and $\overrightarrow{BC} = \begin{pmatrix}2\\-4\end{pmatrix}$. Find $\overrightarrow{AC}$.

b $\mathbf{a} = \begin{pmatrix}-3\\5\end{pmatrix}$ and $\mathbf{b} = \begin{pmatrix}-2\\1\end{pmatrix}$. Find **a** + **b**.

11 $\mathbf{p} = \begin{pmatrix}2\\4\end{pmatrix}$, $\mathbf{q} = \begin{pmatrix}-1\\-3\end{pmatrix}$ and $\mathbf{r} = \begin{pmatrix}0\\-5\end{pmatrix}$.

a Work out

i **p** + **q** ii **q** + **p**

b Work out

i (**p** + **q**) + **r** ii **p** + (**q** + **r**)

12 $\mathbf{p} = \begin{pmatrix}3\\6\end{pmatrix}$ and $\mathbf{q} = \begin{pmatrix}2\\-4\end{pmatrix}$. Work out **p** − **q**.

13 $\mathbf{a} = \begin{pmatrix}2\\5\end{pmatrix}$, $\mathbf{b} = \begin{pmatrix}-1\\-2\end{pmatrix}$ and $\mathbf{c} = \begin{pmatrix}-4\\0\end{pmatrix}$.

Write down the column vector for

a −**a** b **a** + **b** c **a** + **b** + **c**

d **a** − **b** e **b** − **c**

18.3 More vector arithmetic

1 **R** $\overrightarrow{AB} = \begin{pmatrix}4\\2\end{pmatrix}$, $\overrightarrow{BC} = \begin{pmatrix}1\\3\end{pmatrix}$ and $\overrightarrow{CD} = \begin{pmatrix}-5\\-4\end{pmatrix}$.

a Find the column vector for $\overrightarrow{AD}$.
Draw a diagram to show this.

b Show that $\overrightarrow{DB} = \begin{pmatrix}4\\1\end{pmatrix}$.

2 $\mathbf{a} = \begin{pmatrix}2\\5\end{pmatrix}$ and $\mathbf{b} = \begin{pmatrix}4\\4\end{pmatrix}$.

Work out the magnitude of

a **a** b 2**b** c **a** + **b** d **a** − **b**

3 In the quadrilateral ABCD, $\overrightarrow{AB} = \mathbf{a}$, $\overrightarrow{BC} = \mathbf{b}$ and $\overrightarrow{CD} = \mathbf{c}$.

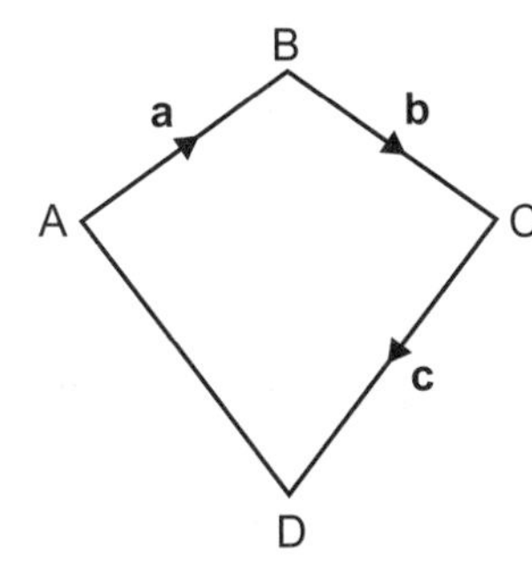

Find in terms of **a**, **b** and **c**

a $\overrightarrow{AC}$ b $\overrightarrow{AD}$

4 $\overrightarrow{OA} = \mathbf{b}$

M is the midpoint of OA.

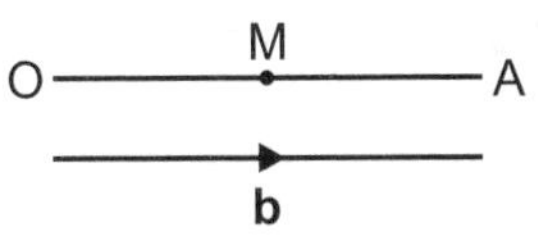

a Write down $\overrightarrow{OM}$ in terms of **b**.

$\overrightarrow{OA} = \begin{pmatrix} 6 \\ 2 \end{pmatrix}$

b Express as a column vector

i $\overrightarrow{AO}$ ii $\overrightarrow{OM}$

5 In the diagram $\overrightarrow{AC} = \mathbf{a}$ and $\overrightarrow{CM} = \mathbf{b}$.

M is the midpoint of CB.

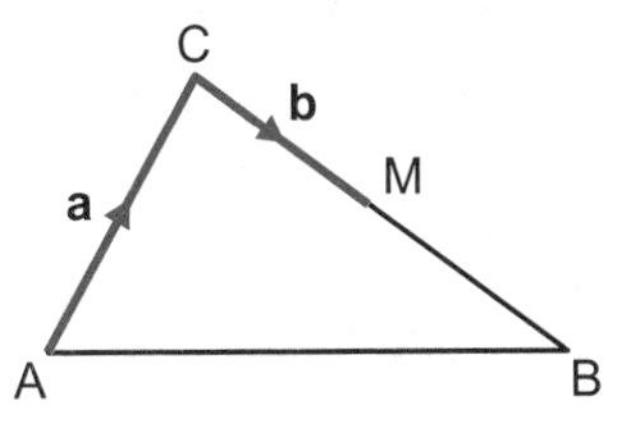

Write down in terms of **a** and/or **b**.

a $\overrightarrow{CB}$ b $\overrightarrow{MA}$ c $\overrightarrow{AB}$

6 JKLM is a parallelogram.

$\overrightarrow{LM} = \mathbf{a}$ and $\overrightarrow{MJ} = \mathbf{b}$.

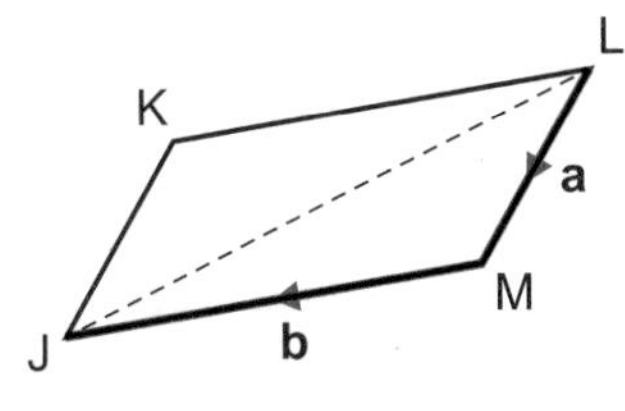

a Explain why $\overrightarrow{KJ} = \mathbf{a}$.

b Find

i $\overrightarrow{LK}$ ii $\overrightarrow{LJ}$

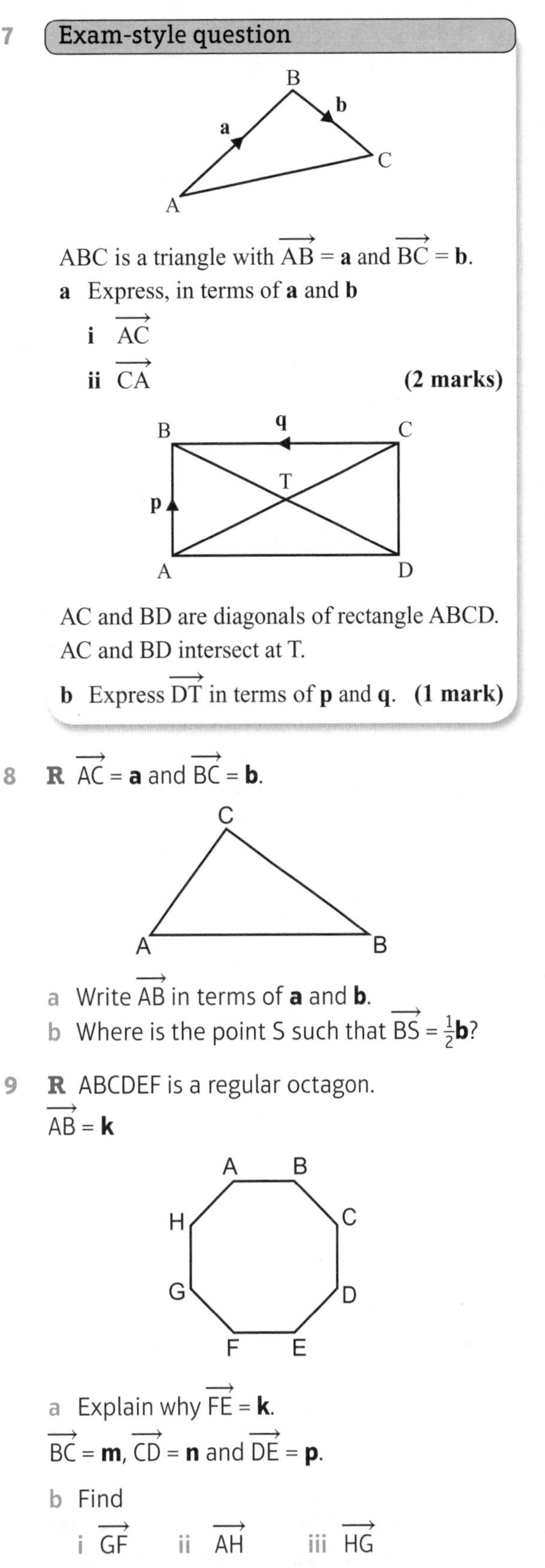

7 **Exam-style question**

ABC is a triangle with $\overrightarrow{AB} = \mathbf{a}$ and $\overrightarrow{BC} = \mathbf{b}$.

a Express, in terms of **a** and **b**

i $\overrightarrow{AC}$

ii $\overrightarrow{CA}$ **(2 marks)**

AC and BD are diagonals of rectangle ABCD.

AC and BD intersect at T.

b Express $\overrightarrow{DT}$ in terms of **p** and **q**. **(1 mark)**

8 **R** $\overrightarrow{AC} = \mathbf{a}$ and $\overrightarrow{BC} = \mathbf{b}$.

a Write $\overrightarrow{AB}$ in terms of **a** and **b**.

b Where is the point S such that $\overrightarrow{BS} = \frac{1}{2}\mathbf{b}$?

9 **R** ABCDEF is a regular octagon.

$\overrightarrow{AB} = \mathbf{k}$

a Explain why $\overrightarrow{FE} = \mathbf{k}$.

$\overrightarrow{BC} = \mathbf{m}$, $\overrightarrow{CD} = \mathbf{n}$ and $\overrightarrow{DE} = \mathbf{p}$.

b Find

i $\overrightarrow{GF}$ ii $\overrightarrow{AH}$ iii $\overrightarrow{HG}$

c Find

i $\overrightarrow{AD}$ ii $\overrightarrow{AE}$

d What is $\overrightarrow{HE}$?

10 **R** ABCD is a rectangle.
M is the midpoint of DC.
$\overrightarrow{AB} = \mathbf{q}$ and $\overrightarrow{DA} = \mathbf{p}$.

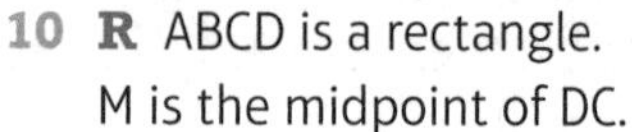

Write in terms of **p** and **q**

a $\overrightarrow{DC}$ b $\overrightarrow{DM}$ c $\overrightarrow{AM}$ d $\overrightarrow{BM}$

11 **R** Here are five vectors.

$\overrightarrow{AB} = 2\mathbf{a} - 8\mathbf{b}$ $\overrightarrow{CD} = \mathbf{a} + 4\mathbf{b}$ $\overrightarrow{EF} = 4\mathbf{a} - 16\mathbf{b}$

$\overrightarrow{GH} = -2\mathbf{a} + 8\mathbf{b}$ $\overrightarrow{IJ} = \mathbf{a} - 7\mathbf{b}$

a Three of these vectors are parallel. Which three?

b Simplify

i $5\mathbf{p} - 6\mathbf{q} + 2\mathbf{p} - 7\mathbf{q}$

ii $3(\mathbf{a} - 2\mathbf{b}) + \frac{1}{2}(3\mathbf{a} + 4\mathbf{b})$

12 **R** In parallelogram ABCD, $\overrightarrow{AB} = \mathbf{a}$ and $\overrightarrow{BC} = \mathbf{b}$.
M is the midpoint of AD.

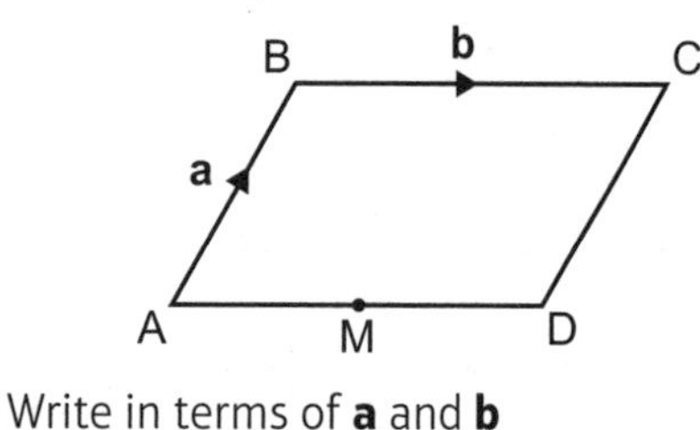

Write in terms of **a** and **b**

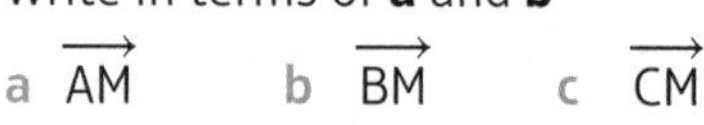

a $\overrightarrow{AM}$ b $\overrightarrow{BM}$ c $\overrightarrow{CM}$

13 In triangle ABC, $\overrightarrow{AB} = \mathbf{a}$ and $\overrightarrow{AC} = \mathbf{b}$.
M is the midpoint of AB.
N is the midpoint of AC.

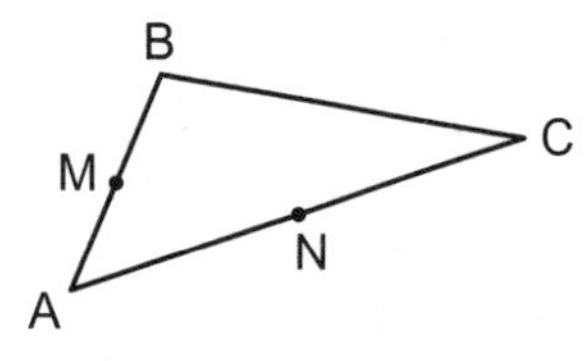

Write in terms of **a** and **b**

a $\overrightarrow{BC}$ b $\overrightarrow{BM}$ c $\overrightarrow{CN}$ d $\overrightarrow{MN}$

18.4 Parallel vectors and collinear points

1 $\mathbf{a} = \begin{pmatrix}-2\\5\end{pmatrix}$, $\mathbf{b} = \begin{pmatrix}-3\\1\end{pmatrix}$ and $\mathbf{a} + \mathbf{c} = \mathbf{b}$.

Calculate **c**.

2 $3\begin{pmatrix}x\\y\end{pmatrix} + \begin{pmatrix}2\\-5\end{pmatrix} = \begin{pmatrix}5\\10\end{pmatrix}$

Find $\begin{pmatrix}x\\y\end{pmatrix}$.

3 $\mathbf{e} = \begin{pmatrix}2\\-3\end{pmatrix}$ and $\mathbf{f} = \begin{pmatrix}4\\-3\end{pmatrix}$.

Calculate **g** given that $2\mathbf{e} - \mathbf{g} = \mathbf{f}$.

4 O is the origin (0, 0). A has coordinates (3, 6) and B has coordinates (5, 1).
Find as column vectors

a $\overrightarrow{OA}$ b $\overrightarrow{AO}$ c $\overrightarrow{OB}$ d $\overrightarrow{AB}$

5 $\overrightarrow{AB} = \mathbf{a}$ and $\overrightarrow{BC} = \mathbf{b}$.

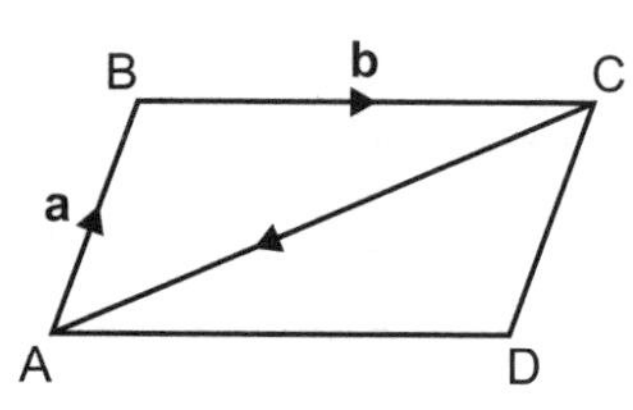

Express $\overrightarrow{CA}$ in terms of **a** and **b**.

6 **R** The points P, Q, R and S have coordinates (1, 3), (7, 5), (−12, −15) and (24, −3) respectively.
O is the origin.

a Write down the position vectors $\overrightarrow{OP}$ and $\overrightarrow{OQ}$.

b Write down as a column vector

i $\overrightarrow{PQ}$ ii $\overrightarrow{RS}$

c What do these results show about the lines PQ and RS?

7 **Exam-style question**

P is the point (2, 9) and Q is the point (−4, 7).

a Find $\overrightarrow{PQ}$ as a column vector. **(1 mark)**

R is the point such that $\overrightarrow{QR} = \begin{pmatrix}5\\7\end{pmatrix}$.

b Write down the coordinates of the point R. **(2 marks)**

X is the midpoint of PQ. O is the origin.

c Find $\overrightarrow{OX}$ as a column vector. **(2 marks)**

Q7 hint In this type of vector question it can be helpful to draw a sketch.

8 **R** Point A has coordinates (2, 7), point B has coordinates (1, 3) and point C has coordinates (−1, −6).

a Write $\overrightarrow{AB}$ as a column vector.

b $\overrightarrow{CD} = 6\overrightarrow{AB}$. Find $\overrightarrow{CD}$.

c Find the coordinates of D.

9 **P** $\mathbf{a} = \begin{pmatrix} -1 \\ 3 \end{pmatrix}$ and $\mathbf{b} = \begin{pmatrix} 4 \\ -2 \end{pmatrix}$.

Find a vector **c** such that **a** + **c** is parallel to **a** − **b**.

10 **R** OABC is a quadrilateral in which $\overrightarrow{OA} = 2\mathbf{a}$, $\overrightarrow{OB} = 2\mathbf{a} + \mathbf{b}$ and $\overrightarrow{OC} = \frac{1}{2}\mathbf{b}$.

a Find $\overrightarrow{AB}$ in terms of **a** and **b**.
What does this tell you about $\overrightarrow{AB}$ and $\overrightarrow{OC}$?

b Find $\overrightarrow{BC}$ in terms of **a** and **b**.
What does this tell you about $\overrightarrow{OA}$ and $\overrightarrow{BC}$?

c What type of quadrilateral is OABC?

11 **P** The points A, B and C have coordinates (1, 5), (3, 12) and (5, 19) respectively.

a Find as column vectors

i $\overrightarrow{AB}$ ii $\overrightarrow{AC}$

b What do these results show you about the points A, B and C?

12 **P** The point P has coordinates (3, 2).
The point Q has coordinates (7, 7).
The point R has coordinates (15, 17).
Show that points P, Q and R are collinear.

18.5 Solving geometric problems

1 **R** In parallelogram ABCD the point M is the midpoint of AB and the point N is the midpoint of BC.
$\overrightarrow{AB} = \mathbf{a}$ and $\overrightarrow{BC} = \mathbf{b}$.

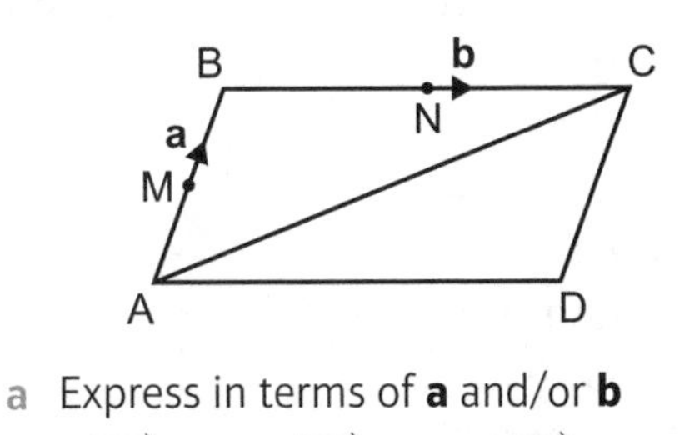

a Express in terms of **a** and/or **b**

i $\overrightarrow{AM}$ ii $\overrightarrow{BN}$ iii $\overrightarrow{MB}$ iv $\overrightarrow{NB}$

b Express $\overrightarrow{AC}$ in terms of **a** and **b**.

c Express $\overrightarrow{MN}$ in terms of **a** and **b**.

d Explain what the answers to parts **b** and **c** show about AC and MN.

2 **R** In triangle ABO, $\overrightarrow{OA} = \mathbf{a}$, and $\overrightarrow{OB} = \mathbf{b}$.
The point X divides AB in the ratio 2 : 3
Express in terms of **a** and **b**

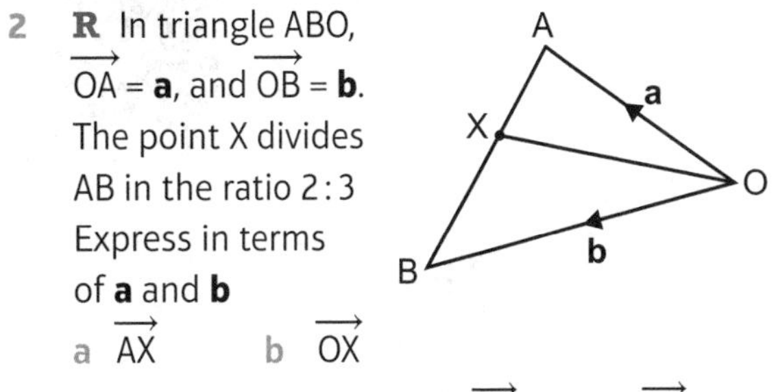

a $\overrightarrow{AX}$ b $\overrightarrow{OX}$

3 **P** In parallelogram ABCD, $\overrightarrow{AB} = \mathbf{a}$ and $\overrightarrow{BC} = \mathbf{b}$.

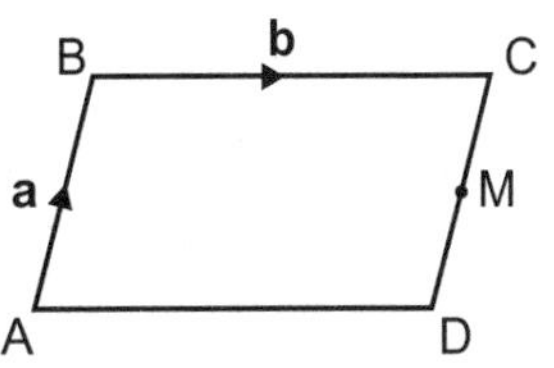

a Find in terms of **a** and/or **b** the vector $\overrightarrow{CD}$.
M is the midpoint of CD.

b Find in terms of **a** and/or **b** the vector $\overrightarrow{CM}$.

c Find in terms of **a** and/or **b** the vector $\overrightarrow{AM}$.

4 **Exam-style question**

$\overrightarrow{OX} = -\mathbf{a} + 2\mathbf{b}$ and $\overrightarrow{OY} = 2\mathbf{a} + 3\mathbf{b}$

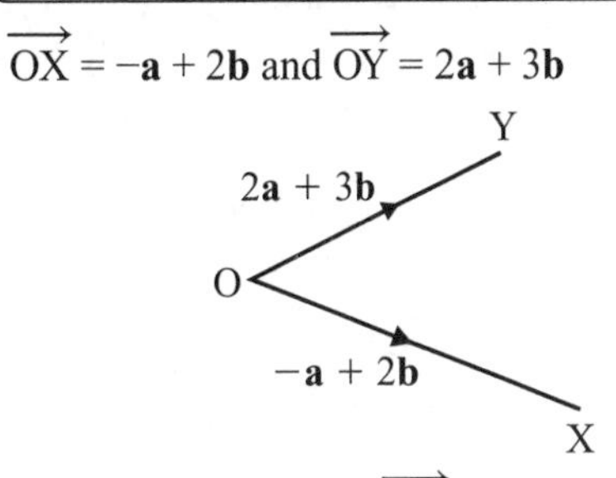

a Express the vector $\overrightarrow{XY}$ in terms of **a** and **b**.
Give your answer in its simplest form.
(2 marks)

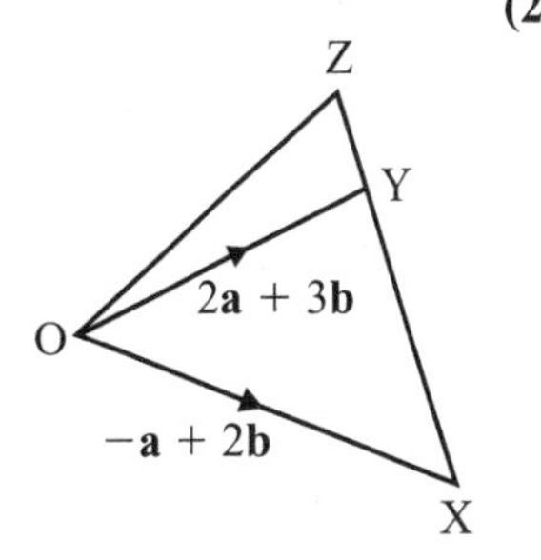

XYZ is a straight line.
XY : YZ = 3 : 1.

b Express the vector $\overrightarrow{OZ}$ in terms of **a** and **b**.
Give your answer in its simplest form.
(3 marks)

Q4b hint XY : YZ = 3 : 1 YZ = $\frac{\square}{\square}$XZ

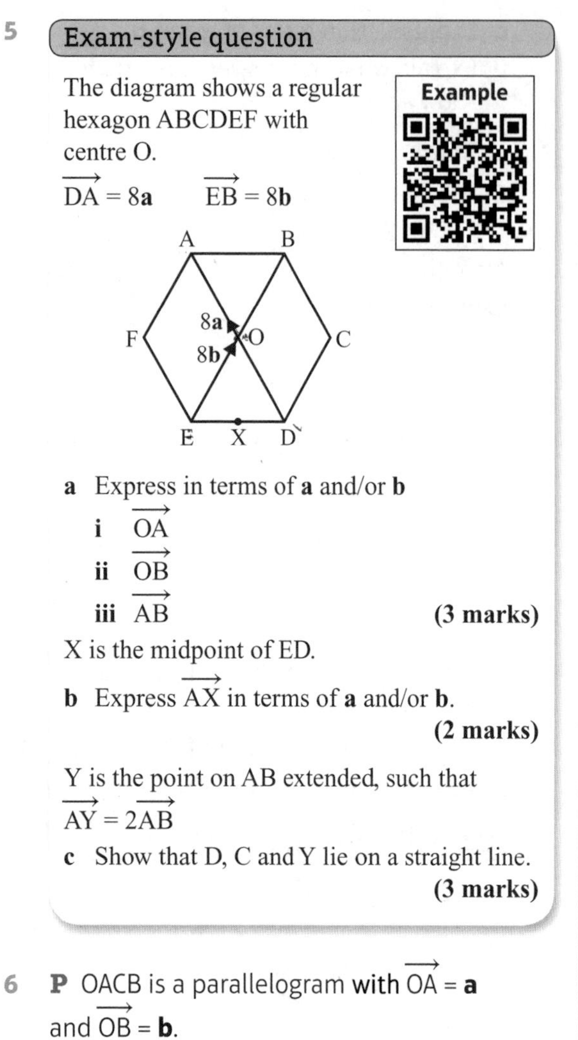

5 **Exam-style question**

The diagram shows a regular hexagon ABCDEF with centre O.

$\overrightarrow{DA} = 8\mathbf{a}$ $\quad \overrightarrow{EB} = 8\mathbf{b}$

a Express in terms of **a** and/or **b**

i $\overrightarrow{OA}$

ii $\overrightarrow{OB}$

iii $\overrightarrow{AB}$ **(3 marks)**

X is the midpoint of ED.

b Express $\overrightarrow{AX}$ in terms of **a** and/or **b**. **(2 marks)**

Y is the point on AB extended, such that $\overrightarrow{AY} = 2\overrightarrow{AB}$

c Show that D, C and Y lie on a straight line. **(3 marks)**

6 **P** OACB is a parallelogram with $\overrightarrow{OA} = \mathbf{a}$ and $\overrightarrow{OB} = \mathbf{b}$.

E is the point on AC such that $AE = \frac{1}{3}AC$.

F is the point on BC such that $BF = \frac{1}{3}BC$.

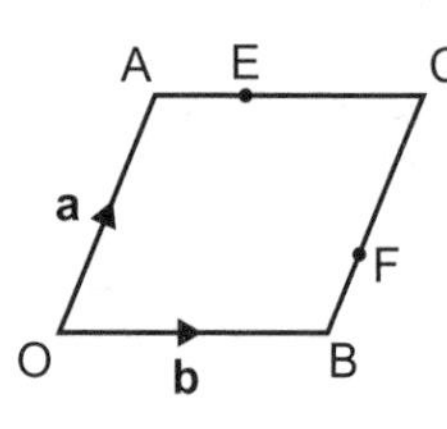

a Find in terms of **a** and/or **b**

i $\overrightarrow{AB}$

ii $\overrightarrow{AE}$

iii $\overrightarrow{OE}$

iv $\overrightarrow{OF}$

b Show that EF is parallel to AB.

7 **P** In triangle ABC, $\overrightarrow{AB} = \mathbf{b}$ and $\overrightarrow{AC} = \mathbf{c}$.

The point M is the midpoint of BC and Q is the point such that $\overrightarrow{AQ} = \frac{2}{3}\overrightarrow{AM}$.

a Find in terms of **b** and **c**

i $\overrightarrow{AM}$ $\quad$ ii $\overrightarrow{AQ}$ $\quad$ iii $\overrightarrow{BQ}$

N is the midpoint of AB.

b Find in terms of **b** and **c** the vector $\overrightarrow{CN}$.

c Show that the points C, Q and N are collinear.

8 **P** In the diagram $\overrightarrow{OR} = 9\mathbf{a}$, $\overrightarrow{OP} = 4\mathbf{b}$ and $\overrightarrow{PQ} = 3\mathbf{a}$.

The point M is on PQ such that $\overrightarrow{PM} = 2\mathbf{a}$.

The point N is on OR such that $\overrightarrow{ON} = \frac{5}{9}\overrightarrow{OR}$.

The point T is on MN such that $\overrightarrow{MT} = \frac{2}{3}\overrightarrow{MN}$.

a Find in terms of **a** and/or **b** the vector $\overrightarrow{NM}$.

b Find in terms of **a** and/or **b** the vector $\overrightarrow{OT}$.

c V is the point such that $\overrightarrow{QV} = 9\mathbf{a}$.
Find in terms of **a** and **b** the vector $\overrightarrow{OV}$.

d Show that T lies on the line OV.

e When $\mathbf{a} = \begin{pmatrix} 1 \\ 5 \end{pmatrix}$ and $\mathbf{b} = \begin{pmatrix} 2 \\ 7 \end{pmatrix}$ find the length of QR.
Give your answer to 3 significant figures.

18 Problem-solving

Solve problems using these strategies where appropriate:

- **Use pictures or lists**
- **Use smaller numbers**
- **Use bar models**
- **Use x for the unknown**
- **Use a flow diagram**
- **Use arrow diagrams**
- **Use geometric sketches**
- **Use graphs**
- **Use logical reasoning**
- **Use problem-solving strategies and then 'explain'.**

1 Write the expression $\frac{y+4}{3} + \frac{5y-2}{8}$ as a single fraction in its simplest form.

2 **R** The diagram shows a circle with centre O. A, B, C, D and E are points on the circumference.
Angle DOC = 112° and OA bisects angle DAC.

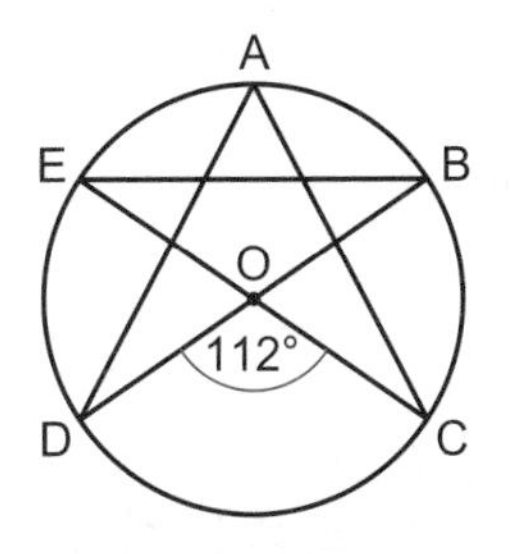

Work out the sizes of the angles at the points of the star shape.

3 **R** WXYZ is a parallelogram.

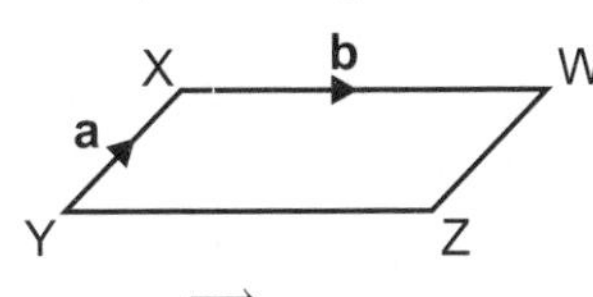

a Explain why $\overrightarrow{YZ} = \mathbf{b}$.

b Describe $\overrightarrow{YW}$ in terms of **a** and **b**.

c C is the point on $\overrightarrow{YZ}$ such that $\overrightarrow{YC} = \frac{1}{2}\overrightarrow{YZ}$.
$\overrightarrow{CD}$ is parallel to $\overrightarrow{YW}$. Describe where point D is on this parallelogram.

4 $\overrightarrow{AB} = \begin{pmatrix} 8 \\ -6 \end{pmatrix}$, $\overrightarrow{CD} = \begin{pmatrix} 12 \\ -8 \end{pmatrix}$, $\overrightarrow{NP} = \begin{pmatrix} 5 \\ 9 \end{pmatrix}$ and $\overrightarrow{XY} = \begin{pmatrix} -4 \\ 7 \end{pmatrix}$.
Put the vectors in descending order of length.

5 **R** Vector **c** is parallel to **a** + **b**.
Vector **d** is parallel to **a** − **b**.
Vector **e** is parallel to vector **d**.
Find vectors **c**, **d** and **e** when $\mathbf{a} = \begin{pmatrix} 3 \\ 4 \end{pmatrix}$ and $\mathbf{b} = \begin{pmatrix} 5 \\ -3 \end{pmatrix}$.

6 In triangle ABC, $\overrightarrow{AB} = \begin{pmatrix} 3 \\ 4 \end{pmatrix}$ and $\overrightarrow{BC} = \begin{pmatrix} -3 \\ -1 \end{pmatrix}$.

a Draw triangle ABC on suitable grid paper.

b Work out $\overrightarrow{AC}$.

c Use vector addition to show that $\overrightarrow{AC}$ is equivalent to $\overrightarrow{AB} + \overrightarrow{BC}$.

Q6 hint Think of the triangle law of vector addition.

7 A curve has the equation
$y = ax^3 + bx^2 + cx + d$
It crosses the x-axis at $x = -3$, $x = -1$ and $x = 1$.
What are the values of a, b, c and d?

8 **Exam-style question**

a Find the resultant of the vectors $\begin{pmatrix} 4 \\ -1 \end{pmatrix}$ and $\begin{pmatrix} -2 \\ 5 \end{pmatrix}$. **(1 mark)**

b OABC is a parallelogram.
M is the midpoint of AB.
N is the midpoint of BC.
$\overrightarrow{OA} = \mathbf{a}$ and $\overrightarrow{OC} = \mathbf{c}$

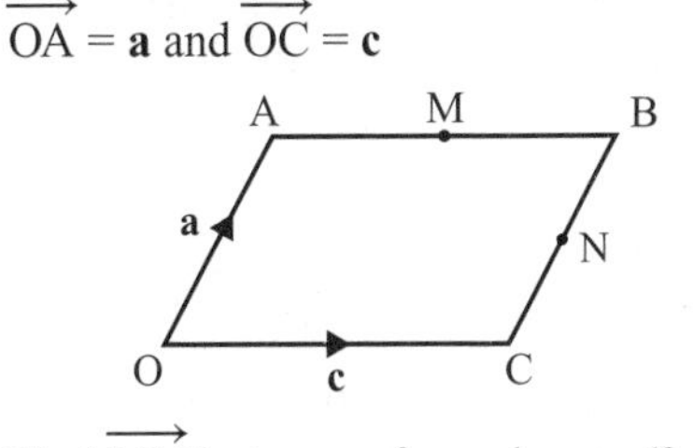

Find $\overrightarrow{MN}$ in terms of **a** and **c**. **(2 marks)**

November 2012, Q12, A502/02

9 **R** In triangle DEF, DE = EF, X is the midpoint of DE and Y is the midpoint of DF.
$\overrightarrow{DE} = 2\mathbf{a}$ and $\overrightarrow{DF} = 3\mathbf{b}$.

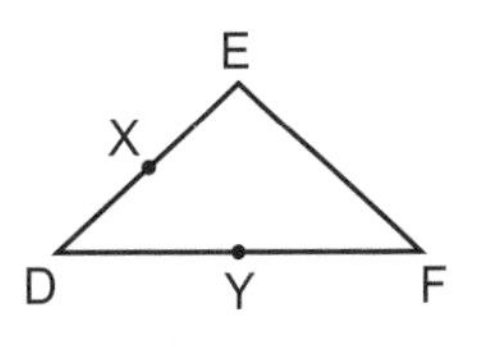

Express in terms of **a** and/or **b**

a $\overrightarrow{EF}$ b $\overrightarrow{DX}$

c $\overrightarrow{DY}$ d $\overrightarrow{DE} + \overrightarrow{EF}$

10 **R** In the diagram, A is the midpoint of OP, C is the midpoint of PQ and B is the point on OQ such that OB : BQ = 1 : 2
$\overrightarrow{OA} = \mathbf{a}$ and $\overrightarrow{OB} = \mathbf{b}$.

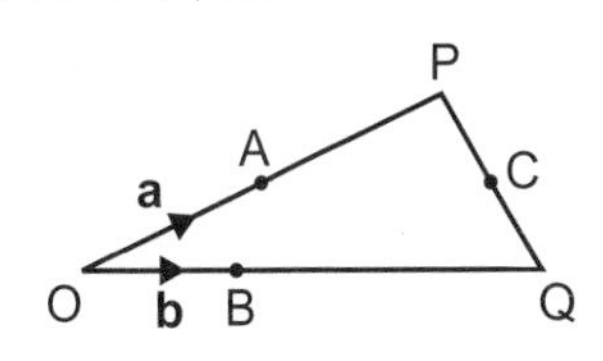

Work out these vectors, in terms of **a** and/or **b**.

a $\overrightarrow{OP}$ b $\overrightarrow{OQ}$ c $\overrightarrow{PQ}$

d $\overrightarrow{AB}$ e $\overrightarrow{AC}$ f $\overrightarrow{BC}$

19 PROPORTION AND GRAPHS

19.1 Direct proportion

1 **R** The tables show the prices paid for different quantities of US dollars from two currency exchange websites during July.

Website A

Sterling (£)	40	180	500	450	160	80	140	320
US dollars ($)	70	300	750	700	255	120	210	510

Website B

Sterling (£)	300	200	280	410	100	200	370	50
US dollars ($)	540	340	480	700	170	350	610	80

a Draw a scatter graph for both sets of information on the same axes.

b Draw a line of best fit for each set of data.

c Write a formula for dollars, D, in terms of sterling, S, for

i Website A

ii Website B.

d Which currency exchange website offers better value for money when buying dollars? Explain your answer.

2 y is directly proportional to x.
$y = 60$ when $x = 12$

a Express y in terms of x.

b Find y when $x = 20$

c Find x when $y = 2.5$

Example

Q2a hint Start with the statement $y \propto x$, then write the equation $y = kx$. Use the values of x and y to find the value of k.

3 y is directly proportional to x.
$y = 46$ when $x = 6$

a Write a formula for y in terms of x.

b Find y when $x = 24$

c Find x when $y = 161$

4 **P** y is directly proportional to x.

a $y = 12$ when $x = 5$
Find x when $y = 7.2$

b $y = 16.2$ when $x = 9$
Find x when $y = 15.3$

c $y = 17.4$ when $x = 0.6$
Find x when $y = 58$

5 **Exam-style question**

y is directly proportional to x.
When $x = 500$, $y = 25$

a Find a formula for y in terms of x. **(3 marks)**

b Calculate the value of y when $x = 360$. **(1 mark)**

19.2 More direct proportion

1 **R** The extension of a spring, d (in metres), is directly proportional to the force, F (in newtons, N), used to extend the spring.
When $F = 20$ N, $d = 5$ m.

a Express F in terms of d.

b Find F when $d = 12.5$ m.

c Find d when $F = 65$ N.

2 **R** The table gives information about the total surface area, A_1, of different 3D objects and the area of their smallest face, A_2.

A_1 (cm²)	26	52	97.5	130
A_2 (cm²)	4	8	15	20

a Show that A_1 is directly proportional to A_2.

b Given that $A_1 = kA_2$, work out the value of k.

c Write a formula for A_2 in terms of A_1.

d Work out

i the value of A_1 when $A_2 = 5$

ii the value of A_2 when $A_1 = 162.5$

3 **R** The distance, d (in km), covered by a long distance runner is directly proportional to the time taken, t (in hours).
The runner covers a distance of 42 km in 4 hours.

a Find a formula for d in terms of t.

b Find the value of d when $t = 8$

c Find the value of t when $d = 7.7$

d What happens to the distance travelled, d, when the time, t, is

i trebled

ii divided by 3?

4 **R** The amount, C (in £), a plumber charges is directly proportional to the time, t (in hours), that the plumber works. A plumber earns £247.50 when she works 5.5 hours.

a Sketch a graph of C against t.

b Write a formula for C in terms of t.

c Use your formula to work out how many hours the plumber has worked when she earns £1035.

5 y is proportional to the square of x.
When $x = 2$, $y = 16$

a Write the statement of proportionality.

b Write an equation using k.

c Work out the value of k.

d Find y when $x = 10$

e Find x when $y = 100$

6 **P** y is proportional to the cube of x.
When $x = 5$, $y = 25$

a Write a formula for y in terms of x.

b Find y when $x = 2$

c Find x when $y = 25\,000$

> **Q6a hint** Write $y = k\square$ and find the value of k.

7 y is proportional to the square root of x.
When $x = 16$, $y = 20$

a Find a formula for y in terms of x.

b Find y when $x = 64$

c Find x when $y = 55$

8 **Exam-style question**

y is directly proportional to the cube of x.
When $x = 2$, $y = 48$
Find the value of y when $x = 5$. **(4 marks)**

9 **R** When an object accelerates steadily from rest, the distance, d (in metres), it moves varies in direct proportion to the square of the time, t (in seconds), it has been moving. An object moves 176.4 m in 6 seconds.

a Write a formula for d in terms of t.

b How far does an object move if it accelerates like this for 10 seconds from rest?

c How many seconds has an object been accelerating for if it has moved 1102.5 m?

d What happens to the distance moved, d, if the time the object has been accelerating for is doubled?

10 **P** The volume, V (in cm³), of a sphere is directly proportional to the cube of its radius, r (in cm). A sphere with a radius of 5 cm has a volume of 523.5 cm³.

a Write a formula for V in terms of r.

b Calculate V when the radius is 20 cm.

11 **R** The y-coordinate of a point on a parabola is directly proportional to the square root of the x-coordinate of that point.
When $x = 36$, $y = 24$

a Write a formula for y in terms of x.

b Find the value of y when $x = 81$.

12 **P** In an experiment, measurements were taken of a and b.

a	45	125	320
b	3	5	8

Which of these laws fits the results?

$a \propto b$ $\quad a \propto b^2$ $\quad a \propto b^3$ $\quad a \propto \sqrt{b}$

19.3 Inverse proportion

1 y is inversely proportional to x.
When $y = 9$, $x = 4$

a Write a formula for y in terms of x.

b Calculate the value of y when $x = 12$

c Calculate the value of x when $y = 2$

Example

2 **R** The average speed of a car, s (in km/h), is inversely proportional to the time, t (in hours), it takes to travel a fixed distance.
$s = 60$ km/h when $t = 0.2$ hours

a Write a formula for s in terms of t.

b Work out the time of a journey when the average speed is 50 km/h.

c Work out the average speed when the time of the journey is 1.5 hours.

d What happens to the average speed when the time of the journey doubles?

3 **P** As a balloon is blown up, the thickness of its walls, t (in mm), decreases and its volume, V (in cm³), increases. V is inversely proportional to t. When V is 15 000 cm³, t is 0.05 mm.

a Write a formula for V in terms of t.

b When the thickness of the wall of the balloon is 0.03 mm, the balloon will pop. Is it possible to blow up this balloon to a volume of 30 000 cm³?

4 **R** y is inversely proportional to x.

x	0.25	0.5	1	2	4	5	10	20
y	48	24	12	6	3	2.4	1.2	0.6

a Draw a graph of y against x.
What type of graph is this?

b $y = \frac{k}{x}$ where k is the constant of proportionality. Find k.

c Work out $x \times y$ for each pair of values in the table. What do you notice?

5 **R** Which of these graphs shows variables in inverse proportion?

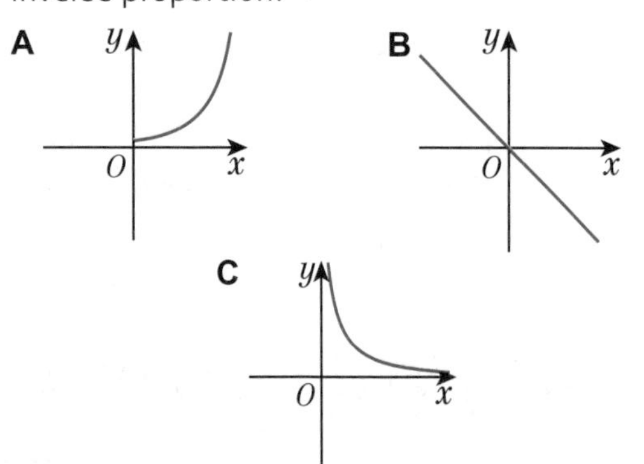

6 **R** The length, l (in cm), and width, w (in cm), of a rectangle with a fixed area are inversely proportional. When the length of the rectangle is 4.5 cm, the width is 2.2 cm.

a Write a formula for l in terms of w.

b Copy and complete the table of values for l and w.

Length, l (cm)	Width, w (cm)
0.3	
	19.8
0.9	
2	
3	3.3
	1.65
15	

c Sketch a graph to show how l varies with w.

7 **P** The graph shows two variables that are inversely proportional to each other.
Find the values of p and q.

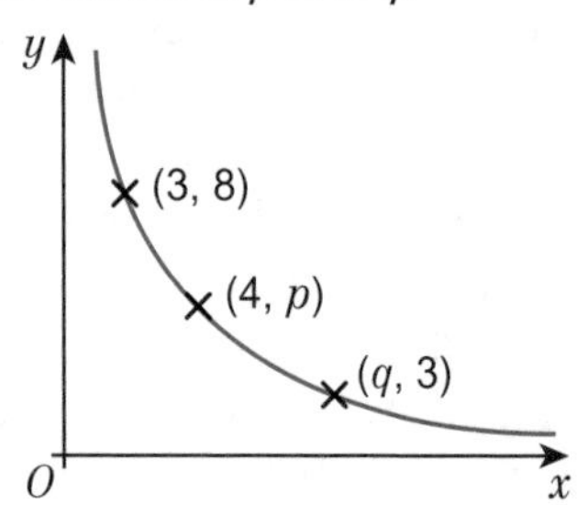

8 **R** A vineyard owner employs teams of people to pick grapes.
Teams have different numbers of people.
The owner records the time it takes different teams to pick the grapes from 12 rows of vines.

Number of people in team, n	Time taken, t (hours)
2	3.5
3	1.8
12	0.5
6	1
4	1.4
5	1.3
8	0.8
9	0.8
10	0.6

a Plot t against n.

b Draw a curve of best fit.

c Write a formula for estimating t in terms of n.

d Use your formula to estimate how long it would take a team of 15 people to pick the grapes from 12 rows of vines.

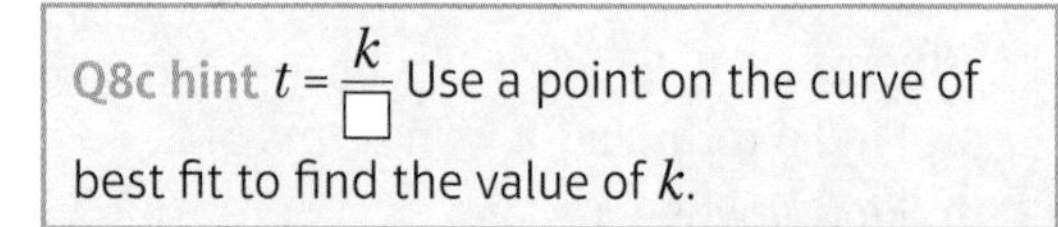

Q8c hint $t = \frac{k}{\square}$ Use a point on the curve of best fit to find the value of k.

9 **Exam-style question**

p is inversely proportional to the square of q.
When $q = 10$, $p = 7.5$
Find the value of p when $q = 5$ **(3 marks)**

10 y is inversely proportional to the cube of x.
When $x = 3$, $y = 6$

a Write a formula for y in terms of x.

b Calculate y when $x = 2$

c Calculate x when $y = 0.75$

11 y is inversely proportional to the square root of x.
When $x = 4$, $y = 5$

a Write a formula for y in terms of x.

b Calculate y when $x = 16$

c Calculate x when $y = 2$

12 **R** When 30 litres of water are poured into any cylinder, the depth, D (in cm), of the water is inversely proportional to the square of the radius, r (in cm), of the cylinder.

When $r = 30$ cm, $D = 10.6$ cm

a Write a formula for D in terms of r.

b Find the depth of the water when the radius of the cylinder is 15 cm.

c Find the radius of the cylinder (to 1 decimal place) when the depth is 60 cm.

d Cylinder P has radius x cm and is filled with water to a depth of d cm. This water is poured into cylinder Q and fills it to a depth of $3d$ cm. What is the radius of cylinder Q?

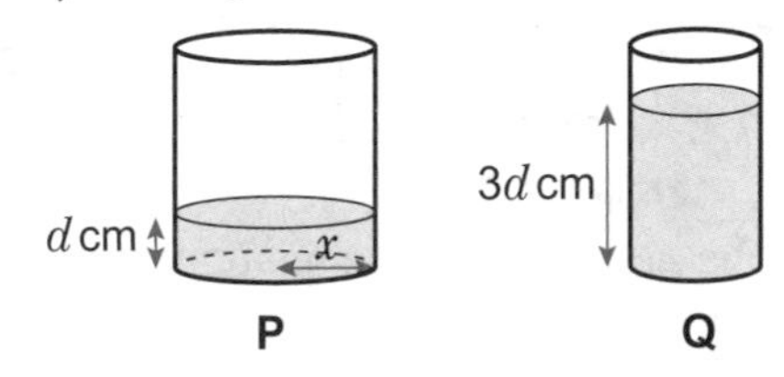

13 **R** The gravitational force between two objects, F (in newtons, N), is inversely proportional to the square of the distance, d (in metres), between them.

A satellite orbiting the Earth is 4.2×10^7 m from the centre of the Earth.

The force between the satellite and the Earth is 60 N.

a Write a formula for F in terms of d.

b The force between two objects is 16 N. What is the value of the force when the distance between the objects doubles?

19.4 Exponential functions

1 Find the value of x for each of these equations.

a $3^x = 27$ b $4^x = 256$ c $10^x = 1\,000\,000$

2 a Copy and complete the table of values for $y = 3^x$.

Give the values correct to 2 decimal places.

x	−4	−3	−2	−1	0	1	2	3	4
y									

b Draw the graph of $y = 3^x$ for $-4 \leqslant x \leqslant 4$

c Use the graph to find an estimate for

i the value of y when $x = 2.5$

ii the value of x when $y = 30$

3 **R** a Draw the graphs of

i $y = 3.5^x$ ii $y = 4.5^x$

b Predict where the graph of $y = 4^x$ would be. Sketch it on the same axes.

c At which point do all the graphs intersect the y-axis?

4 a Copy and complete the table of values for $y = 4^{-x}$. Give the values correct to 2 d.p.

x	−3	−2	−1	0	1	2	3
y							

b Draw the graph of $y = 4^{-x}$ for $-3 \leqslant x \leqslant 3$

c Use the graph to find an estimate for

i the value of y when $x = 1.5$

ii the value of x when $y = 5$

5 **R** The table gives information about the population of the UK.

Year	1751	1801	1851	1901	1951	2001	2014
UK population (millions)	5.8	7.8	15.3	30.1	38.7	49.1	64.1

a Draw a graph of the data.
Plot the year on the horizontal axis and population on the vertical axis.
Draw in a trend curve of best fit.

b Is this an example of exponential growth or exponential decay?

c Use your graph to estimate the number of years it took for the population to double from its value in 1901.

6 **Exam-style question**

The sketch shows a curve with equation $y = ka^x$, where k and a are constants and $a > 0$.

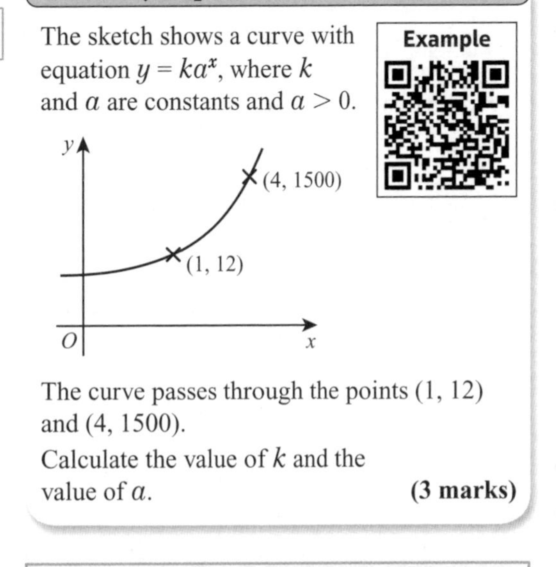

The curve passes through the points (1, 12) and (4, 1500).

Calculate the value of k and the value of a. **(3 marks)**

Q6 hint Start by using the point (1, 12) to express k in terms of a.

7 **R** The value, V (in £), of a motorbike depreciates exponentially over time.

The value of the motorbike on 1 January 2015 is £15 000.

The value of the motorbike on 1 January 2017 is £8400.

The sketch graph shows how the value of the motorbike changes over time.

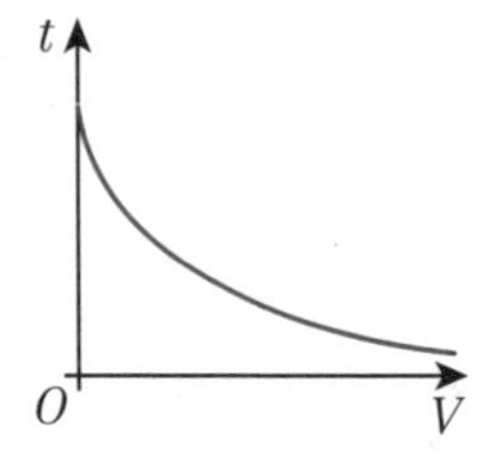

The equation of the graph is $V = pq^{-t}$ where t is the number of years after 1 January 2015, and p and q are positive constants.

a Use the information to find the values of p and q.

b Use your values of q and p in the formula $V = pq^{-t}$ to estimate the value of the motorbike on 1 January 2018.

c By what percentage (to 1 decimal place) does the motorbike depreciate each year from 2015 to 2018?

8 **R** A business is currently valued at £2.4 million, and is growing at a rate of 15% a year.

The expected value, v (in £ millions), in t years' time, is given by the formula $v = 2.4 \times 1.15^t$.

a Use a table of values to draw the graph of v against t for the next 5 years.

b Use your graph to estimate

i the value of the business after 2.5 years

ii the time taken for the value to reach £3 million.

9 **R** £8000 is invested in a savings account paying 5% compound interest a year.

a Write a formula for the value of the savings account (V) and the number of years (t).

b Draw a graph of V against t for the first 8 years.

c Use the graph to estimate when the investment will reach a value of £10 000.

19.5 Non-linear graphs

1 Water is poured into a curved vase at a constant rate. h is the height of water after time t.

a Describe how the rate at which the height increases changes over time.

b Which graph best describes the relationship between h and t?

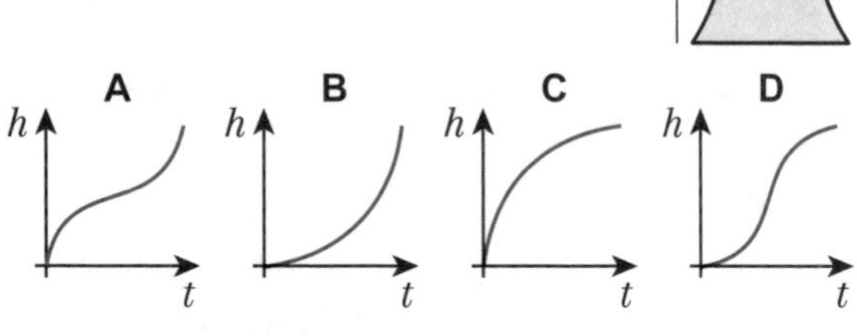

2 **R** The graph shows the relationship between the temperature of cooling water in a kettle, in °C, and the time, in minutes.

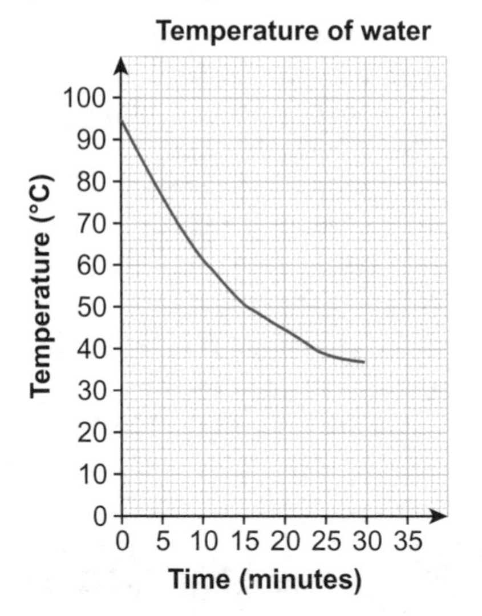

a What is the temperature of the water after 20 minutes?

b Describe the rate at which the water cools down.

c Calculate the drop in temperature between 0 and 5 minutes.

d Calculate the average rate of cooling between 5 and 10 minutes.

e Zain says, 'The water cools at least three times as quickly between 0 and 5 minutes as it does between 20 and 25 minutes.'
Is Zain correct? Explain your answer.

f Compare the average rate of cooling over the first 15 minutes with the rate of cooling at exactly 15 minutes.

3 **R** The distance–time graph shows information about a plane accelerating.

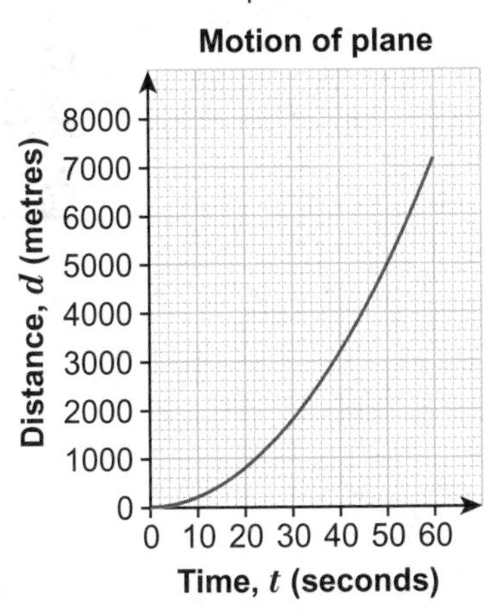

a Estimate the speed of the plane 30 seconds after it starts moving.

b What speed does the plane reach after 60 seconds?

c Will the speed of the plane continue to rise after 60 seconds following the relationship shown by the graph? Explain your answer.

4 **R** A man is cycling at a speed of 12 m/s. The velocity–time graph gives information about the motion of the bicycle as he decelerates to rest in 5 seconds.

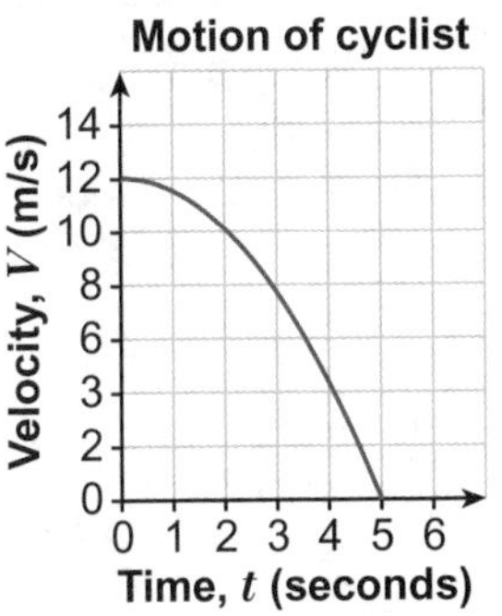

a Copy the graph. Draw a chord from $t = 0$ to $t = 5$.

b Calculate the average deceleration of the bicycle over the 5 seconds.

c Estimate the deceleration at time $t = 4$ seconds.

d Describe how the deceleration changes over the 5 seconds.

e Draw a chord from $t = 0$ to $t = 3$. Use the chord to form a trapezium. Estimate the distance travelled between $t = 0$ and $t = 3$ by finding the area of the trapezium.

5 a Draw the graph of $y = \frac{1}{2}x^2 + 1$ for $0 < x < 4$

b Draw in a chord from $x = 0$ to $x = 1$ and use it to make a trapezium under the graph.

c Repeat with chords from $x = 1$ to $x = 2$, $x = 2$ to $x = 3$ and $x = 3$ to $x = 4$.

d Calculate the areas of your trapezia to estimate the area under the graph of $y = \frac{1}{2}x^2 + 1$ from $x = 0$ to $x = 4$.

6 **R** The distance–time graph shows information about a 20 km bike race between Dawn and Aarti.

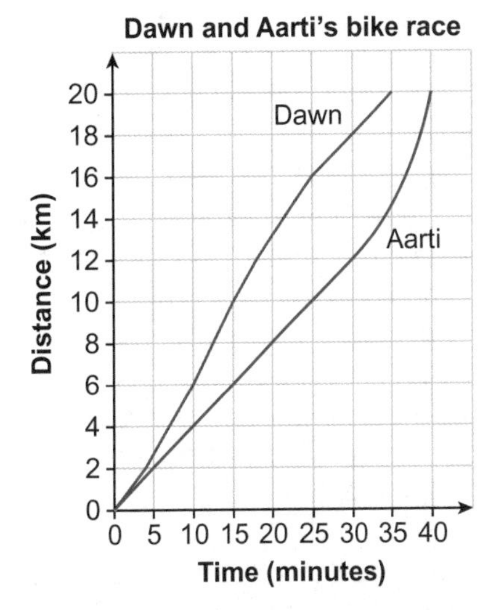

a Describe the race between Dawn and Aarti.

b Compare Dawn's speed for the first and second half of the race.

c Estimate the difference in their speeds 20 minutes into the race.

7 **Exam-style question**

The velocity–time graph describes the acceleration of a truck.

Velocity, v, is measured in metres per second (m/s) and time, t, is measured in seconds.

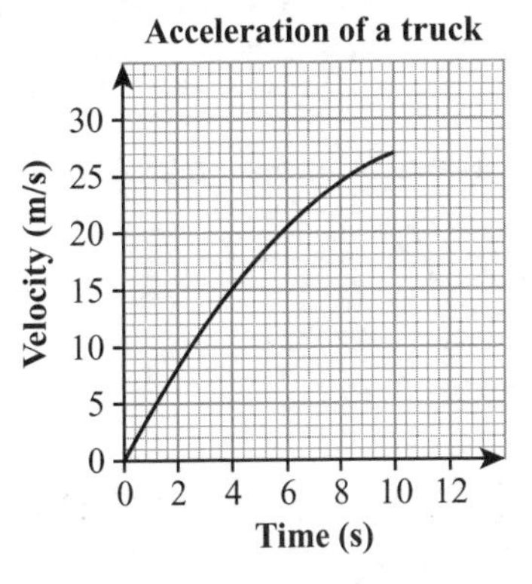

a Estimate the acceleration at $t = 5$. **(2 marks)**

b Estimate the distance travelled between $t = 2$ and $t = 6$. **(3 marks)**

c The instantaneous acceleration at time T is equal to the average acceleration over the first 10 seconds. Find an estimate for the value of T. **(3 marks)**

19.6 Translating graphs of functions

1 Draw a coordinate grid with –5 to 5 on both axes.

a On the same set of axes draw the graphs of

i $y = f(x) = -x^2$

ii $y = f(x) + 3 = -x^2 + 3$

iii $y = f(x - 1) = -(x - 1)^2$

b The maximum point of $y = f(x)$ is (0, 0). Write the coordinates of the maximum point of

i $y = f(x) + 3$

ii $y = f(x - 1)$

c Describe the transformation that maps the graph of $y = f(x)$ onto the graph of

i $y = f(x) + 3$

ii $y = f(x - 1)$

Q1a hint Create a table of values for each graph.

2 Here is the graph of $y = f(x) = -x^2$.

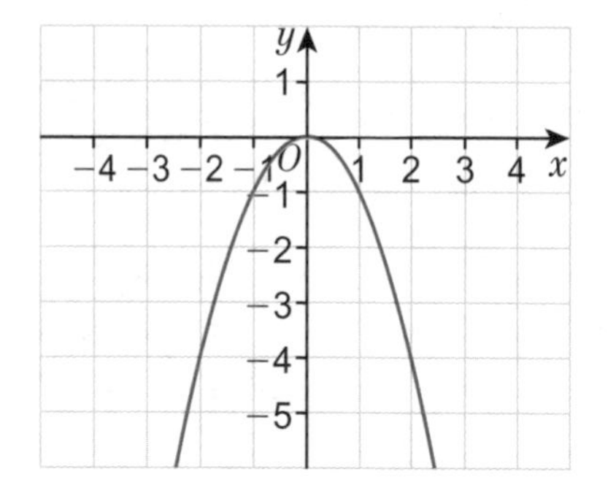

Copy the axes and sketch the graphs of

a $y = f(x) + 2$

b $y = f(x) - 4$

c $y = f(x - 2)$

d $y = f(x + 3)$

3 Write the vector that translates $y = f(x)$ onto

a $y = f(x) + 3$

b $y = f(x) - 5$

c $y = f(x - 5)$

d $y = f(x + 4)$

e $y = f(x - 5) + 3$

4 **Exam-style question**

The graph of $y = f(x)$ is shown on the grid.

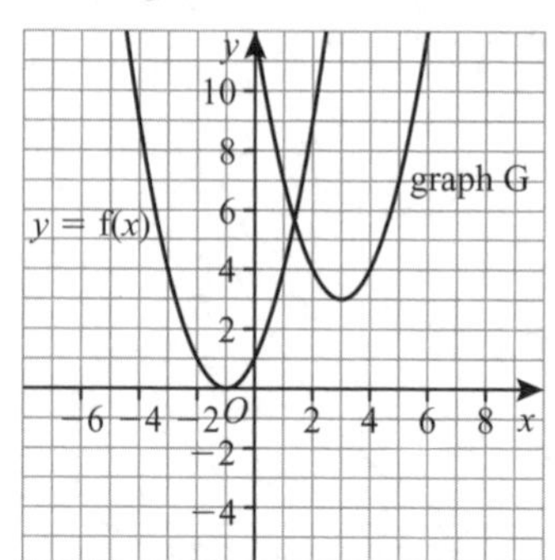

The graph G is a translation of the graph of $y = f(x)$.

Write down the equation of graph G. **(1 mark)**

Q4 hint First see if the translation is to the right or to the left or up or down and by how many squares.

5 **Exam-style question**

The graph of $y = f(x)$ is shown on the grid.

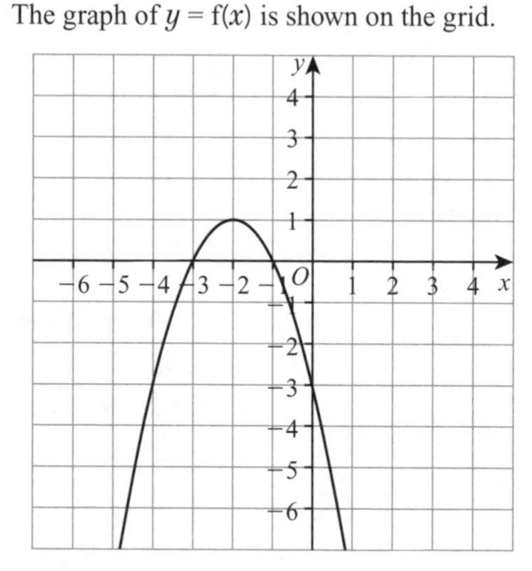

Copy the diagram and sketch the graph of $y = f(x - 4) - 1$. **(2 marks)**

Q5 hint Make sure you translate the points that have integer coordinates such as (–2, 1) and (0, –3) exactly the right number of squares in the correct direction.

6 Here is a sketch of $y = f(x) = -x^3$

a Draw sketches of the graphs

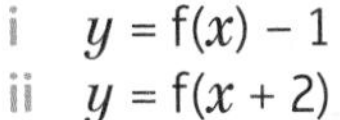

i $y = f(x) - 1$

ii $y = f(x + 2)$

b Write the coordinates of the point which (0, 0) is mapped to for both graphs.

7 **R** $f(x) = 3 - 4x$

a Draw the graph of $y = f(x)$

b Draw the graph of $y = f(x - 1)$

c Write the algebraic equation of $y = f(x - 1)$

8 **P** $f(x) = -\frac{1}{x}$

a Sketch the graph of $y = f(x + 1) - 2$

b Write the equation of each asymptote.

19.7 Reflecting and stretching graphs of functions

1 $f(x) = 5 - 2x$

a Copy and complete the table.

x	−2	−1	0	1	2
f(x)					
−f(x)					
f(−x)					

b On the same set of axes, draw the graphs of

i $y = f(x)$ ii $y = -f(x)$ iii $y = f(-x)$

c Describe the transformation that maps $f(x)$ onto $-f(x)$.

d Describe the transformation that maps $f(x)$ onto $f(-x)$.

2 **R** The diagram shows the graph of $y = f(x)$.

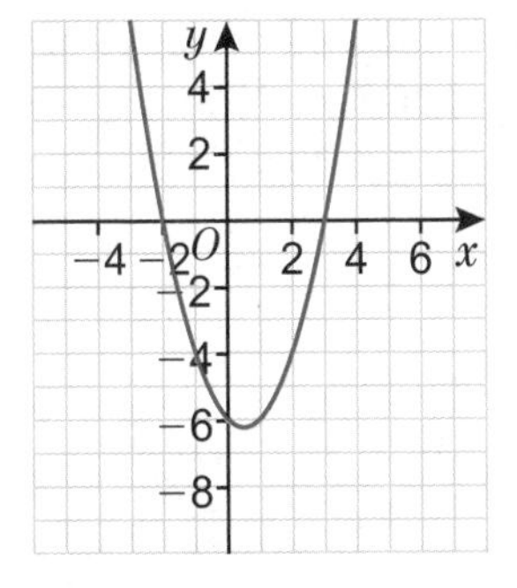

a Sketch a copy of the graph.
On the same axes sketch the graphs of

i $y = -f(x)$ ii $y = f(-x)$

b Caz says, 'The graph of $y = -f(x)$ is a reflection of the graph of $y = f(x)$ in the line $x = 0$.
The graph of $y = f(x)$ is a reflection of the graph of $y = f(-x)$ in the line $y = 0$.'
Is Caz correct?
Explain your answer.

3 **R** The diagram shows the graph of $y = f(x)$.
The turning point of the curve is A(3, −4).

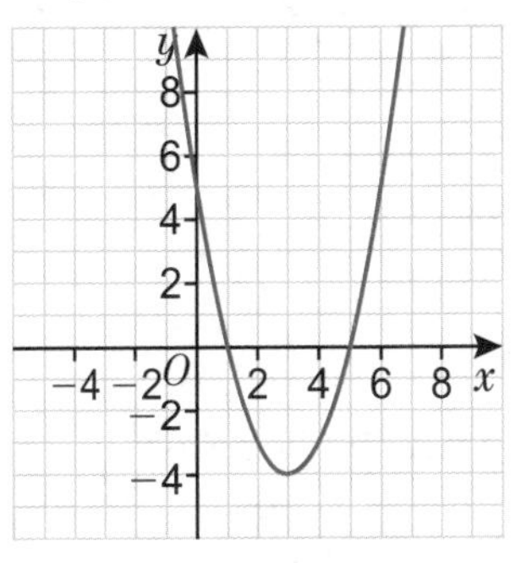

Write the coordinates of the turning point of the curve with equation

a $y = -f(x)$

b $y = f(-x)$

c $y = -f(-x)$

4 **R** Here is the graph $y = f(x) = -x^3 - 1$

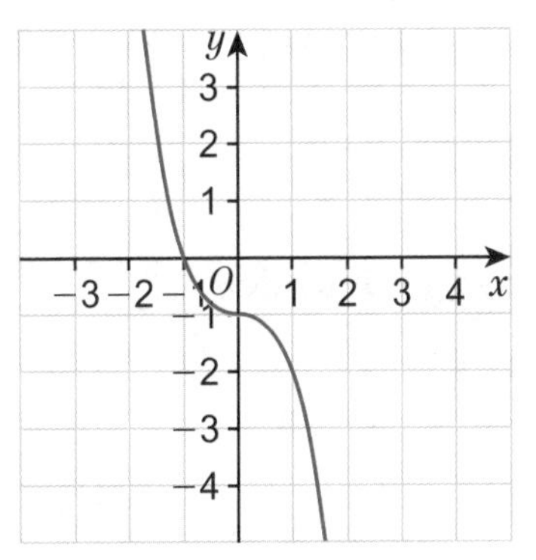

a Sketch a copy of the graph and on the same axes sketch the graphs of

i $-f(x)$ ii $f(-x)$ iii $-f(-x)$

b Describe the transformation that maps $f(x)$ onto $-f(-x)$.

5 $f(x) = 2 - x^2$

a Copy and complete the table.

x	−4	−3	−2	−1	0	1	2	3	4
f(x)									
3f(x)									
f(3x)									

b On the same set of axes draw the graphs of

i $y = f(x)$ ii $y = 3f(x)$ iii $y = f(3x)$

6 Here is the graph of $y = f(x)$

Draw the graphs of

a $y = 2f(x)$

b $y = 3f(x)$

c $y = \frac{1}{2}f(x)$

7 Here is the graph of $y = f(x)$

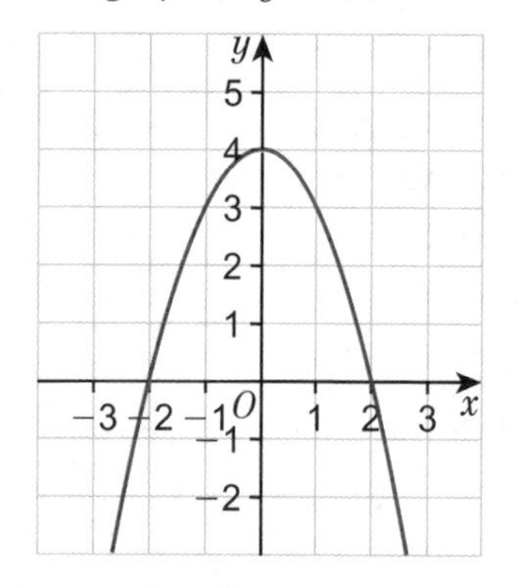

Draw the graphs of

a $y = f(3x)$ b $y = f(\frac{1}{2}x)$

8 **Exam-style question**

The graph of $y = f(x)$ is shown on the grid.

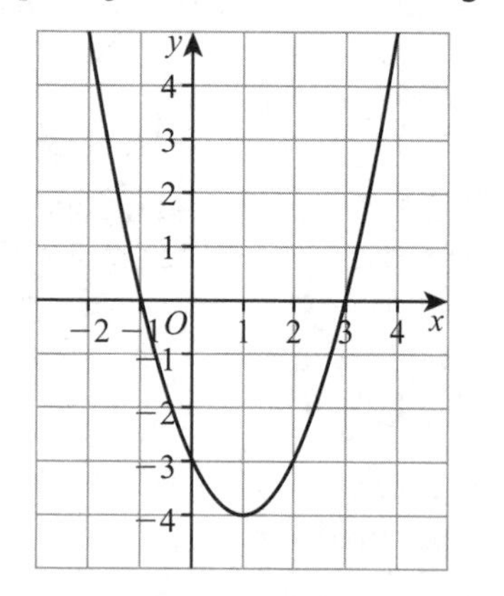

Copy the diagram and sketch the graph of $y = 2f(x)$. **(2 marks)**

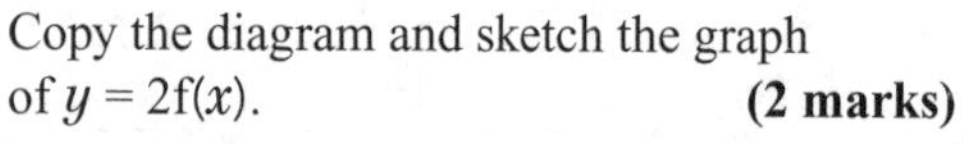

Q8 hint Calculate some y-values for $y = 2f(x)$ to make sure your graph passes through the correct points.

9 **R** The diagram shows the graph of $y = f(x) = 5 - \frac{1}{2}x$ and the graphs of some transformations of $f(x)$

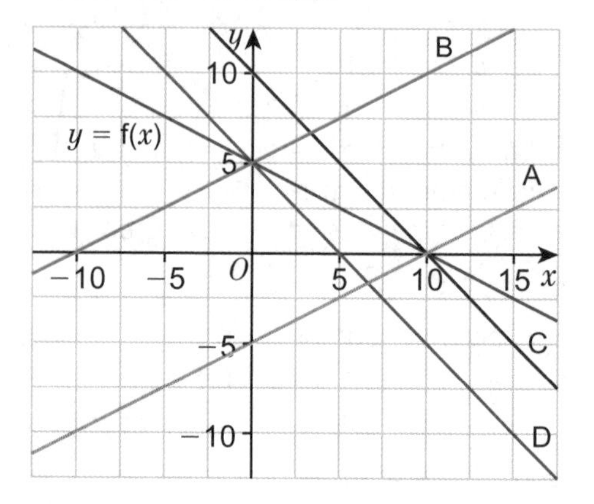

Match the function notation to the graphs.

a $f(2x)$

b $2f(x)$

c $f(-x)$

d $-f(x)$

10 **R** Here is a sketch of $y = f(x) = (x - 3)^2 + 2$

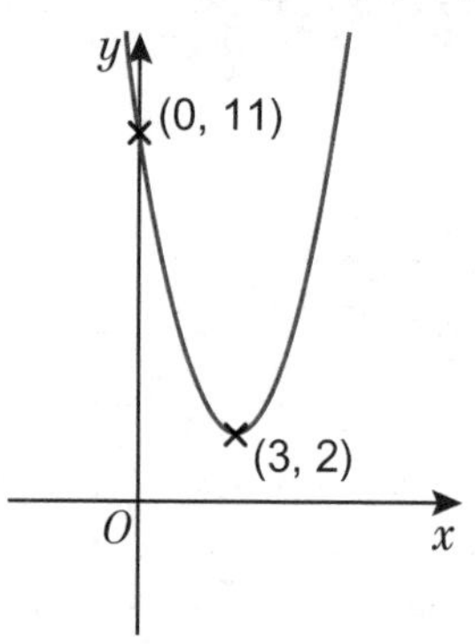

The graph has a minimum point at (3, 2).
It intersects the y-axis at (0, 11).

a Sketch the graph of $y = (\frac{1}{2}x)$

b Write down the minimum value of $f(\frac{1}{2}x)$

c Write down the coordinates of the minimum point of $y = f(2x)$

d **R** Explain why the graphs of $y = f(x)$, $y = f(\frac{1}{2}x)$ and $y = f(2x)$ all intersect the y-axis at the same point.

19 Problem-solving

Solve problems using these strategies where appropriate:

- **Use pictures**
- **Use smaller numbers**
- **Use bar models**
- **Use x for the unknown**
- **Use a flow diagram**
- **Use arrow diagrams**
- **Use geometric sketches**
- **Use graphs**
- **Use logical reasoning**
- **Use problem-solving strategies and then 'explain'.**

1 A 3 inch chord PQ is cut across a circle with centre O. The minimum distance from the centre of the circle to the chord is 1 inch. What is the radius, PO, of the circle? Give your answer to 1 d.p.

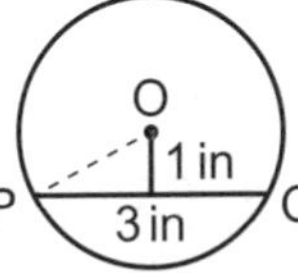

2 **R** A tap is dripping at a steady rate. A container is put underneath to catch the drips. Use the graph of the water depth in the container to sketch the shape of the container.

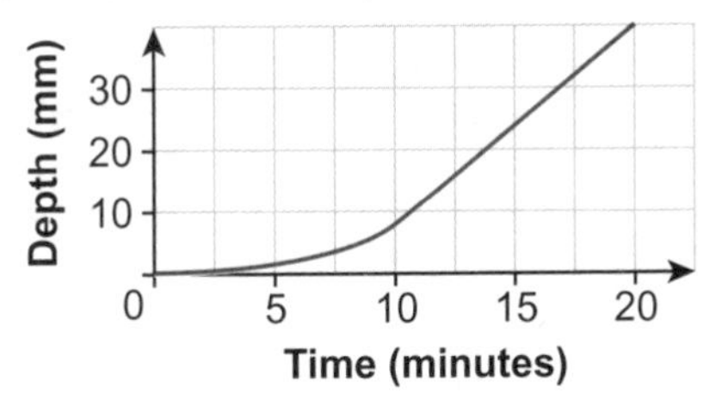

3 **R** a Find a counter example to prove that the statement '$n^2 + 3$ can never be divided by 2, where both n and the answer are integers' is not true.

b What could you say about n to make the statement true?

4 Luke's wages are calculated for every 45 minutes worked and are shown in the table.

Minutes	45	90	135	180
Wages (£)	6.30	12.60	18.90	25.20

The wages earned (w) are directly proportional to the number of minutes worked (m).

a Find a formula for w in terms of m.

b How many 45-minute blocks must Luke work to earn £200?

c How much does Luke earn per hour?

5 Jennifer is building a ramp from her gate to her doorstep.
The height of her step is 28.4 cm and the distance from the gate to the step is 240.6 cm.
Both measurements are to the nearest mm.
What are the upper and lower bounds for angle x, to 3 d.p?

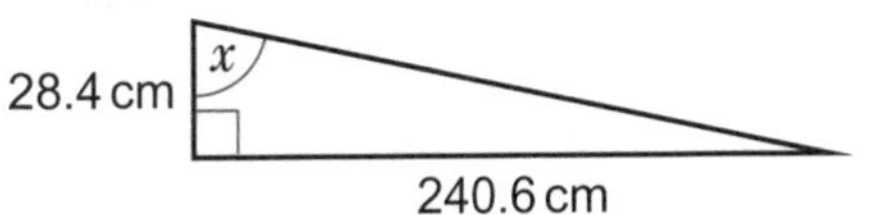

6 **R** Marcia and Judy are both asked to sketch the graph of the exponential function $y = 6^x$
Marcia sketches this graph and explains that the function shows exponential growth.

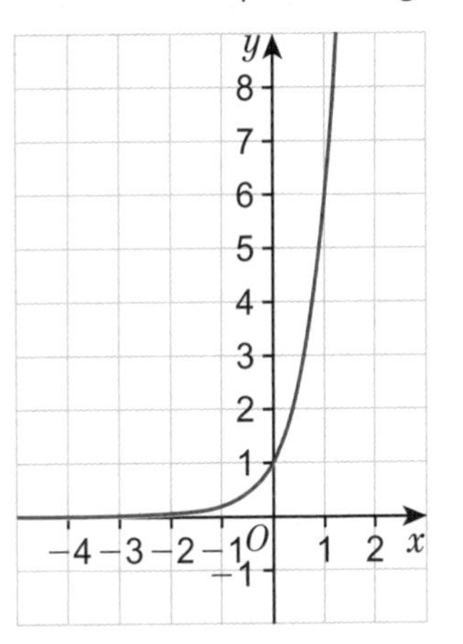

Judy sketches this graph and explains that the function shows exponential decay.

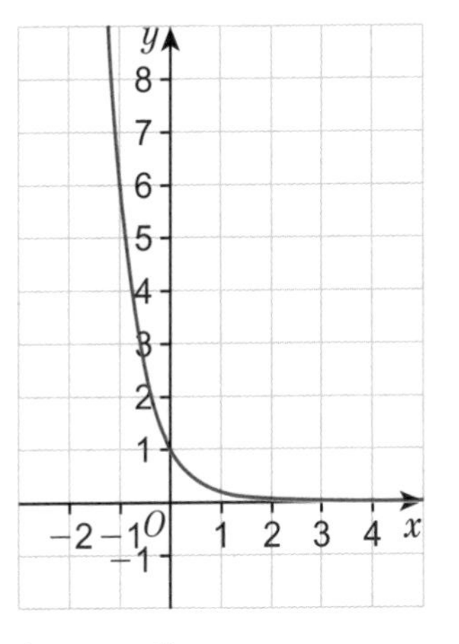

a Who is correct?
Explain why.

b If $x = 5$ what would the value of y be?

7 p is inversely proportional to the cube of q.
When $q = 3$, $p = 5$
Find the value of p when $q = 5$

8 **Exam-style question**

a y is inversely proportional to x^2.
$y = 8$ when $x = 3$.
Work out an equation connecting x and y. **(3 marks)**

b Work out the value of y when $x = 12$.
Give your answer as a fraction in its simplest form. **(2 marks)**

9 **R** The points M, N and P have the coordinates (1, 9), (7, 3) and (5, 5) respectively.
Prove that the three points are collinear.

Q9 hint **Collinear** means 'lie on the same straight line'.

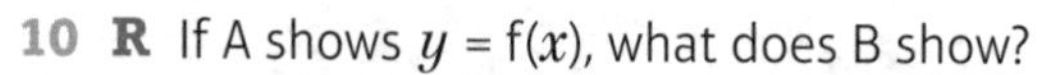

10 **R** If A shows $y = f(x)$, what does B show?

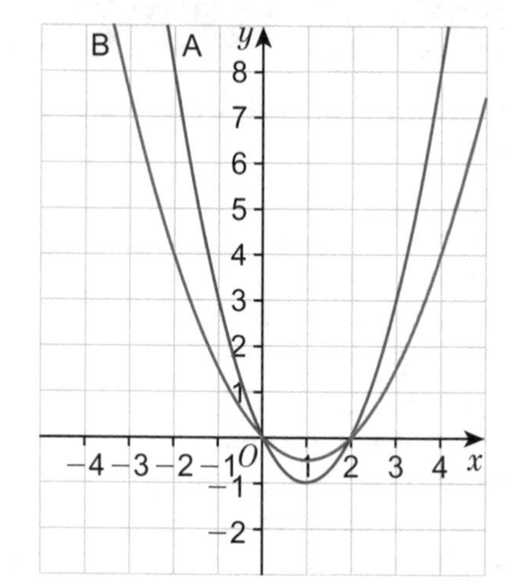